MAVERICK SCIENTIST

ELECTRICAL CIRCUITS

AN ELECTRICAL CIRCUIT IS ANY ARRANGEMENT THAT PER AN ELECTRICAL CURRENT TO FLOW. A CIRCUIT CAN B AS SIMPLE AS A BATTERY CONNECTED TO A LAMP AS COMPLICATED AS A DIGITAL COMPUTER.

☐ A BASIC CIRCUIT — THIS BASIC CIRCUIT CONSISTS A SOURCE OF ELECTRICAL CURR (A BATTERY), A LAMP AND TI CONNECTION WIRES. THE PA OF A CIRCUIT WHICH PERFORMS WORK IS CALLED THE LOAD. HERE THE LOAD IS THE LAM IN OTHER CIRCUITS THE LOAD CAN BE A MOTOR, A HEATIN ELEMENT, AN ELECTROMAGNET,

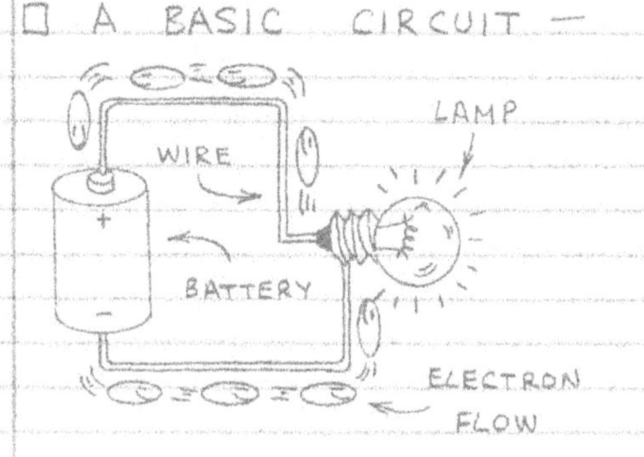

☐ A SERIES CIRCUIT — A CIRCUIT MAY INCLUDE MOR THAN ONE COMPONENT (SWITCH, LAMP, MOTOR, A SERIES CIRCUIT IS FOR WHEN CURRENT FLOWIN THROUGH ONE COMPON FIRST FLOWS THROUG ANOTHER. (ARROWS SHOW DIRECTION OF ELECTRON FL

☐ A PARALLEL CIRCUIT — A PARALLEL CIRCUIT IS FORM WHEN TWO OR MO COMPONENTS ARE C NECTED SO CURREN CAN FLOW THROUG ONE COMPONENT W OUT HAVING FIRST TO FLOW THROUG ANOTHER.

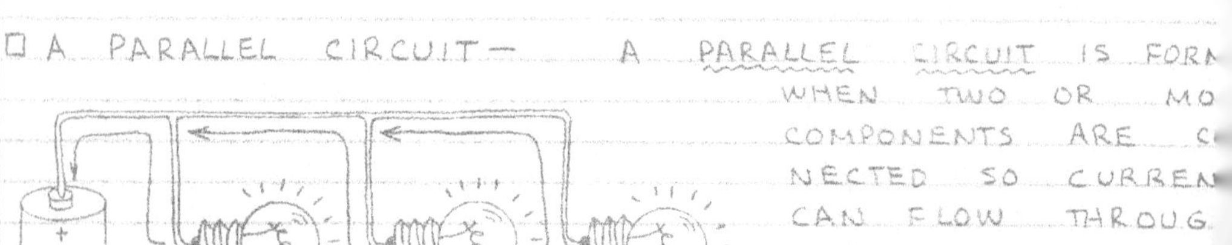

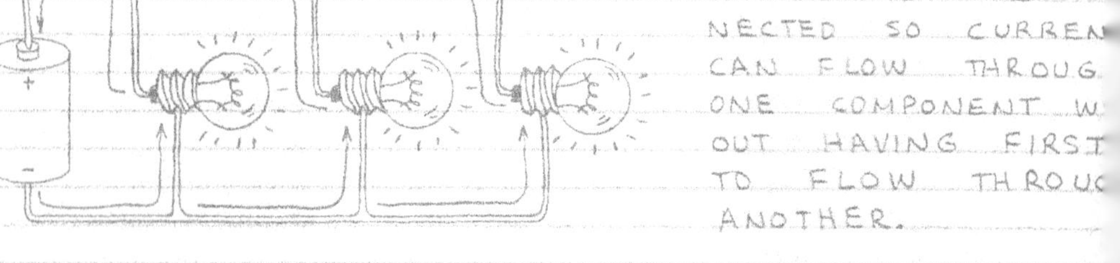

☐ A SERIES-PARALLEL CIRCUIT — MANY ELECTRICA CIRCUITS A BOTH SERIE AND PARALL ALL PROVIDE COMPLETE BETWEEN T CIRCUIT AN POWER SU

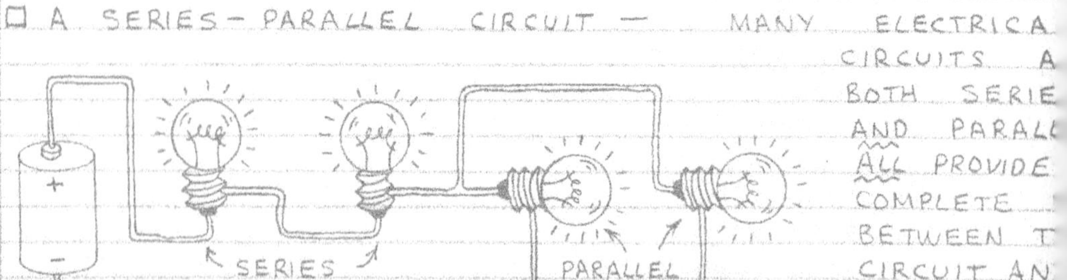

MAVERICK SCIENTIST

My Adventures as an Amateur Scientist

By Forrest M. Mims III

Make:

Make: MAVERICK SCIENTIST

ISBN: 978-1-68045-816-9
March 2024: First Edition
See www.oreilly.com/catalog/errata.csp?isbn=0636920346234 for release details.

Make: Books
President: Dale Dougherty
Creative Director: Juliann Brown
Editor: Kevin Toyama
Designer: Terisa Davis
Copyeditor: Mark Nichol
Proofreader: Sophia Smith
Indexer: BIM Creatives

Make: Community is a growing, global association of makers who are shaping the future of education and democratizing innovation. Through *Make:* magazine, 200+ annual Maker Faires, Make: books, and more, we share the know-how of makers and promote the practice of making in schools, libraries, and homes.

Make: books may be purchased for educational, business, or sales promotional use. Online editions are also available for most titles. For more information, contact our corporate/institutional sales department at 800-998-9938.

Make Community, LLC
150 Todd Road, Suite 100
Santa Rosa, California 95407
www.make.co

(Cover photo: Forrest's experimental rocket launch from his Saigon apartment roof in 1967, which caused military police to think the base was under attack.)

DEDICATION

This book is dedicated to my fabulous wife Minnie and our three children, Eric, Vicki, and Sarah, each of whom assisted me many times during my science career.

TABLE OF CONTENTS

SOLAR CELLS

SOLAR CELLS ARE PN JUNCTION PHOTODIODES WITH AN EXCEPTIONALLY LARGE LIGHT SENSITIVE AREA. A SINGLE SILICON SOLAR CELL GENERATES 0.5 VOLT IN BRIGHT SUNLIGHT.

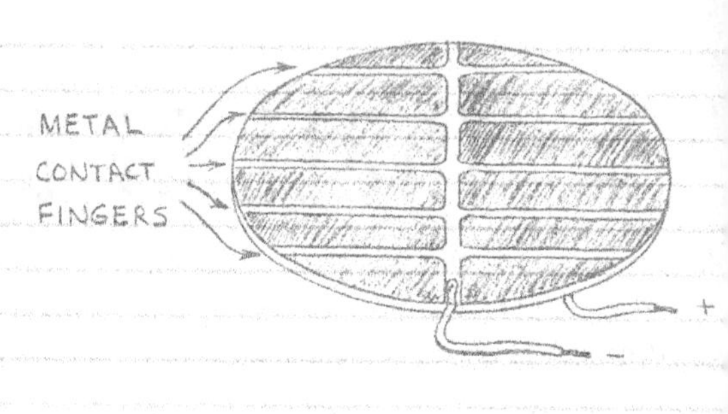

☐ SOLAR CELL OPERATION

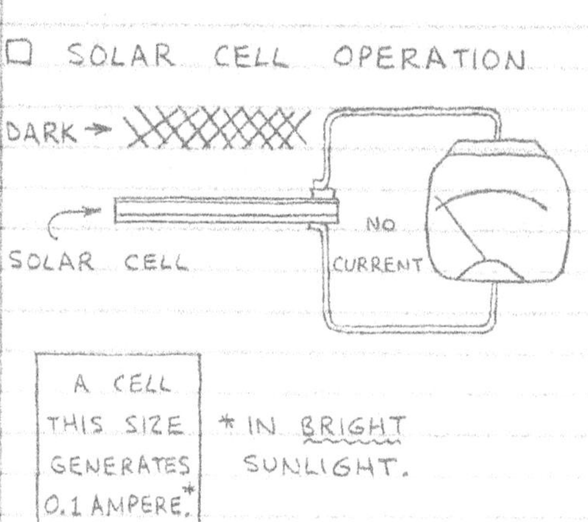

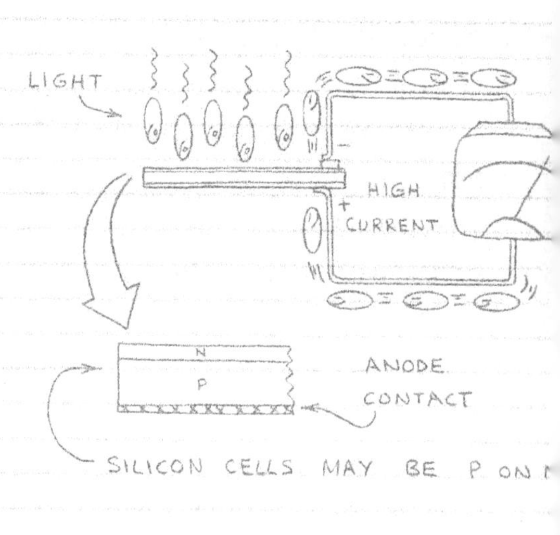

A CELL THIS SIZE GENERATES 0.1 AMPERE.* * IN BRIGHT SUNLIGHT.

SILICON CELLS MAY BE P ON N

☐ KINDS OF SOLAR CELLS.

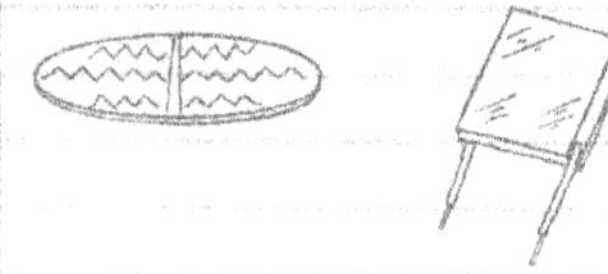

MANY DIFFERENT KINDS OF SILICON SOL CELLS ARE MADE. OFTEN INDIVIDUAL CE ARE CONNECTED IN SERIES OR PARALL

SERIES: OUTPUT VOLTAGE IS SUM OF CELL VOLTAGES.

PARALLEL: OUTPUT CURRENT IS SUM OF CELL CURRENTS.

☐ SOLAR CELL SYMBOL.

REMEMBER, CELLS MAY BE P ON N.

HOW SOLAR CELLS ARE USED

P. 11

ARRAYS OF SOLAR CELLS CAN CHARGE RECHARGEABLE CELLS AND BATTERIES.

SUN

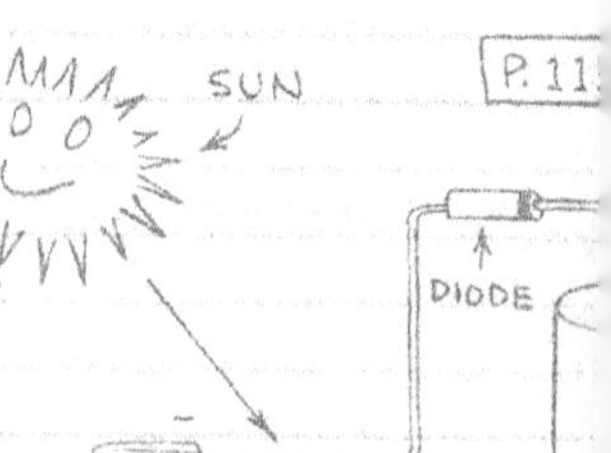

SERIES CONNECTED SOLAR CELL ARRAY

PREFACE

Hawai'i's Mauna Loa Observatory, the world's most famous atmospheric monitoring station, is perched on the barren slope of the world's largest volcano, 11,200 feet above the Pacific. One afternoon, I was there alone analyzing my ozone data when the blaring of a horn suddenly blocked the whirring sound of nearby air pumps. Moments later, a car door slammed shut, and a man began shouting and banging on the steel door of the building where I was working.

As the banging continued, I walked toward the door and heard the man shouting again and again, "Are any scientists in there?" Finally, I shouted back, "How can I help you?"

The banging stopped, and a friendly voice gently replied, "Are you a scientist?" When I slowly opened the door, a cherubic fellow with a short, gray beard smiled broadly and respectfully asked, "Are you a real scientist?"

Though I received a Rolex Award for my homemade instrument that measures the ozone layer, and published my discovery of an error in NASA's ozone satellite data in *Nature*, the world's leading scientific journal, I do not have a science degree. Yet no one had ever asked me whether I was a scientist, much less a real scientist. My life was an ongoing series of science adventures and projects, and I, too, had spent years banging on the doors of science, sometimes successfully. But I did not want to disappoint the visitor by telling him I was self-taught.

Those thoughts spun through my mind before I offered a four-word reply that would need a book to explain. This is that book. Whether you want to join me in doing science or simply accompany me on some of my adventures, like the gentleman who banged his way into the Mauna Loa Observatory, this book is for you.

1

COMING OF AGE IN THE SILICON ERA

When I was growing up in Texas, Alaska, Florida, Colorado, and Alabama, no one told me dreams might never come true; I just assumed they would. Later in life, when people said my science projects and goals were impossible, I kept pursuing them. When scientists said some of the ambitious science fair projects my children planned would not be successful, I told them that meant they would be. Their projects went on to earn many awards and scholarships and were published in books and scientific journals and displayed in major museums.

My father, Forrest M. Mims Jr., influenced this approach to life. My brother, Milo, and I sometimes made life difficult for our parents, but our hero was our dad. In our eyes, Daddy could do anything, for he had emerged from the Great Depression to become a US Air Force officer, jet pilot, and civil engineer. He could repair anything that went wrong with our family car, with Milo or me holding a flashlight when darkness arrived.

He expected us to be as disciplined as he was, and each morning inspected our room to make sure it was clean and that our beds were made with military precision. He taught us to respond to him, our mother, and all other adults with "Sir" and "Ma'am." He expected us to eat everything on our plates, open doors for ladies, shine our shoes before church (I sometimes do), and never lie or use profanity (I still do not).

During Daddy's time away from work, he organized family outings and taught Milo and me basic electricity, how to use tools, and how to bleed the brake lines in our car. He spent any remaining free time as an artist, architect, sculptor, woodworker, fisherman, hunter, and church deacon. Daddy encouraged my early interest in science, which was inspired not by courses taught in school but by the natural world, hobby science magazines, and many trial-and-error experiments. My life became an endless string of science fair projects; it still is.

My poor aptitude in math, however, ruled out majoring in physics at Texas A&M University. But while I was majoring in government and history during the day and sending voice-modulated light beams across the campus at night, I did not realize that my liberal arts diploma would be accompanied by a do-it-yourself degree in electronics.

Back then, I had no idea that walking along a country road with my blind great-grandfather as a young boy would inspire my invention of a handheld travel aid for the blind during my senior year at A&M. Nor did I realize that the travel-aid project and my weekend rocket experiments would lead to top secret research at the Air Force Weapons Laboratory and the founding of Micro Instrumentation and Telemetry Systems, a partnership with my friend Ed Roberts that led to the founding of Microsoft when Bill Gates and Paul Allen moved to Albuquerque to work at MITS.

I did not know that the columns I would someday write for *Scientific American* magazine would be accompanied by an international controversy, and that I would receive a Rolex Award, which fulfilled my dream of doing serious science for the National Aeronautics and Space Administration (NASA), the National Oceanic and Atmospheric Administration (NOAA), the Environmental Protection Agency (EPA), and several universities.

Even today, I never think of myself as a real scientist. While I have designed and built some novel instruments and reported on their discoveries in leading journals of science, I have always thought of myself as a self-taught amateur scientist. After all, professional scientists are a relatively recent development. Though devout Christian Michael Faraday never took a science course, he is widely considered history's greatest amateur scientist. Thomas Edison, the Wright brothers, Benjamin Franklin, and Thomas Jefferson were significant amateur scientists in their day.

Several childhood experiences contributed to my career doing science, and I distinctly remember each of them. The first occurred in January 1949, when I was barely four years old, and 3 inches (8 cm) of snow fell on Houston, Texas, my hometown. That unforgettable experience led to my lifelong interest in weather.

My fascination with spiders and their webs began when I walked through shoulder-high grass into the circular web of a black-and-yellow *Argiope* the size of my palm while trying to observe a large, gray spider on its web. This occurred during the summer of 1950, after we moved to Victoria, Texas, when I was five.

My birthday is in September, which meant I could not begin first grade in the public school system that year. Instead, my parents, who were Baptists, enrolled me in Nazareth Academy, one of the oldest Catholic schools in Texas. There, I learned about critical thinking during a chapel service when the priest checked the bowl of fruit he had earlier left for Jesus in a little room behind the podium. I was seated in the back row, but I could see a table in the room when the priest opened the door. He picked up an empty bowl on the table and showed us that Jesus had eaten the fruit. I wondered how that could be true, for there were no orange peels or apple cores.

During the fall of 1951, we were back in Houston, and migrating robins were prancing across the front lawn. When I asked Mama how to catch one, she told me to sprinkle some salt on its tail and poured some in my open hand. I went outside and began crawling toward the nearest robin. That bird and the next one flew away, but the third paid me no attention as it cocked its head back and forth, listening for activity in the soil. Soon, I was close enough to reach out and grab it without using the salt. The startled bird began squawking and flapping its wings so vigorously that I let it go.

Moving to Alaska

In January 1952, we moved to Elmendorf Air Force Base, near Anchorage, Alaska, where Daddy was assigned as a B-17 pilot. A blizzard occurred just before our arrival, which required that snowplows clear the runway so our plane could land. We walked down the stairs from the plane into air 29 degrees Fahrenheit below zero. My brother, Milo, and I were soon attending school in log buildings and later in corrugated half-cylinder steel structures known as Quonset huts. That is when I won $5 for third prize in a letter-to-Santa-Claus contest, my first freelance sale.

Other Alaskan adventures included learning to fish through holes cut in ice, blizzards that left 6-foot-high drifts, giant two-story icicles, and frozen streams. During summer, we went on hikes and scenic drives. We once watched as a few dozen fishermen caught hundreds of migrating salmon simply by snagging them with treble hooks they dragged across a fast-moving, 5-foot-wide stream that flowed into Cook Inlet. While no major earthquakes occurred during our time in Alaska, we frequently experienced shaking from the minor quakes known as tremors.

Concerns about the Soviet Union were very real back then. Sometimes, Elmendorf AFB and nearby Fort Richardson had blackout drills. No lights were allowed, and we had to practice wearing gas masks. The masks were large for Milo, my sister Sherry, and me, and Alaska-born Karen Lea was too small to wear one. During blackouts, Mama changed Karen's diapers by laying her on the open door of the clothes dryer, which provided enough light to see what she was doing.

Mount Spurr Erupts

During long summer days, we wanted to play until midnight. And during long winter nights, we were mesmerized by the aurora borealis, the northern lights. Those waving streamers of pastel green and pink painted across the black sky were sometimes so stunningly spectacular that we often stared for minutes without uttering a word.

Then there was Thursday, July 9, 1953, when my mother was driving me to a doctor's appointment. Suddenly, a dark cloud began blocking the deep blue sky so quickly that Mama had to switch on the headlights, which illuminated what appeared to be gray snow falling from the sky. Soon, it was pitch black, and Mama explained that Mount Spurr, the highest and most northeastern volcano in the Aleutian Arc, was erupting and dumping a thick cloud of ash over the region. Mount Spurr was 86 miles (138 km) west of our quarters at Fort Richardson, but the ash was so thick that it was difficult to see the road ahead with headlights.

So much volcanic ash coated the area that heavy equipment was required to clear the runways at Elmendorf. Even though Mama taped paper across the windows in our quarters, the superfine ash particles managed to sneak inside and coat everything in a fine film of gray. Daddy missed much of this, for he was among the pilots who hurriedly evacuated B-17s to a safe location before the ash cloud reached Elmendorf.

Milo and I attended Ursa Major Elementary School, a new school built in a bulldozed clearing at the edge of a wilderness loaded with wildlife and sprinkled with abandoned US Army foxholes that we enjoyed exploring while walking home. One day during recess in the spring of 1954, when I was a nine-year-old fourth grader, I participated in an unplanned wildlife demonstration that Milo later described in his fourth-grade assignment: "My Life Story": "One day, a little girl saw a mother moose, a father moose, and a baby moose. The little girl was walking out to them, saying, 'I want to pet the horses.' She started out to pet them. Then the big moose started snorting and pawing the ground. Just as the big moose started for the girl, my brother ran out to save the girl. The moose waited, and Forrest saved her just in time. The whole school had been watching. Later, we heard some machine guns, and that was the MPs [military police] shooting the big moose."

After Daddy was assigned to Eglin Air Force Base in Florida, in June 1954, we flew to Seattle to pick up the family car, which had been shipped there by boat. We then began an unforgettable drive back to Texas via Yellowstone National Park, the Grand Tetons, the Colorado Rockies, and Carlsbad Caverns.

From Alaska to Florida

At my maternal grandparents' house in Houston, we watched television for the first time in three years. We also became reacquainted with bananas, which were unaffordable in Alaska at 15 cents each. My grandfather Pa took us fishing in Galveston Bay after trolling for jumbo shrimp that we used for bait. Today, those shrimp would be far more desirable than the speckled trout and croakers we caught. Back in Houston, we visited the zoo, where one of the animal keepers entered a birdcage to retrieve feathers for Milo and me. On the way to Florida, we passed through a town where huge piles of bright yellow sulfur were being loaded into railroad cars, and we drove through Louisiana swamps.

The contrast between chilly Alaska and tropical Florida was dramatic. In Alaska, we ice-skated, built igloos, followed animal tracks in the snow, and walked through blasts of snow thrown off the streets by motorized plows. In Florida, we walked through clouds of mosquito-killing DDT, and I learned how to swim when my Boy Scout troop was wading along a waist-deep, crystal-clear creek bordered by dense green foliage under a canopy of tall trees.

When we encountered a large tree that had fallen across the stream, I told the scout leader I could not swim. He calmly assured me that staying in the water was much safer than braving the thorns and rattlesnakes in the impenetrable brush and woods along the creek. He told me to take a deep breath, kneel under the water, look straight ahead, and swim like a frog. His swimming lesson worked, and I was never again afraid of water.

During another Boy Scout expedition, we camped by a lake where we learned how to spotlight alligators with our flashlights. We saw as many as three pairs of glowing orange alligator eyes in the lake. Late that night, we were awakened by screams and shouts when a gator poked its head inside one of the tents. We spent the rest of the night in the beds of several pickup trucks.

Daddy took us to church on Sundays and prayed before our meals. He took Milo and me fishing and the entire family to the snow-white beaches at Destin. Back then, Destin was undiscovered, and the beach was wide open. During one beach outing, I hiked to where Choctawhatchee Bay pierced the dunes and entered the Gulf of Mexico. Suddenly, a porpoise poked his upper body out of the bay only 20 feet (6 m) away, looked directly at me, and then dropped back into the water.

Electrical Developments

The many adventures in the natural world we experienced in Florida were soon expanded to include technical topics. This informal technical education began when I nearly electrocuted myself while Milo and I were exploring a house under construction. When I noticed a shiny copper wire emerging from an electrical fixture, I carefully touched the end of the wire with my right index finger. While I suspected there might be a sensation, I did not realize it would occur in the form of a powerful jolt that threw my arm away from the outlet. That experience left a permanent respect for electricity that probably saved me from future trouble.

More hands-on but much safer experiences with electricity occurred a year or so later when Milo and I were building soapbox racers from scrap wood and parts salvaged from broken wagons. I wanted to install a headlight on my racer, and Daddy offered to help. He drilled a hole through the bottom of an empty tin can, inserted a flashlight bulb into the hole, and soldered it neatly in place. He then soldered wires to the lamp and the can. My job was to mount the headlight on the racer and connect its wires to a battery and an on-off switch made from two nails. Instead of placing the switch between one of the lamp wires and the battery, I connected it *across* the battery. I could not understand why the light glowed brightly when the switch was in the Off position. When the switch was in the On position, the light was very dim, the two nails became hot, and the expensive battery was soon depleted.

This experience taught valuable lessons: I learned how to track down and solve a problem. I learned how to draw a diagram on paper before wiring things together from scratch. And I learned to solder almost as neatly as Daddy.

Looking back, Daddy's role was much more significant, for he was mentoring Milo and me as we learned about mechanics, soldering, airplanes, nature, science, and God. For Christmas, we received a chemistry set. Daddy saved the money to buy us the ten-volume *Book of Knowledge*, every page of which we read. He subscribed to *National Geographic*, which became our favorite magazine. He took us to the weather station at Eglin AFB, where we saw all the instruments and were each given a weather balloon. He took us to Eglin's famed Climatic Hanger, where we saw a Russian MiG-15 jet encrusted in ice while undergoing secret tests to determine its winter capabilities.

When Daddy volunteered his time to design church buildings, he let us accompany him when he inspected construction progress. He read the Bible to us, and he and Mom watched as Milo and I were baptized in 1955. Each night, Milo and I would discuss the latest things we had learned, and we often ended the day with a contest of alternately naming military rockets until there were none left to name.

The Crystal Radio

When Daddy told us about the crystal radio he built while growing up in Houston during the Great Depression, we asked if he could build one for us. He built the radio on a board the size of a textbook. A toilet paper tube coated with shiny shellac and wound neatly with orange insulated wire formed the coil. He used sandpaper to expose an arc of polished copper across the upper side of the coil. A copper strip that could be moved back and forth across the coil's exposed wire served as the tuner.

The most important ingredient was a half-inch section of copper tube containing a quartz crystal from my rock collection. The end of a safety pin mounted nearby was pressed against the crystal. We connected the coil to an antenna made from a wire strung across the garage. When we connected Daddy's old pilot's headphones to the radio, we could tune in several stations. We soon learned how to probe the surface of the crystal with the safety pin for the best results.

We were intrigued by the simplicity and neatness of this radio. It required no battery, for it was powered by radio waves. Even more amazing was the magical ability of the crystal to transform the fluctuations in the amplitude of the radio waves modulated by voice and music into audible sound.

That radio taught us about the connection of radio science and engineering. The word science comes from a Latin term that means "knowledge," and engineering is the design and building of things. That made the radio an engineered device that employed science to achieve a goal. This merger of science and engineering is often at the root of discovery.

My career has bounced between science and engineering and has often been a merger of both. For example, the instruments I have designed to measure the ozone, water vapor, and dust in the atmosphere are engineered. The processing of the data they collect is science. A merger of science and engineering occurs when the data are processed by a computer, an engineered device.

National Geographic Magazine

The arrival of a slick magazine with a yellow outline around the cover was the highlight of each month. The July 1954 issue of *National Geographic* was especially memorable for its pair of articles and sixty photographs about the conquest of Mount Everest by Sir Edmund Hillary and Tenzing Norgay. Sir Edmund became one of my boyhood heroes, and I never imagined that someday I would spend quality time with him.

That 1954 issue of *National Geographic* included another article that sharply boosted my interest in electronics: "New Miracles of the Telephone Age," by Robert L. Conly. I read that article repeatedly and stared at its photographs of transistors invented at Bell Laboratories and a miniature radio transmitter powered by a solar battery, another Bell Labs breakthrough. The article also included photos of Nike Ajax ground-to-air missiles being tested at the White Sands Proving Ground in New Mexico. I hoped to someday make transistorized electronic circuits powered by solar cells and become a rocket scientist at White Sands.

A series of events during the fall of 1956 further stimulated my interest in science, and the arrival of Hurricane Flossy on September 25 was especially momentous. Milo and I were impressed by the power of its wind, which sent trash cans rolling down the street. Even more interesting was how the air became perfectly still for twenty minutes when Flossy's eye passed directly over our house in Fort Walton Beach.

After the eye passed, Flossy's wind arrived from the opposite direction, and left me with a lifelong interest in hurricanes that culminated fifty years later, when I came to know Dr. Robert H. Simpson, for whom the Saffir-Simpson hurricane scale is half named. Simpson, a world-class hurricane scientist, also founded Hawai'i's Mauna Loa Observatory, about which I extensively interviewed him while writing *Hawai'i's Mauna Loa Observatory: Fifty Years of Monitoring the Atmosphere* (University of Hawai'i Press, 2011). We became friends, and he wrote the book's foreword.

The October 1956 cover of *Science and Mechanics* magazine featured a tiny radio worn on a man's wrist. The article inside, "Build Your Own Wrist Radio," provided instructions on how to make a one-transistor radio that worked much better than a crystal set. I dreamed of someday being able to afford the parts to build one. This goal was motivated in part by Judith Moore, a sixth-grade classmate, who dedicated the song "Blueberry Hill" to me on a local radio station, an event that excited my mother and sisters but embarrassed me. With a personal radio, I could listen in private out of the range of the rest of the family.

Two assemblies at Westwood Elementary School (now Edwins Elementary School) in Fort Walton Beach also stimulated my interest in science. During one assembly, a meteorologist explained the details of weather balloons. During another, a physicist demonstrated what happens to everyday objects when frozen by liquid nitrogen. It was amazing to see a flower dipped into the liquid shatter like glass when it was tapped against a table. Equally astonishing was how a frozen banana could be used to hammer a nail into a piece of wood.

Those school assemblies might have been organized by the school's elderly guidance counselor and his wife, who spoke with distinction and encouraged academic excellence. Maybe they were retired college professors. After the students were given IQ tests, they told my parents that Milo, Sherry, and I had unusually high scores. They recommended that Sherry be advanced from second to third grade, and our parents agreed. They did not agree that Milo should be advanced to fifth grade with me. I wondered why my supposedly high score did not simplify long division and the mysterious art of diagramming sentences.

The Piltdown Man Hoax

Another development at school was the introduction of human evolution. When the sixth graders presented the musical *The Frog Prince*, everyone knew that the transformation of a frog into a prince was a fairy tale. (Biochemist Dr. Duane Gish popularized this notion in his books and lectures.) But that is not what our science book taught. The book also taught that human evolution had been confirmed by the discovery of Piltdown Man in 1912. It even included a sketch of this putative ancestor that was part ape and part man. The teacher passed over this subject so quickly that she seemed to have doubts.

I was fascinated by fossils, and the fossil skull of Piltdown Man depicted in our textbook was so persuasive, it could not be denied. While I still believed in God, I decided He must have used evolution to create us. I did not yet know that in November 1953, the London *Star* had declared Piltdown Man "The Biggest Scientific Hoax of the Century!"

But I did know that our bus driver looked exactly like the Neanderthal man pictured in our textbook. He had the same prominent brow, broad nose, wide face, and scruffy hair. He was a walking humanoid fossil wearing a modern shirt, trousers, and shoes. How could a prehuman Neanderthal be such a careful driver? Why did he always welcome us human kids to his bus with a friendly smile and greeting instead of a grunt? How could he even exist when his ancestors had gone extinct many thousands of years ago?

A Fan of Invention

In December 1956, all seven of us, including my new baby sister, Eileen, piled into our blue 1954 Mercury station wagon and left Florida for Houston, where we were to stay while Daddy was assigned to Korea for a year.

My seventh-grade classes at Hamilton Junior High School (now Hamilton Middle School) were often interrupted by visions of science projects and inventions, one of which I remember far better than whatever was being taught at the time. One hot and humid afternoon, while I was seated in the third row at the right side of a classroom near a big rotating fan by the door, an idea for controlling the flight of a rocket without the use of movable vanes or fins flashed into my mind.

The idea was for the nose cone of a rocket to have a forward-facing hole to admit oncoming air. The air would then exit through a ring of holes around the base of the nose cone, and the rocket would fly a straight path. To steer the rocket, air would be allowed to exit through only one hole, and this would push the rocket in the opposite direction. I was not able to implement this scheme, which I later called ram-air control, but it stayed on my mind through high school and college and eventually played a leading role in fulfilling my dream to do serious science.

Milo and I often explored Buffalo Bayou, foolishly walked across a railroad trestle, and collected flattened copper disks made from pennies we placed on railroad tracks. I also tried making miniature rockets propelled by powder extracted from dud firecrackers. These experiments came to a halt, at least temporarily, when a homemade rocket exploded in my hand and degraded my hearing.

During my first semester in seventh grade back in Florida the previous fall, an older student known as Gator had showed me a radio that he made from a used hearing aid. I very much wanted to emulate Gator's accomplishment, but where could used hearing aids be found? The answer arrived during the summer of 1957, when Milo and I, along with our cousins Joe and Paul Mims, rode a city bus to downtown Houston to visit coin dealers.

When I noticed a hearing aid store, I split from the others to request used hearing aids. After several downtown trips, I accumulated a dozen or more used hearing aids, some of which were early transistor models. Unfortunately, I then lacked the experience to repair them, much less convert them into radios.

Historic space-related developments occurred during the fall of 1957, especially on October 4, when the Soviet Union launched *Sputnik 1*, the world's first artificial satellite. The International Geophysical Year had begun on July 1, and *Sputnik* stole the show from the many science projects sponsored by the sixty-seven participating nations. The *Sputnik* launch was widely covered by newspapers, and radio news programs played its "*. . . beep . . . beep . . . beep . . .*" signal over and over.

Sputnik triggered a huge spike in interest in science among youth in the US. While the 184-pound satellite was much too small to observe without a telescope, the rocket booster that carried *Sputnik* into space was also orbiting the earth and was sometimes visible with the naked eye during twilight. When a Houston overpass was announced, I went outside around 4:30 a.m. to try to get a glimpse of *Sputnik*'s booster. Soon, I spotted what appeared to be a star rising in the sky. It was the booster racing across the sky at the incomprehensible speed of 18,000 miles (almost 29,000 km) per hour. I still recall how sunlight reflected from the tumbling booster flickered as it passed more than 100 miles (161 km) overhead.

Three weeks later, another major event occurred in the sky over Houston. *Around the World in 80 Days*, named Best Picture of 1956 by the Academy of Motion Picture Arts and Sciences, featured a gas-filled balloon that was later flown across Houston. The balloon's 12-mile (19 km) flight path came within 2 miles of our house, and I had a good view of it as it slowly drifted across the sky.

Several days after the *Around the World in 80 Days* balloon flight, the Soviet Union again stunned the world, with its launch of *Sputnik 2*. This satellite was much larger and more sophisticated than *Sputnik 1*, and it carried a dog named Laika. On December 6, a US Navy attempt to launch a grapefruit-sized satellite aboard a Vanguard rocket failed spectacularly in a huge explosion shown on television and printed in newspapers and magazines. This failure infuriated Wernher von Braun, the German rocket scientist who was the technical director of the US Army Ballistic Missile Agency. Von Braun claimed he could launch a satellite in sixty days if given a go-ahead. On January 31, 1958, von Braun met his goal with the launch of *Explorer 1*.

There were legitimate fears that the Soviet Union was well ahead of the primitive US space program. Leading politicians and scholars urged the nation to vastly improve the science and math education of US students. As for me, I was more motivated than ever to build a transistor radio like the one on the cover of the October 1956 issue of *Science and Mechanics*.

My dream came close to fruition when Mama told the three oldest of us five children that we could each make a list of what we wanted for Christmas so long as it did not exceed $10 ($106.28 in today's dollars). I filled out the order form from a Lafayette Radio catalog with the parts needed to make a transistor radio, which Mom then mailed with a check. The list included a 1N34 diode, a ferrite antenna coil, a capacitor, and, best of all, an RR66 germanium transistor.

One afternoon a few weeks later, when Mama was away shopping, I snuck into the closet in her bedroom and found a box from Lafayette Radio. Inside were all the parts I had ordered, including the transistor, a silvery object the size of a small tooth from which three wires emerged. The 1954 *National Geographic* article contained only photographs of transistors; now I was holding the real thing. It was a moment never to be forgotten.

A few weeks later, Christmas arrived, and several days later, I was listening to Elvis Presley singing "Don't Be Cruel" and Buddy Holly singing "Peggy Sue" with my one-transistor radio and an earphone provided by a hearing aid dealer. Listening to those songs with that homemade radio was nearly as exciting as watching the flickering booster stage of the Soviet Union's *Sputnik 1*.

After Daddy returned from Korea in January 1958, new outdoor experiences awaited us in Colorado, where Daddy was assigned as a project officer for the US Air Force Academy's dramatic chapel. We were back in snow country, and we experienced hailstorms for the first time. Daddy took us on drives through the mountains and on camping and fishing trips, where Milo and I sometimes spent more time searching for snakes than catching fish. When Daddy volunteered to design a new building at Glen Eyrie, the estate of Colorado Springs's founder, Gen. William Jackson Palmer, he took Milo and me on a tour of the estate, which had been acquired by the Navigators, an evangelical Christian organization. Adjacent to the remarkable geological formations in the famous Garden of the Gods, it includes unique formations of its own.

I was enrolled in eighth grade in the spring semester of 1958 at North Junior High School (now North Middle School), exactly 1 mile from our Victorian-style, two-story house on North Nevada Avenue. Snowstorms sometimes slowed the 1-mile walk to school, but we Alaska veterans were fully prepared despite our time in Florida and Texas.

Colorado provided an opportunity to resume rocket experiments using newly developed model rockets designed for hobbyists. G. Harry Stine, a former range safety officer at White Sands Missile Range, cofounded Model Missiles Inc. in Denver with Orville Carlisle, who had developed the single-use Rock-A-Chute rocket engine.

During the fall of 1958, Daddy took me to a rocket presentation by representatives from Model Missiles. That Christmas, my main present was an Aerobee Hi rocket kit, which reached a measured altitude of 671 feet (205 m) when launched at a model-rocket club. This was one of the first launches organized by William S. Roe, who worked with Stine to organize the National Association of Rocketry. We later attended another launch, where John Roe, William's son, launched a rocket that carried a radio transmitter. I was intrigued by that transmitter, especially since I had yet to build one.

The rapid development of model rocketry and the flight of the transmitter-equipped rocket renewed my dream to implement the rocket-control mechanism inspired by the big fan back in the seventh grade. I did not have the money or the skills to begin the project, but I learned as much as I could about model rocketry, which was to eventually play a significant role in my budding science career.

The Language Translator

The June 1958 issue of *Radio-Electronics* magazine included "Mr. Math Analog Computer," an article by Forrest H. Frantz Sr. that described a very simple electronic computer that "adds, subtracts, divides and multiplies."[1] Frantz published a similar article in the July 1958 issue of *Popular Electronics*: "Compute—With Pots: How to Multiply, Divide, Add, and Subtract with Simple Potentiometer Circuits."[2]

Those two articles made a major impact on my thirteen-year-old mind, for they described how to do basic math with a battery, a voltmeter, and some variable

1. Forrest H. Frantz Sr., "Mr. Math Analog Computer," *Radio-Electronics* 29, no. 6 (June 1958), 52.
2. Forrest H. Frantz Sr., "Compute—With Pots: How to Multiply, Divide, Add, and Subtract with Simple Potentiometer Circuits," *Popular Electronics* 9, no. 1 (July 1958), 39–41, 123.

resistors like those used to adjust the volume of a radio. By 1958, anyone could build a transistor radio, but I did not know anyone who had built a computer. Moreover, a computer might enhance my limited math aptitude.

That fall, when I was a ninth grader, I entered my version of Mr. Math at the city science fair. Much more important than the tiny medal the project received was the inspiration it provided. Daddy was then taking an evening course in Russian. Instead of copying someone else's analog computer that could do simple math, why not devise something entirely original? An analog computer that could translate Russian into English was added to my dream of building a guided rocket controlled by ram-air.

The language-translator idea was based on the realization that variable resistors (potentiometers, or pots) that adjust the volume of most radios or TVs can be thought of as memory elements. After you adjust the volume, the control "remembers" where it was set even after the radio or TV is turned off.

Resistance is measured in units called ohms. At one end of its rotation, the volume control might be set to 0 ohms; at the opposite end of its rotation, it might have a resistance of 10,000 or 100,000 ohms. If a pot can be manually adjusted to 100 positions between 0 and 10,000 ohms, I reasoned, it can store up to 100 words to be translated.

I further reasoned that an array of pots, each set to the resistance representing a single word, could form an electronic dictionary. Each pot would be adjusted to a resistance that matched that of a word dialed letter by letter into a row of input pots connected in series. If, for example, dialing each letter of *hello* into the input pots gave a resistance of 1,200 ohms, the memory pot for *hello* would be set for 1,200 ohms. I built a primitive version of the language translator computer, but it was not the automatic machine I envisioned.

The US Air Force Academy

The language-translator project was placed on hold after we moved from Colorado Springs to the US Air Force Academy in June 1959. We lived adjacent to hundreds of acres of wilderness just waiting to be explored, and that is what Milo and I did nearly every day that summer. We investigated magpie nests up close, followed a great horned owl as it was chased through the trees by mobs of birds, and tried to see how close we could get to deer. I visited a beaver dam and studied its construction and the teeth marks on nearby stumps of the trees from which it was built. That fall, Milo and I began delivering the *Colorado Springs Free Press* along a route in Douglass Valley that included half the housing units at the academy. Every day, we arose at 4 a.m. to wait for the route manager to deliver a big bundle of papers. We rolled the papers, secured them with rubber bands, and loaded them into bags mounted on our bicycles.

My section of the paper route covered 4 miles (6 km), with a dozen up-and-down segments having an elevation change of 310 feet (94 m). While riding my bike during the first glow of twilight, I often saw meteors flash overhead. A particularly brilliant fireball left a trail of meteor smoke across the western sky. On an especially frigid day, at the base of a hill, I laid the bike in the snow and walked across yards to deliver papers. Someone had left their garden hose filled with water, which had frozen, and the hose shattered to bits when I stepped on it.

Riding our bikes before sunrise required a headlight powered by a generator pressed against the side of the front tire. The generator was not a problem when coasting down hills but was a major drag when riding uphill with a full load of papers. Learning how much physical labor is required to illuminate a low-power, 6-volt lamp made a lifetime impression, and I am still obsessive about switching off unused lamps in our house and my office.

After a year delivering papers, Milo and I calculated that we were earning 17 cents per hour ($1.79 in 2023) and decided it was time to quit. This provided more time for school activities and hikes through the woods, but our house was so crowded with all seven of us, there was little space for projects. Daddy gave Milo and me small spaces in the ground-floor workroom to build rockets, but there was no space for my project to build a language-translating computer.

One of the few technical things I learned was that dipping the exposed wires from a 120-volt power cord into a cup of lightly salted water creates a maelstrom of noisy bubbles. (Please do not try this.) One summer day, I made a solar cooker from a cardboard box lined with aluminum foil. When Daddy was building a speaker cabinet for his stereo system, Milo and I learned about woodworking as we watched him work while listening to *The Sound of Music* on his reel-to-reel tape recorder.

Teachers I still remember at Air Academy High School taught Latin, US history, algebra, geometry, chemistry, and biology. After reading one of my essays, Mr. Lamb, my English teacher, told me I should become a writer and loaned me a set of his science fiction novels. Colonel Duffner, a retired army officer, taught Latin, which still comes in handy when identifying clouds and classifying plants and animals. Chemistry teacher Miss Godfrey was interested in an ultraviolet source I made from an argon glow lamp. These and other teachers made their classes so interesting that I stopped daydreaming about rockets and computers and made a serious effort to do well in school. After the first semester of tenth grade, Principal John Asbury sent my parents a letter informing them I had made the honor roll.

On many summer days, I hiked through the woods to the swimming pool or the library at the academy's community center. I learned to dive at the pool, and at the library, I explored books on ancient history, religion, and evolution. Although we attended church every Sunday, the evolution and religion books raised doubts in my mind. But those doubts were themselves doubted when I read that the famous Piltdown Man I had studied about in sixth grade had been revealed as a forgery in 1953, three years before I was taught about this supposed link between apes and people.

Why did scientists need forty years to expose what in retrospect was so obviously a hoax? Especially troubling was why my sixth-grade science book did not include a correction slip at the page with the Piltdown Man fable. And I still wondered about the family tree of the kindly, Neanderthal-looking gentleman who had driven my Florida school bus.

Moving Back South

By the summer of 1961, Daddy's work on the academy's chapel was concluded, and he was assigned to the Air Command and Staff College at Maxwell Air Force Base, near Montgomery, Alabama. Our new home was Prattville, Alabama, where we moved into an old, roomy house with plenty of space for science experiments.

While the schools were poorer and less advanced than what we had left behind at the US Air Force Academy, the students at Autauga County High School (now Prattville High School), which Milo, Sherry, and I attended, extended a friendly welcome. Some of the girls were among the prettiest we had ever seen, and some of the teachers were as good as those at Air Academy High School. I still recall how D-Day veteran Mr. Platt made physics comprehensible, Miss Tankersly provided informative civics lectures, and Miss Davis told us, "Someday, y'all will thank me" for the vocabulary we learned from her daily twenty-word spelling tests.

Those were segregation days, and the civil rights movement had yet to achieve its goals. As in Florida, the schools we attended were for whites only. "KKKK" ("Knights of the Ku Klux Klan") was stenciled in big white letters across the main highway that entered Prattville. A shiny KKKK emblem was mounted on a steel post nearby. However, Prattville long ago abandoned those racist relics.

Though Daddy was born in Lufkin, Texas, and Mama in Alexandria, Louisiana, we were raised to avoid the prejudice exhibited by some of our friends and relatives. That is why Milo and I once annoyed the grandmother we called Nanny by drinking from a Colored Only water fountain where she worked at downtown Houston's Foley's Department Store.

The Russian-to-English Language-Translating Computer

Earning good grades back in Colorado had required hours of homework. But thanks to a daily study hall and few homework assignments, it was easy to earn As at the Alabama high school. This provided plenty of time to pursue various experiments and projects, especially the language-translating computer, which had been shelved since ninth grade.

That fall, I began sketching various ways to build the computer. The plan was to dial the letters of the word to be translated into a bank of six potentiometers, each of which could be rotated to select one of the twenty-six letters of the alphabet. These six input pots were connected in series like a string of Christmas lights. Thus, the total resistance of the six pots was the sum of the resistance of each. Since each of the six pots had twenty-six positions, the total possible combinations of letters was 308,915,776. (The number of words was less, due to coincidences in which the sum of the resistances was identical for different words.)

This plan would give each word a unique resistance to be compared with the resistances of the words stored in a memory bank made from twenty miniature pots called trimmer resistors. The resistance of each trimmer would be adjusted with a miniature screwdriver to match the resistance of a word dialed into the six input pots. In operation, the resistance of a word dialed into the six input pots would be sequentially compared with that of each of the twenty trimmers in the memory. If a match were found, the translated word would be indicated by a glowing red light on a panel of twenty lamps.

While the idea was simple, implementing it was not, and, during the fall of 1961, I spent considerable time completing the project. Back in Houston, a neighbor of my grandparents' Nanny and Pa had given me a sharp-looking black box he bought at a surplus sale that would be perfect for the computer. I began by sawing a 4-inch section from the top of the 18-inch high box. This was set aside to become the computer's readout panel.

The memory array of twenty trimmer resistors was installed on a thin sheet of wood mounted inside the box over the battery compartment. The resistance of a word dialed into the input pots was compared with the resistance of each of the twenty trimmers in the memory array by an electric music box mechanism mounted with all the other parts on a second panel inside the box. The rotating cylinder of the music mechanism was wrapped with black electrical tape. Four slots were cut in the tape so that only one of four metal fingers mounted in place of the music prongs would touch the bare metal of the cylinder as it rotated. If there was a match in the resistance of the input pots and a selected trimmer, a meter needle would move to zero.

Determining when the meter indicated zero posed a problem. The solution was to cement a tiny solar cell directly onto the face of the meter's zero position and then cement a sliver of aluminum foil onto the meter's needle. The glass over the meter readout was then replaced with a piece of black plastic in which I had cut a slot over the zero position. A flashlight bulb was mounted over the slot.

When the needle was indicating anything other than zero, light from the bulb illuminated the solar cell, which generated an electrical current that caused a transistor to activate a relay that turned on the electric music box. When a word match occurred, the needle moved to zero, the foil blocked the light from the solar cell, the relay switched off the music box motor, and a single red light on the readout panel glowed to indicate the translated word.

The readout panel was built inside the 4-inch section cut from the top of the box. Twenty flashlight bulbs were installed behind twenty holes bored in a square of black Formica that fit inside the open front of the readout box. A sheet of red plastic film was installed between the lamps and the front panel. A cable of twenty-one wires connected the lamps to the main processing box. The display looked sharp when the red lights were flashing on and off during scans.

After the computer was completed and the memory array was programmed with twenty words, Daddy drove the project, the poster, and me to the regional science fair at a school gymnasium in nearby Montgomery. In those days, students could not be interviewed by the science fair judges, so I made a list of operating instructions and leaned it against the computer. Parents and teachers could observe the judging from the bleachers, while the students had to wait in an adjoining room.

While I was optimistic the language translator would receive an award, it received only a certificate of participation, probably because the instruction list slipped behind the computer during the judging. Daddy later told me that judges who stopped by the project seemed interested in the translator but were unable to operate it.

Embarrassed by Neutrinos

While the poor performance at the science fair was a disappointment, even worse was an embarrassing incident that occurred in physics class one day. Mr. Platt, the physics teacher, was a cordial gentleman who doubled as the school photographer. During a lecture about subatomic particles, I raised my hand to report on a subatomic particle called the neutrino that could pass through the entire earth without being stopped.

Mr. Platt looked puzzled and politely assured the class and me that this would be impossible. The class knew about my interest in science, and I could see their sympathetic looks in response to what they perceived was my error. But the experience was so embarrassing that I did not bother to follow up with news reports about the discovery of neutrinos, which can still travel through the entire earth today. While that was the first time a statement of mine about science was challenged, it was certainly not the last, as we shall soon see.

After graduation from high school on May 30, 1962, I began working as a grocery sacker at the nearby Piggly Wiggly and tinkering with experiments that were considerably simpler and easier to implement than the language translator. At the Prattville swimming pool, I placed a microphone in a plastic bag, connected it to a battery-powered amplifier, and spoke to friends at the surface when I was under the water. At home, I installed an old car ignition coil and a 6-volt battery in a wood box to provide 20,000 volts, enough to produce a crackling arc of plasma nearly an inch long.

My First Discovery

Because a standard magnetic earphone could double as a microphone, I naively reasoned that a detector of light could also emit light, so I connected the spark coil across the two wire leads of a light-sensitive photocell used to trigger nightlights, switched on the power, and watched in amazement as the light-sensitive film on the surface of the photocell emitted bright green flashes. The flashes soon ended, and the film emitted a soft green glow. Since the cadmium sulfide that formed the light-sensitive material of the photocell was most sensitive to green light, I knew that this experiment might have significance, for it demonstrated that a single device could both emit and detect light. But I did not realize how much this simple experiment would impact my future in science.

Texas A&M University

After Daddy completed his one-year assignment at the Air Command and Staff College, the Air Force sent him to Texas A&M University, in College Station, Texas, so he could complete his degree in architecture. We moved to College Station in August 1962, where Daddy and I both enrolled at Texas A&M.

All first-year students had to be in the Texas A&M Corps of Cadets for at least two years. While I did not realize it back then, leadership lessons I learned in the corps have lasted a lifetime.

In those days, all students took the same courses during their first two years. I did reasonably well in English and history and made an A in a botany course. But I nearly failed first-year algebra with a grade of 69.5. Fortunately, the instructor rounded up my score to a minimum passing grade of 70.

While I did well in high school trigonometry and geometry, it was time to face up to my poor aptitude for algebra and change my major from physics to a field that did not require math. I told Daddy that governments need people who understand science, and my new goal would be to major in government while continuing to do science projects. I told him that someday I would earn enough to be able to hire a mathematician to assist with my science projects. Daddy was understandably skeptical, but he supported my decision. He also supported my plan to stay in the Corps of Cadets and graduate with a commission in the Air Force.

During my first three years at Texas A&M, there was little time for science experiments. The busy academic schedule and the military drills and activities left time only for working at a string of part-time jobs, including busing tables at Coach Norton's Pancake House ($0.50 per hour) and campus jobs as a library-book sorter ($0.75 per hour) and painter's helper ($0.90 per hour). The Corps of Cadets taught leadership, not humility, much of which I learned while cleaning dormitory toilets as a custodian ($0.85 per hour). Today, when I see a grim-faced janitor cleaning an airport restroom, I bring a smile to his face by letting him know I once did what he does.

Each of these jobs provided interesting experiences, the most memorable of which occurred while I was sweeping the theater stage at Texas A&M's Guion Hall. Suddenly, the heavy curtain fell across the stage inches behind where I was standing. I thought I was alone; apparently I was not.

Meeting Wernher von Braun and Willy Ley

During my sophomore year, Wernher von Braun and fellow German rocket scientist Willy Ley visited Texas A&M in February 1964 to speak at the Space Fiesta. There was a dark side to von Braun, for during World War II he had directed development of the German V-2 rocket.

After he surrendered to US forces in 1945, von Braun and dozens of his missile coworkers were taken to New Mexico's White Sands Proving Ground (now White Sands Missile Range) during Operation Paperclip. There, they supervised the assembly and launch of dozens of V-2s made from original parts captured in Germany. Later, von Braun's rocket team at the army's Redstone Arsenal in Alabama launched *Explorer 1*, the first US satellite, in January 1958. During my college years, von Braun led the development of the Saturn program, which culminated in the Apollo series of lunar landings.

I met both Von Braun and Ley during a reception, where they signed their coauthored book, *The Exploration of Mars*, for Daddy, who loaned me his copy in case I met its famous authors. von Braun was as handsome and charming as I had read in the press, and even today, I recall looking into the blue eyes that had peered at Adolf Hitler during the war when Von Braun briefed *der Führer* on progress with the V-2.

Crushing Rocks and Setting Dynamite

During the summer before my senior year at Texas A&M, I worked at Capitol Cement in San Antonio, where my parents lived. Every few hours, I lubricated the giant, roaring ball mills that ground limestone to powder. I also had rock-crusher duty. The supervisor showed me how to tie a thick rope around my waist and climb over a safety rail while carrying a heavy, noisy pneumatic hammer to break up large boulders stuck in the crusher. There was no safety helmet, and no one was present when I was splitting boulders.

When it came time to replace the fire bricks in the giant, rotating kiln, the flame, 25 feet (8 m) long, was shut off for several days to allow the kiln to cool enough for us to enter. Only one man at a time could stand under a crude steel shield while dislodging the heavy bricks, which bounced off the shield when they fell.

It was an exciting job, especially when I learned how to operate a bulldozer. During the first lesson, the supervisor directed me to level a patch of ground between two buildings. Suddenly, giant sparks began flying from the steel blade when it struck a buried electrical cable. Much safer was watching the explosives technician attach blasting caps to dynamite sticks that he dropped into rows of drilled holes before filling them with ammonium nitrate and diesel fuel.

White Sands, New Mexico

Near the end of summer break in 1965, I attended Reserve Officers Training Corps training at Biggs Air Force Base, near El Paso. The July 1954 issue of *National Geographic*, which had so influenced me in sixth grade, included photographs of Nike missiles being launched from White Sands, and I was excited to learn that we would witness two Nike launches. We were miles away from the launch site, but we could see the exhaust trails from the missiles.

The training also included several days of survival training in the desert north of Biggs, which featured a lecture on arctic survival by a sergeant drinking a cold soft drink from an ice-filled chest. This was distracting, to say the least, for we were required to find water on our own. Some of us placed slices cut from barrel cactus into handkerchiefs and squeezed out the liquid within. We blamed our headaches on the cactus juice, when they were probably due to dehydration.

Astronaut-Aquanaut Scott Carpenter

Thanks to the income from the cement company job, I began my senior year at Texas A&M in September 1965 without having to work to pay tuition and room-and-board expenses, which totaled $370 for the final spring semester ($3,932 in 2023). Not having to work for the first time during college provided time to resume electronics experimenting and become involved in various extracurricular activities, including serving on the Cadet Court.

Fellow corps member Thomas "Tommy" Tyree chaired Texas A&M's Great Issues Committee, which hosted presentations by visiting dignitaries. During my senior year, Tommy appointed me to escort Brenda Lee at one of her appearances at Texas A&M. *The Battalion*, the campus newspaper, described her as "the most popular female vocalist of the 1960s."

When Tommy appointed me to chair the Great Issues Speakers Series, I met a series of famous personalities, including journalist and author Harrison Salisbury and Scott Carpenter, the second US astronaut to orbit the earth. It was Carpenter who, serving as flight controller, said, "Godspeed, John Glenn" as his fellow astronaut was launched into orbit in 1962 atop an Atlas rocket propelled by 75,000 gallons of kerosene and liquid oxygen. As reported by the Associated Press, on learning of Carpenter's passing in 2013, Glenn issued a statement saying, "Godspeed, Scott Carpenter—Great Friend. You are missed."

During his keynote speech for the Hydrospace Fiesta, Carpenter spoke about "The World Ocean: Challenge to Man's Future." Being on the conference staff, I met Carpenter and asked whether I could ride with him to Houston after his speech. He consented, and there followed a two-hour opportunity to spend quality time with an authentic American hero.

What I remember most about that night was Carpenter's careful attention to the highway. He said he was glad I was along to help keep him awake, but he did not need me. He was highly observant as he switched his gaze to whatever caught his attention along the road ahead. His night-driving style very much reminded me of my jet-pilot father's.

Carpenter dropped me off in downtown Houston around midnight. Still wearing my Corps of Cadets uniform, I immediately began hitchhiking back to Texas A&M, where I arrived just in time for class the next morning with a memory I have never forgotten.

A Vietnam Hint

Senior Air Force ROTC cadets were taken on a field trip to Walker Air Force Base, near Roswell, New Mexico, where we were given a ride in a Boeing KC-135

Stratotanker aerial-refueling plane and allowed to manipulate the big refueling boom hanging out the rear of the aircraft. During a second trip, we visited Eglin Air Force Base, where my father had been assigned when I was in elementary school. We were shown some of the latest armaments and bombs being tested at Eglin, including inert bomblets in an open cluster bomblet unit (CBU).

A loaded CBU contains hundreds of orange-size bomblets, each of which is packed with a high-explosive charge designed to explode on impact and produce a deadly spray of steel balls. The CBU is designed to open at an altitude that would allow its bomblets to spread out over a uniform circle inside of which any vehicles would be damaged and any enemy forces killed or wounded. I wish I had known then what I learned about CBUs a year later in Vietnam.

Laser Hijinks and Talking Over a Beam of Light

Not having to work during my senior year provided time to pursue electronics projects and experiment with a helium-neon laser borrowed from Texas A&M's Physics Lab Center. Aggies enjoy practical jokes, so I decided to introduce some students to the remarkably intense beam of the laser by pointing it at notes students were making on a blackboard in another building a few hundred feet (60 m) away. Back then, lasers were only six years old, and very few people were aware of them, so those students always jumped back or even fell when the bright red dot followed their chalk or made patterns on the blackboard. Ron Schappaugh, commander of Squadron 9, to which I was assigned, enjoyed watching as I pointed the laser beam from the top floor of our dorm at a sidewalk along which people were walking. Near panic ensued as people jumped back or even ran from the bright red spot.

About this time, I devised a simple apparatus for automatically igniting the fuses of a string of fifty firecrackers at thirty-second intervals. When this gadget was concealed high in the branches of a tree, students passing by were startled when every half-minute, a firecracker would explode in midair as a group of us sat in chairs and watched their reactions. Soon, a small crowd gathered, and so did two campus police. Their questions were interrupted every half-minute by an explosion without a detectable source. Since we could honestly say we had not ignited those firecrackers, the puzzled police eventually left, but the crowd stayed.

That fall, I designed a system that would transmit voice and music over the bright beam of light from a 6-volt Burgess Dolphin lantern light, the kind with a handle on top. The receiver was a second lantern light with a pair of silicon solar cells installed in place of the lamp. (It still works, and it is resting on my desk as this is typed.) An inexpensive amplifier was installed inside the lamp housing along with a speaker and a battery.

One night, I parked my 1950 Buick along a dark rural road, connected a radio to the light transmitter, and pointed the beam up the road. I then walked along the dark road with the light receiver pointed back at the car and received the radio's voice and music nearly 1,000 feet (305 m) away. Later, I repeated this experiment with the modulated light source in my dormitory window and received the signal at the drill field across from the Memorial Student Center, 1,200 feet (366 m) away.

I next used an old hearing aid to make a tiny radio that clipped onto my uniform pocket. One day, I was walking across the campus when a newspaper reporter approached and asked what was attached to my uniform. She was quite surprised to learn it was a radio and asked if she could write a story about it. A few days later, a photo of me showing the radio to a Texas A&M secretary accompanied a front-page article about "possibly the smallest radio in the world."[3]

When I applied for a part-time job repairing hearing aids, I was immediately hired. The main problem was dirty on-off switch contacts and volume controls caused by perspiration. After I repaired several of the expensive aids during the first afternoon, the dealer politely informed me he no longer needed me. He explained that sales of new aids would be jeopardized if I repaired all the defective aids customers returned for service. I left the position with experience repairing the latest hearing aids and new knowledge about human nature.

I next prepared a proposal for a magazine article about how to make miniature radios from old hearing aids, which I sent to Larry Steckler at *Popular Mechanics*. Larry rejected the proposal in a kindly written note in which he explained that it would be difficult for readers to find used hearing aids. Years later, I wrote articles for Larry's *Radio-Electronics* magazine, and he made me editor of *Science Probe!*.

Soon afterward, a simple electronic gadget I purchased for $0.98 ($9.21 in 2023) from Bryan Radio and TV Service had a major impact on my electronic projects. The device was a small, two-transistor module amateur radio operators used to practice Morse code. When connected to a battery, a telegraph key, and a speaker, the module produced a tone each time the key was pressed.

I reasoned that the circuit could be made sensitive to light by replacing its single resistor with a light-sensitive photocell like the one I used in high school to produce green light when excited by a spark coil. I modified the circuit accordingly, and the result was a surprisingly sensitive audible light sensor.

3. Shary Brown, "Aggie Senior Builds Mim(s)iature Radio," *Bryan Daily Eagle*, November 1, 1965.

One evening, I asked some freshmen to switch off the lights in our dorm's first-floor hallway while I pointed the circuit toward them from the opposite end of the hall, 175 feet (53 m) away. The hall was dark, and the speaker was silent. I then asked the students to light a match, and, to everyone's amazement, the circuit responded with a rapid series of clicks. When they switched on the hall lights, the speaker emitted a high-pitched hum.

That circuit played the key role in a travel aid for the blind, my first major electronics project since building the language-translating computer three years before. Three years later, it led directly to the founding of MITS, the company that introduced the Altair 8800 microcomputer, which pioneered the personal computer era. It is a story worth telling.

2

THE TRAVEL AID FOR THE BLIND

Inspiration beyond the profit motive often plays a role in science and technology. It certainly did in the case of my next project, which was inspired by my blind great-grandfather, George Edgar Hardy Myer. He lost his sight in a black powder explosion on May 3, 1906, 13 miles (21 km) south of Sweetwater, Texas, while working on the track bed for the Kansas City, Mexico, and Orient Railway about 100 feet (30 m) north of mile marker 654. His watch stopped at 4:20 p.m.

Papaw, as we called him, was tall, with broad shoulders, rugged features, and bushy white hair. He was a humble, hardworking man who was well known and liked around Lufkin, Texas, where he was guided during the early years of his blindness by my grandmother Nanaw and a wooden cane when he sold peanuts on the street. His cane, which is leaning on my knee as this is typed, also accompanied him on daily walks along the rural road where he and my great-grandmother lived in a sparsely furnished house.

When I was around twelve, I joined Papaw on one of his walks and was amazed when he said he could hear the presence of power line poles as we walked by them. As I grew older, I sometimes wondered how modern electronics could supplement the clues provided by his cane.

Designing the Seeing Aid

During the fall of 1965, I began thinking about how to design an object detector for the blind. The idea was to send electrical pulses to a light source, such as a recently invented solid-state laser, which would project bright flashes of light in a narrow beam pointed away from the blind user. If the beam struck an object, some of the light would be reflected to a solar cell connected to a hearing aid amplifier. The blind person would be notified about the object by a tone from an earphone. This idea became the first entry in my *Notebook of Circuit Diagrams and Designs*, begun on December 6, 1965. The entry was titled "Object detector for blind."

During my last two years at Texas A&M, I often visited the electronics reading room in a long-since demolished building near the Cushing Library. The cover story for the November 15, 1965, issue of *Electronics* magazine was "What's New in Semiconductor Emitters and Sensors," by Dr. Edward L. Bonin and Dr. J. R. Biard of Texas Instruments (TI), which I read in early 1966, during the spring semester of my senior year. The article described various kinds of light-emitting diodes (LEDs) developed at TI, including powerful diodes that emitted at 900 nanometers (nm) in the near-infrared just beyond the rainbow of colors that comprise visible light, from 400 nm (violet) to 700 nm (red).[1]

I reasoned that the new diode would be an ideal light source for a travel aid for the blind, especially since the invisible light would not attract attention. On February 12, 1966, I sketched a more detailed outline for the "Object detector for blind" idea under the heading "Seeing Aid" on page seven of my notebook. The entry suggested: "Use light-emitting diode for source." I also made an appointment to visit Bonin at TI.

The Perils of Hitchhiking

Journalism major Larry Jerden and I were among a diminishing number of Aggies whose principal mode of transportation was hitchhiking, which led to some close calls and several dangerous experiences. I do not advise readers to try this means of travel.

One afternoon, I was hitchhiking near the approach to an old-fashioned steel girder bridge when no one would stop for me. When there was a gap in traffic, I began walking across the bridge, when cars suddenly began arriving from both directions. I immediately climbed up into the bridge girders to avoid being hit. A large truck in my lane suddenly slowed to a stop and the driver shouted for me to climb into his cab through the open window. I tossed my plastic briefcase through the window and climbed inside.

One night, a car slowed down to pick me up north of Corpus Christi. The car behind did not approve and started passing on the shoulder with me in his headlights. The first car realized what was happening and sped up to make way for the second car. Within a second, both cars flew by either side of me at high speed only inches away.

---

1. J. R. Biard and E. L. Bonin, "What's New in Semiconductor Emitters and Sensors," *Electronics* 38, no. 23 (November 1965), 98–104.

Then there was the night a driver decided he was in love, grabbed my left thigh, and turned onto a gravel road off Highway 6 just north of Hempstead. I was not hurt when I jumped from his car, but I had to hide in the woods along the road while he slowly drove back, looking for me.

Another driver tried the same trick while we were cruising along Highway 59 near Rosenberg. After he grabbed my leg, I opened the door, and he slowed down enough for me to bail out and roll down a grassy slope. The driver then stopped for three young men farther down the highway. When I shouted a warning, they also rolled down the slope. The four of us soon ended up in the back of a pickup that stopped for us.

When my ancient Buick developed major engine trouble, I sold it for $25 ($235 in 2023), and hitchhiking was the only way I would be able to visit Texas Instruments. Hitchhiking through Dallas during winter on the busy North Central Expressway to get to TI in Richardson was a scary experience back in 1966. It would be impossible today.

Texas Instruments

Bonin made the trip to Texas Instruments worthwhile when he reached into his desk drawer and pulled out a penlight that emitted a distinct red glow. That was the first time I had seen a visible LED. Bonin and another scientist then gave me a tour of their labs. They seemed surprised that I went to such trouble to reach TI, and that may help explain Bonin's generous offer to give me one of their very expensive PEX1201 infrared LEDs. His condition was that I must first send him the receiver that would detect a rapid series of infrared flashes from the LED in the proposed travel aid for the blind.

After a second scary hitchhiking experience along the North Central Expressway, I made it back to my dorm late that evening and stayed up all night making a miniature version of the lightwave receiver used to send voice and music across the Texas A&M campus. The receiver used a solar cell for the light detector and an amplifier salvaged from a hearing aid a dealer gave me in Houston in 1957. Two days after returning from TI, I sent the receiver to Bonin via special delivery.

I planned to use the $0.98 ($9.35 in 2023) code oscillator module from the light-detection experiments to send a stream of pulses to an LED, which would respond by emitting flashes of invisible infrared. Since I did not have an LED to connect to the module, I had no way of knowing whether the receiver would work properly. My concern was alleviated when Bonin responded in a letter dated February 9, 1966:

Dear Mr. Mims:

I received your light receiver, and it certainly worked as well as you described. . . . I will locate emitting diodes for you as soon as the device engineer returns from a trip. . . . When I collect these items I will send them to you, along with your receiver.

Sincerely,
Edward L. Bonin[2]

Building the Seeing Aid

On February 19, a package from Bonin arrived. Inside was my light receiver, a glass infrared filter, and three Texas Instruments PEX1201 infrared LEDs valued at $365 each ($3,432 in 2023 dollars). I immediately began experimenting with those diodes, and they soon expanded my electronics experience far beyond the high school language translator. I began by installing the pulse module, an on-off switch, and some batteries in a small plastic box with a phone jack. I then soldered wire leads to each of the two terminals of one of the new LEDs. The ends of the leads were soldered to a phone plug, which was inserted into the phone jack. Pressing the switch caused the LED to emit powerful pulses of infrared light.

When the pulsed LED system was held in one hand and the receiver in the other, the receiver's earphone emitted a distinct tone when the pulsing beam of infrared struck nearby objects. This development was much more exciting than anything being taught in my government classes, so I told fellow government major Lani Presswood about the project. Lani was a reporter for the *Battalion*, the Texas A&M newspaper, and he asked whether he could write a story about the project.

I visited the *Battalion* office, where Lani blindfolded me and a photographer took a sequence of four photographs of me using the travel aid to detect and walk around a table. Lani's article appeared on page one of the March 9, 1966, issue of the *Battalion* under the title "Aggie Edison Revealed—Invents Object Detector for Blind." As we shall soon see, Lani's article initiated what became my science career.

While I did not know him at the time, decades later, I learned that Jerry Cooper watched that first demonstration of the travel aid. Jerry edited the *Texas Aggie* magazine for thirty-one years and wrote several articles about my career.

By March 21, all the electronics were squeezed into a clear plastic box 3.5 inches (9 cm) long, 1 inch (2.5 cm) high, and 1 inch deep. The LED was installed in a small plastic

2. Edward L. Bonin, letter to author, February 9, 1966.

tube mounted on top of the box. A small lens inserted over the open end of the tube focused the invisible light from the LED into a narrow beam. I made a miniature copy of the two-transistor oscillator to send a rapid stream of pulses to the LED.

When the pulsating flashes of invisible light from the LED struck an object, some of the light was reflected back to the travel aid, where it was detected by a solar cell that transformed the pulsating light flashes into a series of electrical pulses. These pulses were amplified by a tiny hearing aid amplifier and sent to an earphone that converted them to an audible tone.

The device, which I named the Seeing Aid, could detect objects as far as 12 feet (4 m) away, which would give a blind user ample time to avoid an obstacle. The user would still need to use a cane, since the Seeing Aid could not detect holes or other such impediments near the ground. But it would help the blind person avoid overhanging branches and signs and even inform the user when to advance while standing in line.

It was time to test the Seeing Aid with blind subjects. Several agreed to assist, and all of them said the device had potential. I then filled nine pages in my notebook with detailed notes and sketches of various versions of the travel aid. On March 23, I replaced the light bulb in a Burgess Dolphin lantern light with one of the TI LEDs. One of the two-transistor pulse generators was installed inside the lamp's battery compartment. This setup allowed the Seeing Aid to detect objects up to 40 feet away.

Meanwhile, William B. Wood of the Central Intelligence Agency (CIA) visited Texas A&M when the *Battalion* article about the travel aid was published. The article mentioned my interest in intelligence as a career, and Wood arranged for an interview in which he asked whether I would like to work at the CIA after graduation. When I explained that I would soon be commissioned into the Air Force, he asked me to contact the CIA after completing my duty.

The Texas Medical Association Convention

The Texas Medical Association held its annual convention three weeks after the travel aid was completed, and on April 13, 1966, I hitchhiked to Austin with the device and a poster. I had never been to a convention beyond the symposiums held at Texas A&M, and knew nothing about fees for exhibitors. I just assumed the medical convention would be a science fair for adults, and for me it was. The organizers were so intrigued by the Seeing Aid that they placed a table at the main entrance for the device and the poster.

Dozens of physicians stopped to ask questions, and many more watched when I gave demonstrations. A newspaper reported, "Although a political science major at A&M, Mims's second interest obviously is 'science and inventing things.'"[6] The travel aid generated so much interest that I contacted several potential manufacturers, all hearing aid companies, and paid $81.60 ($767.28 in 2023) for a patent search. That was virtually all my savings.

While in Austin, I asked permission to try the Seeing Aid with students at the Texas School for the Blind. I hitchhiked there, and several students used the device with excellent results. This, and a reminder of my blind great-grandfather hearing utility poles as he walked, motivated me to continue the project. The reminder was a young blind boy who was finding his way along an empty hall by snapping his fingers as he walked. The classroom doors were kept open, and the absence of an echo meant he was passing an open door.

Rediscovering Two-Way Emission and Detection

Tinkering with the Texas Instruments infrared LEDs led to several experiments beyond the Seeing Aid project. Of special interest was that I found that some standard diodes and even transistors could both emit and detect infrared. A notebook entry dated March 14, 1966, described how a silicon solar cell was connected to the same pulse circuit used in the Seeing Aid. When this solar cell was placed near the solar cell in the light receiver, a tone could be heard.

This method worked when the solar cells were up to 6.5 inches (16.5 cm) apart. It also worked with various diodes used as both emitters and detectors, though over much shorter distances. I wrote, "This experiment reminiscent of 1962 or 1963 [when cadmium sulfide] CdS photoresistive device and spark coil produced bright green flashes . . ."

Since I had never heard of this application for silicon solar cells and diodes, I scheduled a meeting with a chemistry professor. He looked over my notes and politely suggested that the signal was being transmitted electromagnetically rather than as light.

This was an early example of the many times a professional academic or scientist has expressed polite skepticism about my do-it-yourself research. In this case, those experiments demonstrated that a solar cell could double as both a sensor and an emitter of infrared. I ended the notebook entry by noting that these experiments were reminiscent of the 1962 high school experiment in which a cadmium sulfide light sensor emitted bright green flashes of light when connected to a high voltage.

Until then, my lab notebook consisted of standard ruled pages in a three-ring binder. On April 2, I began entering notes into a thick spiral notebook. The first

---

3. "New Device Helps the Blind," *San Antonio Light*, April 15, 1966, p. 2.

seven pages were copied from the previous loose pages in the binder, and new material followed.

The dreaded but mandatory Graduate Record Examination, for which I was woefully unprepared, was scheduled for April 15. It is impossible to study for this exam, and the day before, I continued my do-it-yourself science degree plan by entering into the new notebook ideas for a miniature laser diode communicator, a simple laser radar, a gamma-ray maser, and a combined laser radar and communication system for military use. The GRE the following day yielded predictably mediocre results, though I was surprised that the math score was much higher than the verbal and government scores.

Texas A&M Officials React

Meanwhile, newspaper stories about the travel aid and the miniature radio attracted the interest of Texas A&M officials, including Andrew D. Suttle Jr., the school's vice president for research, who asked me to meet with him. Suttle was an agreeable fellow with a big smile and a friendly disposition. He asked lots of questions and expressed serious interest in my projects. But he was puzzled and even concerned about why I was majoring in government instead of electrical engineering.

I explained my weakness in math, told him that I was working with the very latest LED technology, and asked whether any of this research might qualify for an alternate degree. He politely said maybe someday, but not then. After our meeting, Suttle sent me a paper titled "A Direct Translation Reading Aid for the Blind," to which he had attached a typed memo: "I thought you might wish to look at an interesting article related to your present work . . . ."[4] The paper was interesting, but even more so was the fact that Suttle thought enough of my research to send it. Soon, he would play a key role in my science career; meanwhile, he suggested I meet with renowned physicist Dr. Clarence Zener, dean of the College of Science and inventor of the Zener diode.

Zener was a famous theoretical physicist known for his practical semiconductor inventions, so I was surprised to find him in an office instead of a laboratory. Zener was not impressed by my projects. Nevertheless, he suggested that I reconsider my major, which would mean staying on at A&M beyond graduation.

Meaningless (to Me) Equations

The meetings with Suttle and Zener caused me to rethink the academic strategy I had told Daddy about four years before, so I visited Bolton Hall, the electrical

4. Andrew D. Suttle Jr., memo to Forrest M. Mims, April 14, 1966.

engineering building, to see a sample of the formal electronics education I had missed. I entered the old brick building, turned right, and noticed an open door on the left side of the hall. I looked in and saw a scene I shall never forget: Two dozen students were tinkering with vacuum tube circuits while seated at several rows of workbenches. A large blackboard across the front of the room to my left was completely covered with mathematical equations that were meaningless to me.

That brief stop by Bolton Hall was a key turning point in my budding electronics career. I left the building reassured about my strategy, returned to my dorm room, and happily resumed experimenting with transistors, silicon solar cells, and state-of-the-art infrared LEDs while worrying about the future of those electrical engineering students being taught yesterday's technology. I was so confident about my decision that I sent Texas Instruments another request for a PEX1201 infrared LED. On April 22, 1966, Elwin Aston sent two more of those state-of-the-art LEDs, which were described on a "Sample Request" form as "electrical rejects" to be sent by air ASAP. While they might not have met TI's strict standards, they more than met mine.

Extracurricular Science

Shortly before graduation, Daddy cosigned the note for a new 1966 Chevy Malibu, and I was able to return to Dallas without having to hitchhike. I again met with scientists and engineers at Texas Instruments, and I met William Holt, president of Varo Inc. Varo manufactured handheld infrared image converters that transformed a scene illuminated by infrared into a visible image. An image converter would be the ideal tool for experiments with the Seeing Aid, which I eventually began calling a travel aid, for it would allow the invisible beam to be easily adjusted and viewed during tests. Holt gave me one of the expensive image converters for the project, which I still use today.

Holt also gave me a professional laboratory notebook with alternate yellow pages that provide a carbon copy of whatever was written or drawn on the preceding white page. He suggested I keep abundant notes and that notebook entries did not have to be as neat as those in the spiral notebook I showed him. I followed the former advice but not always the latter. Keeping neatly drawn entries in that notebook turned out to be an especially important decision years later, when an editor at Radio Shack saw one of my notebooks.

3

VIETNAM

On May 28, 1966, I graduated from Texas A&M and the next day was commissioned into the Air Force as a second lieutenant. A week later, orders arrived for me to report to the Armed Forces Air Intelligence Training Center, at Lowry Air Force Base, Colorado.

The primary role of the AFAITC was to train military personnel to interpret images acquired by airborne cameras, infrared sensors, and radar. The photographic images came from cameras flown from a variety of aircraft, including the high-flying Lockheed U-2. We learned to carefully inspect aerial images to find and measure the dimensions of everything from huts and oil drums to aircraft, landing strips, power plants, and factories.

Some of the training materials and images we studied were classified "Secret," including U-2 images of Soviet missiles installed in Cuba four years earlier. We would need a security clearance to view these images and for whatever assignment we received after completing the school. Soon, the Air Force Office of Special Investigations went to work conducting background investigations on those of us who lacked a clearance. Six weeks after arriving at the intelligence school, a document titled "Record of Personnel Security Clearance" arrived. It stated that I had been assigned a clearance of **TOP SECRET**.

Rocket Projects

At about that time, I resumed my spare-time experimentation, this time with a rocket equipped with a pulsating infrared LED. Both the LED and the circuit that caused it to emit flashes of invisible infrared were nearly identical to those in the travel aid for the blind—with one major change: The frequency of the pulses was altered by the intensity of sunlight detected by a cadmium sulfide photocell. The idea was to use my light receiver to pick up an infrared signal from a rocket in flight. Eight test flights were conducted on August 3 and 4, 1966, but the pulsing signal from the LED was briefly detected during only two flights.

My next project was a miniature device that would allow a blind person to determine the position of the needle on a panel meter. That circuit and the one that flashed the infrared LED in the rocket project were both based on the two-transistor code practice oscillator circuit I used to pulse the LED in the travel aid. I built two of the meter readers, one of which included a tiny hearing aid earphone and was the size of a jelly bean.

I also built a prototype of a urinalysis analyzer for blind diabetics that attracted serious interest from a petroleum engineer who was looking for a method of detecting oil in shale. I told him that the urinalysis circuit could detect changes in the reflectance of rock through which a drill was moving, which I confirmed with a sample of crude oil he provided. I then sketched in my notebook an "Oil Shale Detector" that would transmit the signal to the surface through the drill pipe.

Meanwhile, I began work on the guided-rocket project I had conceived during seventh grade. I built and launched several model rocket with primitive control systems based on the principle I called ram-air control: Air entering a forward-facing hole in the rocket's nose cone rotated a homemade copper turbine that directed the incoming air out the side of the rocket's nose through a series of open ports along the base of the guidance section. The air was ejected through one port at a time in rapid sequence, and this, I reasoned, should cause the rockets to follow a tight helical path during its ascent.

The smoke trail left by these rockets clearly showed the expected corkscrew-like patterns, which I photographed using a Kodak Instamatic camera. The goal for the guidance system was to steer the rocket toward the sun. This would require a brake that would stop the turbine from spinning when a small solar cell that looked through the forward-facing port lost direct sunlight.

One afternoon, I visited the University of Colorado Boulder to acquire some cadmium sulfide crystals to repeat the high school experiment with my spark coil. Back at the apartment, I mounted the crystals either on aluminum foil, on a glass microscope-slide cover slip, or inside a glass tube with a wire contact at one end and a mercury contact at the other.

When stimulated by 20,000 volts from the high school spark coil, all the translucent yellow crystals emitted green flashes reminiscent of the cadmium sulfide photocell light-sensor experiment four years earlier. I then repeated the high school experiment with a discarded cadmium sulfide photocell from a crashed rocket and recorded the results in my notebook on November 6, 1966: "Results astounding! Complete crystalline face [of photocell] turns green with moving, wandering spots and flashes." William Kroes, my Air Force roommate, witnessed some of the experiments and signed the page.

During the visit to the university, one of the staff allowed me to photograph the beam of a helium-neon red laser. I placed my $11 ($100 in 2023) Kodak Instamatic camera directly in the beam and captured a spectacular starburst image that became a cover photo for *Science Digest* after I became a freelance writer. The magazine paid $400 ($3,761 in 2023) for the photo and an accompanying article I wrote about lasers.

Mysterious Orders to Vietnam

Halfway through the twenty-eight-week air intelligence course, Daddy called from Randolph Air Force Base, in San Antonio, to say he would be flying a Convair T-29C to Lowry Air Force Base in a few days. While I was visiting with him at base operations on September 15, 1966, his copilot, a captain, asked, "Where would you like to be assigned?" I replied, "Vietnam." He said, "OK." Several weeks later, I was surprised to receive orders to report to Travis Air Force Base, in California, on January 14, 1967, to catch a military charter flight to Vietnam, where I was being assigned to an air intelligence unit.

The staff at the intelligence school informed me there must have been a mistake: The orders would require them to release me from the course six weeks early, and Air Force policy at the time did not allow second lieutenants in Vietnam. They made a series of phone calls to the Air Force assignments headquarters at Randolph AFB and were informed that the orders had been properly issued.

What the intelligence school staff and I did not know was that Daddy's copilot worked in the assignments HQ. What I did not realize until reviewing Daddy's flight records five decades later was that his eight-hour flight that day was his last before he was reassigned to Vietnam. His influence on my interest in nature, radio, electricity, and model rocketry helped lead to the projects in which I developed the travel aid for the blind and the guided rocket. Now, he was playing a behind-the-scenes role in my new military career.

The orders to Vietnam provided one week to return to San Antonio, place my car in storage, and conduct one more launch of the ram-air control test rocket. Daddy was back in Vietnam on his second assignment, so Mama and Nanny watched the flight after driving me to the launch site. I then had two days to pack for the trip to Saigon.

Because I planned to continue experimenting during off-duty hours in Vietnam, selecting what to pack left little room for clothing. Airport security in those days was minimal, so no one at the San Antonio or Travis AFB's airports noticed that I had packed 40 pounds of electronics, batteries, rockets, and rocket motors and only one spare uniform.

Tan Son Nhut Air Force Base

On January 14, 1967, I walked off the airplane at Tan Son Nhut Air Force Base, near Saigon, and saluted the colonel waiting for me. It was Daddy, and he was soon to end his second tour in Vietnam. Daddy explained the basics of living in Saigon and let me stay at his place near his office in Saigon while I was searching for a room near the base.

My job was to supervise the Immediate Exploitation Branch—Out Country (North Vietnam and Laos), a crew of eighteen photo interpreters (PIs) on the night shift of the 13th Reconnaissance Technical Squadron (RTS). Our building, which was just 60 feet (18 m) from 7th Air Force headquarters, was air-conditioned to protect the expensive film processing equipment and the film itself from excessive humidity.

Most of the PIs had positive attitudes and at least some college education, and all of them were surprisingly good at analyzing negative imagery on rolls of film spooled across light tables. Each night, our shift analyzed thousands of feet of imagery before the film was shipped elsewhere for detailed analysis. We scanned around 3 million feet (914 km) each month using equipment that was primitive by today's standards but the best technology could then provide. The aircraft cameras were high quality, but they used conventional film instead of modern video sensors, and this added a few hours of processing time before we could examine the film.

Our assignment was to find and identify new military targets, which were designated as "Hot Items," provide preliminary assessments of damage from bombing runs, and plot the geographic coordinates covered by the imagery. Our final product was called an Immediate Photo Interpretation Report (IPIR).

Operation Rolling Thunder was in effect, and the air war was very active. January was the driest month of North Vietnam's dry season, and US Air Force and US Navy planes were flying hundreds of missions over North Vietnam and Laos. Some of the air attacks were dramatically successful, but I was surprised by the number of strikes required to take out most of the significant targets. Laser-guided bombs were just being developed, and very few conventional bombs made direct hits on fragged (planned) targets.

There was less work for the PIs during the wet season from May to September, when bombers and reconnaissance aircraft flew fewer missions. During one slow night, I found a large portfolio of disturbing photographs of Vietnamese civilians who had been mutilated and dismembered by the Vietcong. The next morning, I asked a superior officer why the photos were classified and not allowed to be published. He said that the higher-ups had decided they were too graphic for the public.

The Colonel's SAM Site

North Vietnam was heavily defended by antiaircraft guns and Russian surface-to-air missiles (SAMs), the highest-priority targets. On April 1, 1967, the afternoon shift created major excitement when they reported that a suspect SAM site had been discovered just north of the demilitarized zone (DMZ) between North Vietnam and South Vietnam. Dozens of aircraft were assigned to strike the target the next morning.

When my team arrived that night, I asked Airman First Class Beardsley, one of our most experienced photo interpreters, to carefully inspect the suspect SAM site. Instead of camouflaged missiles lying flat on launchers, Beardsley saw what appeared to be trenches. He was right: When viewed at an angle, the trenches appeared to be raised above the ground like camouflaged SAM launchers, but that was an optical illusion.

My supervisor, Capt. Jack Lundgard, was sleeping at his apartment, and I broke curfew to ride my motorbike to his place after midnight and advise him what we had discovered. He agreed that the morning air strike should be canceled, but we had no authority to do so. Lundgard told me to prepare a morning briefing about the new finding for him and his boss, Maj. Harold Clement.

Clement agreed with our findings when he saw the evidence we had prepared, but the new commander did not. The previous afternoon, he had persuaded the generals that the April Fools' Day trenches were SAMs. He said he had no intention of trying to change the minds of the 7th Air Force generals at the top of our chain of command. When we showed him fuel drums we had discovered inside one of the trenches, which proved they were trenches and not camouflaged missiles, he said SAM transporter trucks need fuel. As he spoke, Clement was counting the fuel drums aloud while staring at them through a stereoscope.

Later that morning and the following two mornings, the suspect SAM site was pulverized by some 200 craters left behind by 750-pound bombs. Eventually, new reconnaissance film and pilot reports confirmed the absence of any missiles. Yet we were ordered by our commander to continue reporting the target as a suspect SAM site until his flawed campaign became obvious to everyone.

A sidelight to this sad episode was that *Lover 01*, a Republic F-105 piloted by Capt. John Dramesi, was shot down by ground fire while flying to a secondary target after bombing the suspect SAM site the first day. His aircraft was the 500th to be shot down over North Vietnam. Dramesi became the only US prisoner of war to twice escape his captors and for this was nearly tortured to death.

The fiasco surrounding the suspect SAM site was only one of several troubling incidents. Another sprang from our responsibility to immediately report hot items to the 7th Air Force Command Post, the unit that assigned targets for the next day's missions. When one of the PIs found a significant hot item during our night shift, my job was to verify it, order enlarged photographic prints, and send a report and prints to the command post as soon as possible.

This procedure greatly annoyed the new commander, who became quite angry when his morning intelligence briefings to the generals were scooped during the night—the generals had already seen the night's hot items and sometimes had already fragged missions to strike them.

The 7th Air Force headquarters had ordered the photo interpreter in charge to "report in person to the 7AF Command Post within fifteen minutes . . . to deal with unexpected development of lucrative targets." The new commander overruled this higher-level requirement by ordering us to hold all hot items until he arrived in the morning, when it might be too late to schedule a prompt attack against a newly found significant target. When I suggested that 7th Air Force Ops (Operations) should at least be informed when we found significant enemy targets during the night, the colonel angrily replied—and these were his exact words—"Ops is the enemy, you fool!"

While our colonel's absurd declaration certainly surprised us, he was simply confirming his role in the intense rivalry that had developed behind the scenes between the Air Force generals in charge of intelligence and operations.

We reluctantly followed the colonel's order to withhold information from the command post until one night, a young officer became so concerned about an especially serious hot item that he walked the report to the command post on his own without waiting for the colonel to arrive the next morning. That officer was me. Though I expected a lecture from the colonel, nothing happened.

The rivalry in which our commander was engaged was reminiscent of the much higher-level rivalries between the Joint Chiefs of Staff and the civilians in Washington who decided when and which major targets could be struck: There was universal disdain among the intelligence people with whom I worked for President Lyndon B. Johnson's micromanagement of the air war against North Vietnam.

I learned about another kind of rivalry when I was sent to Phù Cát Air Force Base to supervise a crew of US Army photo interpreters. Before arriving at Phù Cát, I had not fully appreciated the professionalism of the Air Force photo interpreters I supervised. The army photo interpreters were poorly disciplined draftees who resented working under an Air Force lieutenant, much less the airman who accompanied me, who was better trained and much more intelligent than the draftees. A sharp Australian officer assigned to the Phù Cát reconnaissance unit more than once attempted to diplomatically resolve these interservice tensions.

These wartime rivalries are included here since they illustrate how human nature and narcissistic personalities can destroy cooperation, damage morale, and lead to unexpected consequences. I have since learned that scientists and those who rely on their theories, models, and advice are not immune to such behavior.

Malfunctioning Canister Bombs

Unexpected consequences also result from wartime science and engineering that lead to the development of newer and more deadly munitions. This was especially evident when newly developed air-to-ground missiles failed to perform properly, thereby wasting missions that subjected aircraft and their crews to the most concentrated antiaircraft fire in history.

A related problem were the times when antipersonnel cluster bombs (CBUs) of the type I had first seen at Eglin Air Force Base failed to work properly. As explained earlier, a CBU is a bomb-shaped canister packed with hundreds of 1-pound bomblets the size of a baseball. CBUs were designed to open well above the ground so that the bomblets would spread over a large circular area and detonate on impact.

Aerial photos of exploding CBUs revealed that some of the canisters were opening prematurely. As a result, instead of distributing themselves in the desired circular pattern, the bomblets were spreading into a halo around the pattern's perimeter.

One day, a sequence of aerial photos showed that a CBU dropped over an antiaircraft artillery site had deposited a halo of bomblets that detonated across a nearby cluster of houses and an empty field instead of the desired solid circle over the AAA site. The pilot had risked his life in a wasted attempt to take out an AAA site that could down his aircraft and planes in the main flight of bombers behind him. Moreover, the defective bomb had likely killed or wounded noncombatants.

While majoring in government at Texas A&M, I often engaged in vigorous debates with fellow students and professors. We posted essays on bulletin boards that stated our positions on candidates for public office, Supreme Court decisions, and so forth. Other students would pencil in their pro or con comments. In Vietnam, I naively failed to realize that such debates were frowned on by the military when I prepared a report titled *Conduct of the War*.

The report cited four specific examples of issues I suggested were harming the progress of the war, including the debacle regarding the suspect SAM site, the failure of some CBUs, and the withholding of hot items until the commander's arrival. The lieutenant colonel and the major who stood between the commander and me in the chain of command shared my frustrations, and after they read the report, both privately told me its points were valid. But they both said I was on my own should I send it to a higher level, and that is what I decided to do on July 21, 1967.

My father had once faced a major decision when he was a project officer for the US Air Force Academy chapel when he discovered serious errors in the dimensions of many of the structure's aluminum panels. He knew water leaks would occur if the panels were installed. But his supervisor's only concern was his impending retirement, and he refused to act on the panel discovery. When Daddy persisted, his supervisor stormed into his office with a pistol and shot a hole through his desk.

Surely, nothing like that would happen to me, especially since one of my extra duties was to write articles about our air intelligence findings for the 7th Air Force's contribution to the *Weekly Report of the Secretary of Defense to the President*. And I had not volunteered for Vietnam duty to acquiesce in the falsification of target reports, withholding of intelligence, and use of defective ordnance. I simply assumed that my report would be taken seriously if it were sent to the 7th Air Force's inspector general.

I was right. The report was taken seriously, and I was quickly summoned to the office of the cigar-smoking general who was the director of intelligence for the 7th Air Force. He ordered me to stand at attention against a wall while he loudly lectured me about the lowly role of a second lieutenant in a combat zone. But I had been well prepared for being shouted at during my time in the Corps of Cadets at Texas A&M, and the general's lecture failed to intimidate me.

I even had to suppress a smile when an elderly Vietnamese woman sweeping the floor *swish-swished* her way between the shouting general and me. While the general was loudly lecturing, I was wondering why a Vietnamese civilian was allowed to be so near the many top-secret charts, maps, and papers in his office.

I soon filed a formal suggestion about the CBU problem and on August 24 received "Suggestion Acknowledgment TAN-34-68" that "careful consideration would be given" to my suggestions for correcting the problem. Later, I was asked for more details by a colonel at Eglin AFB assigned to resolve the issue of malfunctioning CBUs.

When I was transferred back to my old job at Tan Son Nhut before returning to the States, I learned that high priority had been given to reporting malfunctioning CBUs. While the premature opening of CBUs was eventually solved, back then I did not know that many cluster bomblets failed to explode on impact and that enormous numbers of them remain scattered across Vietnam and Laos half a century after they were dropped. They have killed or maimed thousands of farmers and children, and the US has helped fund a program to find and dispose of unexploded cluster bomblets and other ordnance.

Despite the negative aspects of the Vietnam assignment, I was impressed by the pilots who risked their lives flying unarmed reconnaissance aircraft over heavily defended targets. And I justified my support of the war after meeting several Vietnamese people whose families had to flee North Vietnam when the communists took over and began persecuting and even killing Catholics and others. The thick folder of photos showing corpses of civilians slaughtered by the Vietcong also kept me loyal to what eventually became a hopeless cause back home.

I found the work challenging, and I especially enjoyed working with the two sergeants on the photo-interpretation team, including Sergeant Payne, who had previously held the rank of major and had served as a navigator on Boeing B-47 bombers. When B-47s were phased out, Payne had stayed on active duty at a significant reduction in rank. When facing important decisions, I often turned to him for advice.

As for the photo interpreters who spent long hours inspecting film through magnifying loupes and stereo viewers, I was their paymaster. Each month, I reported to the disbursing station, where I counted $82,223 in military payment certificates and signed a pay-agent agreement that I assumed "pecuniary responsibility" for the money. They then placed the money in an empty ammunition box and handed me a belt with a holster that contained a .45 pistol. I then had to walk several hundred feet back to the 13th RTS while being teased by every officer along the way about what I would do if they or someone else tried to steal the cash-filled ammo box.

On paydays, my sergeants set up a pair of tables, and I sat at one end and doled out the cash to each airman as he passed by. Then the sergeants took out barracks fees and advances the men had received. Some left with little or no cash.

What I remember best after all these years is that I can still hear one of them calling out, "Lieutenant Mims, I've found something unusual. Can you take a look?"

Off-Duty Science in Vietnam

Nine days after arriving in Vietnam, I received the color slides of the two flights of the ram-air rocket launched twice near Denver. Several slides of the smoke trail of the second flight, which reached an altitude of 1,400 feet (427 m), showed the bumpy pattern expected for a rocket following a helical path. The same rocket had been launched again from near San Antonio on January 12, 1967, two days before I had departed for Saigon. The smoke trail from this flight also showed distinctive jerks that also indicated that the ram-air control idea was properly working.

More research flights were needed, so during off-duty time in Vietnam, I continued the rocket experiments. The goal was to concentrate on two categories of experiments based on those previously conducted. Type-one experiments employed rockets with slightly tilted fins that caused the rockets to spin in flight. These rockets had no turbine or electronics. They simply had a forward-facing opening that led to a single output port at the base of the nose cone. Because a sideways force of ejected air from the top of the rocket was present during the entire upward flight, the rocket should follow a helical path and leave behind a smoke trail that resembled a corkscrew.

As with previous flights, the expectation was that the smoke trail would hang in the air a few seconds. This would be long enough for it to be photographed with the new Pentax Spotmatic 35mm SLR camera like Daddy's that I bought through the Base Exchange at Tan Son Nhut. Photographs of these corkscrew patterns would allow estimates of the magnitude of the sideways motion of the rocket that resulted from different nose-cone openings.

Type-two experiments involved a major change from the turbine guidance mechanism that scanned the sky for the target (the sun) while diverting the incoming air out each of seven holes around the nose of the rocket. The turbine idea was attractive because it needed no power and weighed less than half an ounce. But I was unable to design a brake to stop its rotation when the target was lost, thereby providing a corrective force from only one of the air output ports.

It would be easier, so I thought, to design a brake for a miniature electric motor that could be controlled with a radio transmitter. Ace Radio Control, a mail-order model-aircraft company in the States, sold a miniature electric brake designed to slow or stop the rotation of the wheels on a radio-controlled plane, and I ordered one and a miniature motor from Ace.

The Gunship Incident

After preparing a set of type-one rockets, I made a box from scrap wood, painted it red, loaded the launch gear in the box, and strapped it to the back of my Mobylette motorbike for each of several trips to the abandoned Phú Thọ racetrack, which served as my launch site. The brightly painted red-and-yellow rockets were too long to fit in the box, so I mounted them on top. Saigon residents launched rockets of their own while celebrating the Lunar New Year, but it never occurred to me that they were unfamiliar with large model rockets, a conclusion I arrived at after noticing many concerned-looking stares as I rode 3 miles to the racetrack.

The rocket launches at the racetrack always attracted an audience, including young men and boys who were glad to serve as the recovery crew. One of the young men would press the firing switch so I could photograph both the launch and the spiral smoke trail.

The third visit to the racetrack was on April 24, 1967. Five young Vietnamese boys watched and assisted while I launched and photographed a series of four type-one spin rockets. While I was preparing the fifth flight, a US Army UH-1 helicopter gunship arrived and landed around 1,200 feet (366 m) away near the north end of the racetrack with its rotor blades still spinning. After a minute or so, I assumed that the chopper crew had dropped in to watch my experiments and proceeded with the launch. The rocket reached an altitude of around 750 feet (229 m) before arcing over and ejecting its parachute. The five boys ran out to retrieve the bright yellow rocket as it floated down under its parachute.

Just as the boys returned, the green gunship suddenly rose over a plume of orange dust and sped toward us at an altitude of around 50 feet (15 m). When it was several hundred feet (100 m) away, the chopper descended and began a gradual turn around us. That is when I saw the helmeted and goggled door gunner staring straight at me over the barrel of an M-60 machine gun. I immediately began waving my arms while shouting at the boys to wave. The boys could not hear me over the gunship's deafening clatter, but when I pulled off my shirt and began waving it, they got the message and began jumping and waving.

During what seemed like forever, the gunner kept his M-60 trained on me from 75 feet (23 m) away as the gunship slowly turned between us and the south end of the racetrack while we, the rocket equipment, and my new Pentax camera were pelted with stinging clouds of sand.

I wondered why the gunner did not shoot. Maybe one of the gunship's crew was familiar with model rockets and knew we were harmless. Or maybe we were saved because the boys were so young. In the end, after a noisy, dusty half minute, the gunship's tail lifted, and the chopper rapidly flew off toward the southwest.

After the gunship cleared the racetrack, it turned north and headed straight to Tan Son Nhut AFB. The Vietnamese boys pointed at the departing gunship while chattering excitedly with one another. I had no idea what they were saying, but their body language and nervous laughter meant fear in any language. I was thankful that the older teenagers and young men who usually watched the rocket launches had not made an appearance that day.

The next day, I reported the gunship encounter to the flight controllers at the air base, who told me that the Vietcong had once used the racetrack to launch rockets at the base. They also told me to inform them before conducting future rocket launches. Four days later, I devoted a page in my notebook to the five flights that day and included a paragraph about the gunship incident and a diagram of the racetrack with the chopper's path marked with blue arrows. Later, there was good news when the photos of the rocket's smoke trails arrived from the processor, and I reported this in my notebook: "Photo evaluation of 24 April launches confirms spiral tracks of spin-stabilized rockets. Very significant."

Unfortunately, a detailed analysis of the smoke trails in the photos was made difficult by their poor contrast against the hazy, cloudy sky. Another method was needed to measure the course changes caused by the ram-air principle, and the best option would be to launch the rockets at night. The dark sky would provide ideal contrast with the rocket's flame trail, which could be photographed by using the Pentax camera in its time-exposure mode.

"The Base Is Under Rocket Attack!"

A month after the gunship incident at the racetrack, the night launches began on my night off from the roof of the apartment building where I lived. The first flight was by one of the ram-air test rockets that had been launched from the racetrack on April 24. A small light was attached to the side of the rocket, which I hoped would be captured as a string of dots by the camera as the rocket spun in flight. I taped a reward notice with a map to my place on the side of the rocket. After I installed the rocket engine's igniter and slipped the rocket onto the launch rod, I applied power to the light by twisting the two wires between the light bulb and a battery installed inside the rocket body. I then stepped back, opened the camera shutter, and pressed the firing switch.

The rocket rose into the dark sky atop a bright yellow flame trailing a string of orange sparks. After the engine burned out, the tracking light was visible as a series of flashing dots across the sky caused by the rotation of the rocket. But suddenly, bright, crisscrossing beams from searchlights near the main gate of the air base, which was only about 500 feet (152 m) from my apartment, swept across the roofs of my place and nearby buildings. I quickly grabbed the camera and launching equipment, rushed downstairs to my room, and stashed everything in my footlocker. Because I had seen the general direction in which the rocket had parachuted down, I raced downstairs and ran to the adjacent street to look for the rocket. It was well after curfew, and I was dressed only in a swimsuit and flip flops, so I became concerned when a Jeep with four heavily armed US military police arrived.

"The base is under rocket attack!" the helmeted driver shouted. "Get off the street!"

Before I could explain what happened, they sped off to join other heavily armed vehicles also looking for the enemy. I quickly returned to my room, where I remembered a map with my address was taped to the rocket. It was difficult to sleep knowing that at any moment, armed soldiers with my rocket in hand would be pounding on the door. Future night flights were impossible, and I decided to abandon the rocket project for the time being.

The notebook entry for May 22, 1967, records that while the flight was "spectacular" and the light flashes were captured by the camera, the time exposed photograph was difficult to evaluate. My notebook was usually filled with details, but, to avoid confessing an attack on the base, I included nothing more.

Saigon Schools for Blind Boys and Girls

For several weeks, the rocket project was placed on hold while I shifted to testing the travel aid for the blind at two schools for the blind in Saigon. A Vietnamese woman who worked for the US Army sometimes served as an interpreter during these visits, which continued to affirm that the travel aid could indeed supplement a cane in assisting blind children and adults.

These trials caught the attention of Air Force publicity people, who visited the Saigon School for Blind Girls with me and wrote a story for the military newspaper *Stars and Stripes*. This attracted the attention of Black Star, a photojournalism agency that sent Robert Ellison to photograph one of my sessions at the Saigon School for Blind Boys. Ellison, who was my age, expressed a genuine interest in the travel-aid project and even allowed some of the blind boys to climb all over him and handle his expensive camera gear. He later sent me a selection of the photos he took.

A CBS television reporter asked to film one of the test sessions with blind children. We met at the School for Blind Girls on June 5, and the camera operator exposed 1,200 feet (366 m) of film. To demonstrate how the travel aid worked, we connected the camera's microphone to the travel aid while I pointed it at a picket fence. When I walked along the fence, the aid responded with a distinct beep each time its beam was reflected by one of the fence slats.

While the travel aid impressed all those who used it or saw it in action, I was frustrated by the lack of funds to improve the device and make it small enough to mount on eyeglass frames. A week after the session with CBS, I wrote about the interview in my lab notebook and concluded, "[The] problem with this project is no money. Can't go further without money!" I also wrote about starting a formal paper about the project that I planned to submit to the *Journal of the American Optometric Society*.

Back to the Rocket Project

Plenty of affordable supplies for the rocket project were available from Ace Radio Control, so I returned to the rocket project, and on June 14, an improved version of the sunlight-seeking rocket was completed. This was a type-two rocket with a miniature electric motor and gearbox to slow the rotation of the motor and a modified air deflector. Three days later, I completed the rocket, which would need to be launched from the racetrack.

This rocket was larger than any of the previous ones, and I was concerned about the attention it would attract while taking it to the racetrack with the motorbike. At about this time, however, an Air Force public relations team asked to photograph one of my rocket launches. The timing was perfect.

Like the small type-one test rockets, the new and much larger rocket had an air-intake port at the tip of its nose cone and a single outlet port at the nose cone's base. A radio control receiver from Ace Radio Control was connected to a mechanical actuator that could open or close the outlet port to cause controlled movements of the rocket during flight. To carry the weight of the electronics, this rocket was powered by a cluster of three rocket engines.

The ground-control system included a radio-control transmitter activated by a homemade switch that detected holes punched in a portable tape recorder's magnetic tape that passed by the switch. Each time a hole in the tape triggered the switch, a signal was sent to the rocket to open and close the outlet port. The idea was to send a series of commands to cause the nose of the rocket to be pushed to one side by the air ejected from the outlet port. These movements would cause obvious changes in the rocket's smoke trail, which would be photographed.

An Air Force photographer drove the rocket gear and me to the Phú Thọ racetrack in a Jeep, so I did not have to worry about transporting the rocket on my motorbike. The photographer then took pictures of the rocket and the control electronics before the launch.

One of his photographs was published in *Stars and Stripes* and years later was posted on my Wikipedia page. The radio-control system and tape recorder worked as designed, and a series of programmed on-off pulses of air ejected from the side of the rocket's nose left distinct signatures in the rocket's smoke trail that I photographed. No armed helicopters arrived during this approved test flight.

Farewell to Vietnam and Robert Ellison

I left Saigon on January 11, 1968, and the Tet Offensive began on January 30, when large-scale attacks by both North Vietnamese and Vietcong forces caught the US military and the South Vietnamese off guard. Major battles occurred at the racetrack, where I had launched rockets.

Robert Ellison, who had photographed the travel aid being used by students at the Saigon School for Blind Boys a year earlier, was among the combat photographers who recorded the Tet Offensive. I remembered this twenty-three-year-old man for his friendly, laughing response to the blind boys who were clustered around him, handling his camera gear, and for his genuine interest in the travel-aid project. I did not know that he was later considered among the bravest and most influential combat photographers of the Vietnam War for his stunning images of the siege of Khe Sanh published in *Newsweek*.

The Khe Sanh photos were Ellison's last. On March 6, 1968, he and more than 30 US marines and airmen were killed when the C-123 carrying them was shot down at Khe Sanh. Even though the North Vietnamese suffered severe losses and US forces eventually recaptured lost cities and territory, the Tet Offensive and Ellison's final photographs marked a major turning point for the war in the eyes of the American public.

Returning Home

Daddy had been assigned to NATO in Oslo, Norway, and I planned to spend two weeks there with my family before returning to the States. I arranged for visas to reach Europe on the daily embassy flight that made stops in India, Pakistan, Saudi Arabia, and Spain. After spending some tourist time in Madrid, I took a commercial flight to Oslo. I was the only man on the flight not wearing a suit and tie.

Traveling from tropical Vietnam to 3 feet (1 m) of snow in Norway was reminiscent of moving from Texas to Alaska and from Florida to Colorado. After ten relaxing days with my family unwinding in snowy Oslo, I took a ferry to Copenhagen, rode a train to Germany, and caught a military flight back to the US. During a brief stop in Missouri, I had dinner with some college students who had no idea what was going on in a war halfway around the world. All I could think about as I ate pizza with them was the scheduling of reconnaissance missions, the analysis of film, the reporting of hot items, and the pilots who never made it back.

Upon returning to San Antonio, I realized that I had flown around the world, a fact that came in handy twenty years later, when I was accused of being a flat-earther. I picked up my 1966 Chevy at the storage garage and retrieved papers, patent searches, and rocket electronics from a safe deposit box. I then headed for the Air Force Weapons Laboratory (AFWL) at Kirtland Air Force Base, in Albuquerque, New Mexico.

4

THE AIR FORCE WEAPONS LABORATORY

Since visiting Carlsbad Caverns as a ten-year-old, reading about the White Sands Proving Ground in *National Geographic*, and watching Nike missile launches from White Sands during desert-survival training, I had always been fascinated by New Mexico. I spent the first night on the way to Albuquerque at a Travelodge in Roswell, where Robert Goddard had moved to conduct his pioneering liquid-fuel rocket experiments during the 1930s.

I visited the Roswell museum that features Goddard's workshop and his rocket launch tower at the front entrance. I left Roswell the next morning fantasizing that someday, I would marry a pretty New Mexican girl and fulfill my childhood dream of working at White Sands Missile Range. On February 11, 1968, I reported for duty at the Air Force Weapons Laboratory.

The Weapons Lab was known as a blue-suit lab, because most of the scientists and staff were Air Force officers required to hold an advanced degree in science or engineering. How I was assigned there with only a bachelor's degree in government requires an explanation.

Shortly before graduating from Texas A&M, I had showed school official Andrew Suttle the infrared viewer provided by William Holt at Varo. I suggested that it had important military applications, including assisting with the rescue of downed pilots equipped with invisible infrared sources. Suttle was highly interested in this idea and said it could be immediately placed in use rescuing downed pilots. Suttle knew Dr. William McMillan Jr., science adviser to Gen. William Westmoreland, commander of the US Military Assistance Command in Vietnam. He sent him a letter about my science background and the pilot-rescue idea.

After I arrived in Vietnam, McMillan contacted me and asked me to provide periodic science briefings for him and his staff. During the first briefing session,

McMillan was intrigued by my projects and asked me for my ideas about several specific military issues. McMillan also said he wanted me to meet an Air Force colonel from the Defense Intelligence Agency who was in Vietnam looking for technical problems that might be solvable by the laboratory he directed.

I arrived at McMillan's office on February 12 and was introduced to Col. David R. Jones, a bright, enthusiastic officer who looked through my notebook, carefully inspected my travel-aid and rocket electronics, and asked many technical questions. He then asked if I would like to be assigned to the Air Force Weapons Laboratory.

The Weapons Lab would provide me with an unexpected opportunity to do serious science at one of the military's most prestigious laboratories. Albuquerque is 150 miles (241 km) north of White Sands, where I had imagined I would someday work, and that was close enough. When I agreed, Jones said he would have me immediately transferred to the lab. When I replied that I wanted to complete a year in Vietnam, he reluctantly agreed and printed his name and both his work and home addresses and phone numbers on page 32 of my lab notebook.

Later, after learning that I lacked the academic credentials for a Weapons Lab assignment, I had second thoughts, so I requested and received an extension of my Vietnam duty from twelve to eighteen months. When Jones learned of this, he wrote:

> *I believe your motives for wanting to stay over in Vietnam for another year are thoroughly admirable. However, I think it is about time that you realized you have a much larger role to play, one that can only be satisfied in an R&D [research and development] environment. Accordingly, I have asked the Air Force Personnel Center to set up procedures whereby you can be rotated back to the United States immediately. . . . What I have in mind for you is an assignment to the Air Force Weapons Laboratory.*[1]

Though Jones eventually relented and allowed me to stay in Vietnam for the original full year, he canceled the six-month extension that had been granted. To make sure I was qualified to be assigned to the Weapons Lab, he had my Air Force Specialty Code changed from 8041 Photo-Radar Intelligence Officer to 2891A Engineer, Specialist (Technical Intelligence).

One of the unique features of the Weapons Lab was how newly arriving officers were required to tour the entire lab before being assigned to one of its branches. After completing this weeklong process, the incoming officer listed the top three choices for where he would prefer to be assigned. Jones then reviewed the choices

---

1. David R. Jones, letter to Forrest M. Mims, March 1967.

and made the final decision. Officers usually received their first choice.

My first choice was the Solid State Group in the Effects Branch, and that is where I was assigned. The Effects Branch had been established under the lab's Research Division to study the electromagnetic pulse effects of a nuclear explosion (EMP) on the electronics in military aircraft. The Solid State Group pursued several major projects using powerful glass and ruby lasers. A parallel group was responsible for developing exotic high-energy lasers that might eventually be able to shoot down planes and missiles.

Jones knew that the Solid State Group would be my first choice, for he had interviewed me when I arrived, and Maj. John Edgar of the Effects Branch was assigned as my reporting officer. Nevertheless, Jones wanted me to tour the entire lab. Most of the scientists I met during the tour were curious about my Vietnam experience, for I was the only person at the lab who had been to Vietnam.

Minnie Chavez

The Weapons Lab tour provided an introduction to Minnie Chavez, an especially attractive secretary in the Bioastronautics Group of the Biophysics Division. The introduction required planning, for Minnie was busy typing while I was being interviewed by her boss. When I walked by later, she was looking at her hands after placing carbon paper between some pages. I reasoned she would soon visit the restroom to wash her hands, and that is what she did. I was waiting by the water fountain and introduced myself when she arrived, for I knew she was the girl in my Roswell premonition when I first saw her.

I drove Minnie to Sandia Crest the evening of our first date. While headed east on Route 66 near the fairgrounds, she asked where I was from. When I replied Texas, she said, "Oh, a Tejano!" When she told me about her background, she said she had been raised Baptist like me. Her grandfather was an Apache who had ridden with Pancho Villa as a young man. After becoming disenchanted with Mexico, he had taken his wife and children to New Mexico, locally known as the Land of Enchantment, where they became US citizens and active members of Roswell's Iglesia Bautista El Calvario (Calvary Baptist Church).

From Sandia Crest, the view of Albuquerque far below was almost as spectacular as Minnie. The Roswell premonition did not reveal that the girl I would marry would be from the same city, but Roswell was where Minnie was born and raised. On June 15, Minnie and I were married by her childhood pastor, Andres Viera, and we have been loyal partners and best friends ever since.

Captain Roger G. Mark, MD, PhD

In Vietnam, Colonel Jones had told me I would be working with exceptionally talented, highly educated scientists at the Weapons Lab. I was concerned they would look down on me because of my lack of qualifications, but that did not happen when I was assigned to report to Capt. Roger G. Mark, who held a doctorate in electrical engineering from the Massachusetts Institute of Technology and a medical degree from Harvard Medical School.

Roger's project was an elaborate experiment he had designed to answer important questions raised by the new generation of powerful lasers. For example, laser-guided bombs and laser distance-measuring devices were being developed. What if a pilot's eyes were accidentally struck by the beam from a friendly laser or intentionally by the beam of an enemy laser? Could a pilot temporarily blinded by a laser continue to fly a high-performance aircraft? Trained primates had preceded astronauts into space, and in Roger's experiment, monkeys trained to respond to shapes on a screen would substitute for pilots.

Having spent two years designing travel aids for the blind and testing them with many blind subjects, I was troubled by an experiment that exposed the eye of a trained monkey to a laser beam. But I was also concerned about the fate of pilots who might be exposed to a laser beam. I rationalized that the monkeys were being used for a greater good.

The rhesus monkeys slated to become subjects for Roger's project were being trained at the Aeromedical Research Laboratory, at Holloman Air Force Base, 150 miles (241 km) south of Albuquerque. The program was under the direction of Dr. Donald N. Farrer, a civilian research psychologist. At the Weapons Lab, Roger was working on the laser and the complex optics and electronics he designed to temporarily blind the monkeys. Over the next several months, I learned far more from Roger about planning and implementing a sophisticated experiment than from all my prior projects.

My Vietnam assignments had often required twelve-hour days. That was good preparation for working with Roger, who often worked long hours with few breaks. His attitude was consistently cheerful, and he always took time to answer technical questions. When I did not know how to align the experiment's laser, he showed me how to tweak its mirrors. When I asked why an expensive pellicle beam splitter was more desirable than a glass plate, he explained that a pellicle is so ultrathin that it reflects what amounts to a single image instead of the double image reflected from the front and back surfaces of a glass plate. As I later wrote in my book *Siliconnections: Coming of Age in the Electronic Era*, "It was difficult to believe that the hardworking, good-natured guy with a screwdriver in his pocket and sawdust in his hair was as close to being a genius as anyone I'd ever known. Roger Mark was certainly not the aloof, ivory-tower type one might have expected."[2]

Ollie Westfall, a civil service technician, was building the plywood chamber in which the experiment would be installed. My first assignment was to spend several days assisting Ollie in constructing the test chamber. As I described him in *Siliconnections*:

> *Those concerned about the rapid pace of technology had little to fear from Ollie, for years of work at the Bureau of Standards had instilled in him a deep respect for the principle of inertia in Newton's second law of motion: the rate of change in the velocity of a body is directly proportional to the force exerted on it. Though Ollie seemed to walk, talk, and think in slow motion, no one could criticize his skills as a machinist, model maker, glassblower, scrounger of hard-to-find parts, and general-purpose lab technician. So long as time was not of the essence, Ollie could single-handedly construct from scratch virtually anything mechanical Roger or I asked him to make. Though Ollie's manner was maddeningly methodical, he was not lazy. Shrouded in his soiled lab coat, gray hair sprouting from his head like springs from an old sofa, Ollie could almost always be found muttering under his breath as he melted solder onto the end of a copper tube, blew molten glass into a glistening sphere, or bent over a whining lathe.*[3]

Ollie was not a genius like Roger, but neither was I. Even worse, my college transcript did not list a single course in electronics, physics, or calculus, and I had never visited a science lab, much less worked in one. My most sophisticated electronic equipment was the $65 ($611 in 2023) oscilloscope I had assembled from a Heathkit in Vietnam, and I was both impressed and intimidated by the expensive Tektronix scopes in Roger's lab. Somehow, Roger managed to make use of me while we busily prepared the monkey test chamber and its accompanying laser and electronic systems.

Weapons Lab Tales

Most of the Weapons Lab staff were friendly, and some had an unusual sense of humor. One airman told me how he and a colleague had lured a third airman to walk through a darkened lab; they had then pointed the powerful green beam of an argon laser at him and set his lab coat on fire. Then there was the Weapons Lab

2. Forrest M. Mims, *Siliconnections: Coming of Age in the Electronic Era* (New York: McGraw-Hill), 72.
3. *Siliconnections*, 72-73.

picnic Minnie and I attended where I caught a large bull snake. When I showed the crowd, a panicked sergeant backed away and loudly proclaimed he would kill me if I took one step closer. Ordinarily, a death threat against an officer would ruin a sergeant's career, but he was genuinely terrified by the snake, so I took it away and released it.

My first major challenge at the Weapons Lab occurred when Roger assigned me to build a monostable multivibrator, an electronic circuit that delivers an electrical pulse of a specific duration when a switch is pressed. Today, that circuit seems simple, but back then, it was much more complex than the circuit that pulsed the LED in the travel aid for the blind. While I quickly learned to align and operate the project's laser, I was genuinely concerned that I might be unable to build the multivibrator.

After finding a suitable design in my *General Electric Transistor Manual*, I walked through the aisles of the Weapons Lab stockroom to select the parts that were needed and placed them on the checkout table for their values to be totaled by the sergeant in charge. The total was around $120 ($1,128 in 2023) for a handful of parts that could have been purchased from Radio Shack for under $15. When I asked why the parts were so expensive, the sergeant explained that they were all MIL-SPEC, meaning they were designed to meet demanding military specifications. But the monkey project would be operated in dry, air-conditioned facilities that did not require MIL-SPEC parts. It seemed to me that the stockroom should either offer both MIL-SPEC and standard parts or allow us to open an account at a nearby Radio Shack.

Nothing like Roger's experimental apparatus had ever been designed. It had reached a level of sophistication worthy of what scientists designate as elegant, and it worked perfectly. While my role was only that of a basic technician, I had learned much about the design and implementation of a complex experiment. Two very busy months after I arrived at the Weapons Lab, Roger and I delivered the monkey chamber and associated equipment to the Aeromedical Research Laboratory at Holloman AFB for testing with the rhesus monkeys being trained by Farrer's staff.

Our travel orders described us as "cargo couriers," and our trip was aboard a Douglas C-54 piloted by Major Edgar. The flight was uneventful until some excitement occurred as we passed over the north end of the White Sands Missile Range. Suddenly, a North American F-100 fighter jet appeared only a hundred feet (30 m) off the right wing of our aircraft, and the pilot waved his wings up and down and motioned with his arm. We had inadvertently entered restricted air space, probably due to an impending missile test. Edgar quickly banked the plane to the east and continued south toward Holloman AFB.

Our trips to the ARL fulfilled the second half of the premonition about marrying a New Mexico girl, which was that I would someday work at White Sands Missile Range. Holloman AFB is adjacent to the missile range and its dunes of glistening gypsum.

The ARL was famous for its high-speed rocket sled and high-altitude balloon flights, and for training chimpanzees for the early program in manned space flight. The ARL's desert location seemed out of place for its Asian rhesus monkeys and its colony of 200 chimpanzees imported from Equatorial Africa.

There, some chimps were busily shoving levers and poking push-button switches they were trained to activate in response to various lights, shapes, and other stimuli. The associated equipment stored the results of whatever the chimps were trained to do and triggered an apparatus that dropped a food pellet into a feeding cup as a reward for correct responses.

The sound of falling food pellets and clicking relays and solenoids in the electronics racks made the busy animals seem even busier. All this activity contrasted with the animal handlers, who seemed bored as they waited for their animals to complete their test session.

When I asked about the ethical protocols under which the lab operated, the staff assured me that the monkeys were used only for serious studies having a direct impact on human safety. Their protocols for both animal subjects and human volunteers can be seen today on YouTube in a 1967 official Air Force film about the ARL, which begins with a disclaimer: "The animals used in this study were handled in accordance with the 'Guide for Laboratory Animal Facilities and Care' as promulgated by the National Academy of Sciences—National Research Council."[4]

This film shows the spotless, neatly organized facilities and its chimps much as I remember the ARL. All that is missing is a row of cages housing chimps that some of the technicians encouraged visitors to check out before the daily cleaning of their cages. When those unsuspecting visitors walked within a few yards of those cages, they were subjected to a barrage of unmentionables thrown at them through the bars by laughing chimps.

The First Monkey Experiment

The chattering monkeys used in Roger's experiment required three months of patient training at the ARL before being moved to a special trailer equipped with cages parked behind our lab at the Weapons Lab. The 7-pound animals looked cute, but they did not take kindly to being handled. To avoid bites and scratches, their handlers wore thick gloves and exercised great care when removing a monkey

---

4. *Pioneers of the Vertical Frontier: Aeromedical Research Laboratory (1967)*, YouTube, https://www.youtube.com/watch?v=Pq_C26JG6FQ.

from its cage and gently restraining it in the plastic-and-steel seat inside Roger's experimental enclosure.

Once in the chair, the monkeys were ready to work. Their task was to stare at a screen and pull a lever on the right side of the chair when a triangle appeared. If a square appeared, they were supposed to pull a lever on the left side of the chair. A correct response earned the monkey a food pellet, which is why 95 percent of responses were correct. They responded to the test as fast as the symbols on the screen were changed, but they did not work for free. As Roger wrote in his summary report, "Subjects worked for about 250 food pellets per day."

Roger's apparatus included several major subsections. The most crucial was a carefully designed arrangement of optics that permitted only the left or right eye of the monkey to see the symbol screen through a hollow tube. This section also included an infrared light source and a photodiode that detected when infrared from the light was reflected from the retina of a monkey's eye when it was staring directly at the target projected on the screen. The second major section was a camera equipped with an infrared-sensitive plate that photographed the subject's eye at the moment the laser was fired. The third major subsection was a powerful neodymium-doped glass laser system.

All this was coordinated by a clever circuit Roger designed around two early integrated circuits called comparators. When a Fire switch was manually pressed, the two ICs and several additional parts waited for the signal generated by the photodiode when it detected infrared reflected from the eye of the monkey. When the eye was exactly centered in the viewing tube, the trigger circuit automatically triggered the camera shutter, waited until it was fully open, and fired the laser.

After the monkeys were allowed time to become used to us and to working inside the chamber, the experiment was begun. While Roger watched the monkey at work through a television link and I made sure the laser was properly aligned and ready to fire, one of the test monkey's eyes was struck by a laser pulse of near-infrared lasting only one hundred billionths of a second. Since the laser was fired only when the animal was staring directly at the target on the screen, the laser beam was focused precisely on the retina's center of vision.

At low energy levels, the energy in the focused laser beam caused no visible damage to the monkey's retina. This made it impossible to determine whether the system was successfully striking the region of central vision called the macula. When the energy level was increased, the laser triggered a small blob of blood from a tiny wound called a hemorrhagic lesion. This pinpointed the exact location of the laser beam on the retina. While the blood would eventually dissipate, the objective of the experiment was to determine whether the animal could continue to identify the objects on the screen immediately after exposure to the laser. Most of the subjects immediately stopped working for a minute or so after the laser exposure. They then began using their unexposed eye and continued working.

This created an unanticipated problem, for the primary goal of the experiment was to discover the ability of the monkeys to perform immediately after being exposed to various levels of laser light, and the delays after the laser exposures were unacceptable. It was reluctantly decided to replace the positive reward of food pellets with a mild electrical shock that would be delivered when the monkeys refused to work. This new protocol would require a fourteen-month period to train a new batch of monkeys.

Other Weapons Lab Projects

While the ARL staff was training the rhesus monkeys, there was plenty of time to pursue other projects. In March 1968, Roger and I visited Fort Belvoir, Virginia, and laboratories of IBM in New York and RCA in New Jersey to learn as much as possible about semiconductor lasers. We then joined a contingent from the Weapons Lab's Laser Division that flew aboard an Air Force transport piloted by two of our officers to the 1968 Quantum Electronics Conference in Miami. The conference included presentations by some of the world's leading laser scientists.

Some of the Weapons Lab scientists were beginning to earn reputations for their pioneering high-energy laser research, and they were warned not to talk to the Soviet delegation. The problem with this order was that the Soviet attendees included Dr. Nicolay G. Basov, who had shared the 1964 Nobel Prize in Physics for his pioneering work that helped lay the groundwork for the invention of the laser. I wondered how our experts could avoid chatting with him. They could not, and they quickly formed a cluster around the famous Nobel Laureate.

Most of the reports presented at the Quantum Electronics Conference were well above my level of understanding, and a talk by a leading laser expert during a presentation Roger and I attended completely baffled me. I did not understand a word he said. I felt much better as we were walking out of the auditorium when Roger exclaimed, "I didn't understand a word he said!"

In August, I visited Dr. Farrer at the ARL to discuss the monkey project and then traveled with Col. David R. Jones, commander of the Weapons Lab, on a visit to Fort Belvoir and the Pentagon to discuss semiconductor lasers and observe a field test. Jones was well known in the Pentagon. Several generals and colonels spotted him while we were walking along a corridor and stopped to chat. In September, I was sent to the Liquid Crystal Institute, at Kent State University in Ohio, to discuss various ways to use heat-sensitive liquid crystal screens for thermal-vision devices.

They were also painting women's breasts with liquid crystals in an experiment to detect breast cancer.

The trips to learn more about semiconductor laser research culminated in November 1968 when I visited IBM in Gaithersburg, Maryland, to see the high-power semiconductor laser system for which I was the contracting officer. The laser was designed to illuminate objects a considerable distance away with a powerful, invisible beam of near-infrared light that could be observed through a night-vision scope. Back at the Weapons Lab, Roger told me I would need to validate the new laser by measuring the power of its beam. He explained how this was done, and I promptly went to work. The laser met specifications and was formally accepted.

My first year at the Weapons Lab was transformational. With no academic training in engineering and a nearly failed algebra class and only freshman-level courses in biology and chemistry in college, I had become deeply involved with the monkey project, the laser-illuminator contract, and a variety of other projects. I also learned much about aligning high-power lasers and testing semiconductor laser systems.

Meanwhile, the Air Force personnel bureaucracy twice sent notice that my liberal arts degree made me ineligible for an assignment to the Weapons Lab, much less to manage contracts and participate in the monkey experiment. But Colonel Jones intervened both times, and I was allowed to stay. Another bureaucratic snarl occurred when I was promoted from captain to full colonel. My father's name is Forrest M. Mims Jr., and I am Forrest M. Mims III. My promotion lasted only a few hours after I contacted Daddy about the mix-up, and he became a full colonel.

Another unexpected development occurred when I was assigned to visit Eglin Air Force Base, where Daddy had been assigned while I was in the fifth and sixth grades. The orders stated I would depart on July 8, 1969, to "visit AF Armament Laboratory for four days including travel. CBU discussions. Captain Mims." This visit amounted to an official follow-up to one of the major concerns in *Conduct of the War*, the report I had prepared while working as an air intelligence officer in Vietnam.

As explained in chapter 3, defective cluster bombs were opening prematurely and failing to strike their intended target. I had received a thank-you note from a colonel at Eglin AFB while I was in Vietnam, and visiting the technical staff at Eglin was a very different response to my concern than the shouted lecture I had received from a general in Vietnam.

Farewell to Dr. Roger Mark

The second trial of the monkey experiment took place at the ARL from May 26 to June 13, 1969. Rod Chambers and Stan Lupo, the two airmen who assisted us at the Weapons Lab, accompanied Roger and me. The new negative-reinforcement strategy was a success, and the monkeys continued to work after being exposed to the laser.

When July arrived, it was time to say goodbye to Roger, who completed his military service and returned to Massachusetts. Roger's departure was a major loss to both the Laser Division and to me, for he had become a friend as well as a mentor. His wife, Dottie, and Minnie had also become friends. But a much bigger role lay ahead for Roger, for he joined the faculty of the Department of Medicine at the Harvard Medical School and practiced medicine on the side. He also became a professor of electrical engineering at MIT, where he was later named Distinguished Professor of Health Sciences and Technology.

After Roger's departure, I continued building electronic circuits, aligning and repairing lasers, monitoring contracts, conducting outdoor field tests, and designing and conducting a classified experiment I designed. I also spent time discussing technical topics with Ed Roberts, a lieutenant who was assigned to the Solid State Group shortly after I arrived. Ed was an electrical engineer with a strong interest in both computers and medicine who was soon to play a key role in my budding technology career.

The Laser-Safety Guy

Laser research and development soon became the dominant role of the Weapons Laboratory, and during late 1969, the laser-research programs of the Effects Branch were consolidated under a new Laser Division. My group continued to work with semiconductor lasers and both glass and yttrium-aluminum-garnet (YAG) high-power lasers, while the High Energy Branch was responsible for high-energy lasers that melted holes through steel bars and fire bricks. A new building was built to house most of the laboratories of the Laser Division. The building included a long, narrow basement for testing lasers.

Care must be exercised when operating high-power glass and YAG lasers, for the pulse of laser light, while very brief, can have an intensity exceeding a million watts. This is enough to blow chips off the front surface of the laser rod. One time, I cranked up the energy of a glass laser so high that there was a loud pop and the end of the laser rod shot across the lab like a bullet. This was not unusual, and people in our branch who were assigned elsewhere were given a going-away present of a wood plaque on which was mounted a fractured portion of a synthetic-ruby laser rod.

Because most of the lasers we used posed serious hazards, a safety official was appointed to keep an eye on us after the monkey project was well underway. This well-meaning gentleman made sure that we wore the proper safety gloves and exercised designated precautions when handling the monkeys. He also insisted that we wear laser-safety goggles when our high-power lasers were being aligned and test-fired. This posed a safety problem of its own, however, for the safety goggles sharply reduced peripheral vision.

Aligning a solid-state laser requires very careful adjustments of the output mirror, which must be perfectly parallel with the rear mirror. Adjusting the mirror required working adjacent to wires, cables, and exposed connections carrying 20,000-volt pulses that triggered the flash lamp and 600 volts from the powerful bank of capacitors that created the flash.

The 20,000 volts were in the form of a fast pulse that might cause an arm or leg to jerk away. Discharging the 600-volt capacitor bank across one's body was far more dangerous. Placing the steel blade of a screwdriver across the capacitor bank's terminals produced a sound as loud as a gunshot and melted the screwdriver's tip. Touching those terminals was potentially fatal.

When everything was working properly, I usually wore safety goggles when firing the laser. But since their tunnel vision limited my view when working around high voltages, I did not pull the goggles over my eyes when aligning the laser's output mirror. Instead, I looked away and closed my eyes when the laser was fired. This annoyed the well-intentioned safety guy, so I sometimes posted my safety guy to warn me when the official safety guy was sneaking down the hallway to peek into our lab.

The Laser Surgeon

During March 1969, my concerns about laser-safety goggles were reaffirmed when the Laser Division designated part of my work as "biological investigation of laser hazards" and sent me to the Second International Laser Safety Conference in Cincinnati, Ohio. During the conference, the attendees were invited to a hospital to observe an early effort by Dr. Leon Goldman, a pioneering laser surgeon, to treat skin cancer with a high-powered ruby laser. While we watched through a green safety screen, Goldman pointed the laser at a skin cancer on the exposed shoulder of an elderly man. Following the usual "3 . . . 2 . . . 1 . . ." countdown, the laser was fired, and the room was momentarily brightened by the brilliant light from the laser's flash tube. A plume of white smoke curled into a question mark over the patient, followed by the odor of burned flesh. Goldman looked closely at the patient's shoulder and said, "Darn! I missed. Let's try again." His second attempt was on target.

While this incident caused a good deal of murmuring among those of us watching, Goldman was not to be blamed. Based on my experience, I assumed that the visibility limitation caused by the laser goggles he was wearing was responsible for the target being missed during the first try.

Protecting Secrets

Dealing with highly classified information had been a daily responsibility when I was an intelligence officer in Vietnam. That's because every reconnaissance and attack mission was classified Top Secret until the mission was flown. Intelligence reports were also classified. While there were severe penalties for disclosing classified information to those without a security clearance or a "need to know," the rules were often bent, and more than once, I was lectured by a higher-ranking officer when I refused to give classified information over an insecure telephone link.

Security regulations also applied to everyone at the Weapons Lab, especially those assigned to some of our projects that were so sensitive they were marked Top Secret—Special Access Required. But security was difficult to impose on people working in an environment that resembled a university more than a military laboratory.

Scientists and engineers sometimes placed highly classified documents in desk drawers when they went to lunch or ran an errand. Classified briefings were sometimes held in a special meeting room with no sound-suppressing insulation. Anyone in the adjacent hall could hear much of what was said.

Everyone knew that the Air Force's Office of Special Investigations was authorized to listen to our phone calls to make sure classified information was not discussed, so most people avoided mentioning any secrets during phone calls. But the few who did provided the OSI with enough information to understand some of the highly classified programs.

Six months after arriving at the Weapons Lab, I was assigned the extra duty of Top Secret Control Officer for the Solid State Group. This duty required me to remind the others about the need to always place discarded classified papers in specially marked burn bags. This was awkward when the offender outranked me, which occurred more than once after a new major was assigned to our group. The new major, who twice ignited the contents of a trash can in our office with his cigarettes, was our branch's second West Point graduate.

The Air Force benefited when Capt. Glenn Doughty chose the service over the army when he graduated from West Point. He was one of the brightest officers in our group. But the army benefited when our second West Point graduate elected to switch to the Air Force.

Security protocols were sometimes defeated by New Mexico's spring windstorms. Several times, an officer walking between buildings with a batch of top secret files lost everything to the wind. When a captain dropped a briefcase full of classified papers on a windy day, he retrieved all but three of the documents, which meant a search of the base was required.

After several hundred scientists and engineers were organized into a row around a thousand feet (305 m) long, we walked along with the blowing tumbleweeds from where the papers were lost to the barbed wire–topped fence that surrounded Kirtland AFB. After three hours of searching, only two of the lost documents were found, and the third was still missing. Instead, several classified papers that had not been in the captain's briefcase and had never been reported as lost were found among old newspaper pages blown against the perimeter fence. The search ended after the Weapons Lab senior staff held an emergency meeting and concluded that the lost document was overly classified.

Eighth Card Secrets

When I was leaving the office late one afternoon, a courier arrived with a classified report for the High Energy Branch of the Laser Division. The airman needed to get to the airport to catch his return flight and asked whether I would sign for the report, which was marked TOP SECRET—EIGHTH CARD SPECIAL ACCESS. I did not have Eighth Card access but decided it would be best to sign for and lock up the report rather than having the courier miss his flight and spend the night in a motel with a highly classified report.

Before I could sign for the report, however, it was necessary to make sure all the pages were present. While I did not read any of the classified text, it was impossible to avoid looking at illustrations showing an aircraft equipped with a hypothetical gas dynamic laser designed to shoot down enemy aircraft or missiles. Those of us in the Solid State Group always assumed that's what our Eighth Card coworkers were developing, but we were not supposed to ask. The next morning, I retrieved the report from the safe and delivered it to the Laser Division commander's office. They were more embarrassed that none of their staff were present to sign for the report than the fact that I was now in on their top secret program.

Eighth Card was then highly classified because it was one of the military's most important projects to develop a laser weapon. It was the first in a series of programs that eventually evolved into the Airborne Laser Laboratory, a modified KC-135 equipped with a high-energy laser that successfully shot down a series of small rockets and drones in 1983. The Eighth Card program was headed by Col. Donald Lamberson, a no-nonsense officer who expected the best of his team of scientists and engineers.

After the High Energy and Solid State Groups were consolidated into the Laser

Division, Lamberson completed work on his PhD in aerospace engineering. One day, a young officer who was more impressed by academic credentials than rank passed Lamberson in the hallway and said, "Congratulations, Dr. Lamberson!" Lamberson turned and gruffly said, "That's Colonel Lamberson to you, lieutenant!"

While I felt bad for the young officer, I understood Lamberson's reaction: Military rank is based not on academic achievement but on leadership, management skills, general knowledge, and productivity. Long before he earned his doctorate, Lamberson had been promoted to the rank of colonel and placed in charge of the Air Force's top secret Eighth Card program because he possessed and demonstrated those attributes. His new doctorate was merely the icing on a cake.

Off-Duty Research

The part-time science projects I began during high school and continued through college and in Vietnam resumed during off-duty hours after the assignment to the Air Force Weapons Laboratory. Two months after arriving, I listed in my lab notebook thirty-two project ideas, none of which related to my assignment to the Solid State Group. On page 63 of the notebook, I sketched a "Highly Miniaturized Semiconductor Pumped Solid State Laser" consisting of a tiny neodymium YAG or glass laser rod stimulated by a row of four diode lasers. The entire laser would be smaller than a thimble. On April 25, 1968, Captain Doughty witnessed this notebook page, which described a kind of laser that I never developed but that is today widely used in green laser pointers.

Two and a half months after arriving at the Weapons Lab, I bought an RCA laser diode and built a simple circuit to operate it as described in an RCA brochure. When powered by a 300-volt battery, the circuit applied pulses to the laser only 50 nanoseconds (0.00000005 second) wide, with an amplitude of around 20 amps. The invisible infrared beam from this laser was easily seen using the Varo infrared viewer.

This demonstration seems simple in retrospect, but it was a personal breakthrough. During the week, I was learning to operate and repair high-power glass and YAG lasers the size of a desk. During evenings at my apartment, I was experimenting with a low-power laser diode system built inside a matchbox.

During July and August 1969, I filled twenty-three notebook pages with detailed drawings of various ways to build a ram-air guided rocket. Several of these pages were witnessed and signed by Roger Mark and Ed Roberts.

That winter, Roger and Ed met me at a large tract of vacant land on the east side of town, where we launched three rockets in ten flights. Those tests were a more carefully planned version of the tests conducted in Vietnam in 1967. One rocket had no input hole in its nose cone. This was the control rocket. The remaining two rockets had an input port and four output ports. Tape was placed over three of the ports for five of the flights. Roger pressed the firing switch for each launch, and two of Ed's sons retrieved the rockets after they parachuted down. Thanks to their help, I was able to take thirty-eight photos and six movies of the rocket smoke trails. The photos nicely recorded the spiral smoke trails we observed. Roger and Ed witnessed the notebook page that described the flights and four other pages.

The Strap-On Wind Tunnel

While the spiral smoke trails of the test rockets confirmed the validity of ram-air control, a better method was needed to quantify the effect, so I designed and built an inexpensive wind tunnel that could be attached to the passenger side of my 1966 Chevy. The wind tunnel was made from a 29-inch (74 cm) length of sheet metal tubing 6 inches (15 cm) in diameter I bought at a hardware store. The tube was cut in half, and an 8-inch (20 cm) section of Plexiglas tubing was installed between the two metal sections. A 6-to-8-inch expansion segment was installed over each end of the tube. The overall length of the wind tunnel was 4 feet (1.2 m) when the wood bracket for the motor and propeller that measured its airspeed was included. The total cost was $11.15 ($105 in 2023).

The wind tunnel was strapped to my Chevy with seven suction cups and nylon rope. When the car was driven at 70 miles per hour (113 kph), the air through the tunnel reached a speed of 90 miles per hour (145 kph). Ron Schappaugh, my former classmate at Texas A&M, was also assigned to the Weapons Lab, and he drove the car, while I tested various ram-air nose cones in the wind tunnel. The strap-on wind tunnel was sufficiently successful to show that tests with a professional wind tunnel were warranted.

Minnie was much more concerned about another kind of labor after the wind tunnel was strapped onto the Chevy. She was pregnant, her due date was less than a week away, and she was concerned that the passenger door of the Chevy was blocked by the wind tunnel. I removed the tunnel the day before Minnie entered labor, and our son, Eric, was born the next morning.

The Model-Rocket Light Flasher

The Vietnam night launch that was mistaken for a rocket attack against Tan Son Nhut Air Base showed that photographing the flame trail of night-launched rockets would provide the best way to measure the ram-air effects on test rockets. Therefore, I began work on a miniature transistorized circuit that would cause a small light bulb to emit very bright flashes. The circuit was identical to the one that pulsed the infrared LED in the travel aid for the blind.

The light flasher produced much brighter flashes than expected. After testing it, I coated it with Silastic, a clear, rubbery silicone, to protect it. I then installed the flasher and a battery inside the clear plastic section of a rocket payload section. The upper end of the payload section was designed to accept various nose cones with a ram-air hole bored through their center.

When the flasher rocket was ready, I drove to the rocket test site where Roger and Ed had earlier helped me launch and recover ten daylight flights. I brought the single light-flasher payload section and five booster rockets. After I mounted my Pentax Spotmatic camera on a tripod, the light-flasher payload was flown sixteen times atop the five booster rockets. The notebook page that described these flights in detail noted that the light flasher worked properly on all flights, even though it experienced two hard landings when the parachute failed. Under "Notes," I wrote, "Flashing light visible only after burnout to chute ejection; occasional flashes until about 100' above ground. Recoveries very easy. Time to construct rockets and conduct launches = 12 hours."

The Southwestern Model Rocket Conference

Earlier in 1969, I had organized a model rocketry club at the Albuquerque Academy, a private school built on a large tract of land with plenty of space for rocket launches. Patrick Miller, a math major at the University of New Mexico and one of the founders of the ARC-Polaris Rocket Club in Portales, New Mexico, learned about our club and attended some of our Saturday rocket launches.

The connection with Patrick resulted in an invitation for the Albuquerque Model Rocketry Club to participate in the Southwestern Model Rocket Conference, organized by Patrick and Don Stone. The conference was held at Eastern New Mexico University in Portales from July 27–29, and Ford Davis, Mike Vinyard, and I attended from the school's club. Also in attendance was George Flynn, then editor of *Model Rocketry* magazine. Earlier, I had sent Flynn a proposed article about a neater version of the light flasher. The new flasher measured ⅝ inch (1.6 cm) by 1½ inches (3.8 cm) and weighed 0.15 ounces (4.2 g). The article included step-by-step construction details for the flasher and three hand-drawn illustrations. The article was titled "The Theory and Construction of a Transistorized Tracking Light for Night Launched Model Rockets." On April 11, 1969, Flynn sent a signed acknowledgment form for the article that stated that is was being held for publication, at which time the magazine would send an unspecified payment.

I gave a talk on the model-rocket light flasher at the July conference, and on the second evening I conducted a demonstration launch. Flynn expressed considerable interest in the light-flasher flight and said he would expedite the publication of my article along with a report about the meeting.

Origin of the Light-Flasher Circuit

Since 1966, I had wondered who had developed the two-transistor code practice oscillator that I had modified to pulse the LED in my travel aids for the blind back in college and later modified for the model-rocket light flasher. On March 12, 2019, Dr. Michael Covington of the University of Georgia sent me "'p-n-p'—'n-p-n' Oscillators," an article by Louis Garner (under the pseudonym E. G. Louis) in *Radio & Television News*.[5] This article provided a detailed account of how to experiment with the circuit that began my writing career.

Later, I learned that Garner's circuit also played a role in Covington's early electronics experimenting. While I have never met Covington in person, I have learned much from his online posts about amateur astronomy, photography, computers, electronics, his commitment to Christianity, and a host of other topics. While Covington's doctorate is in linguistics, he is a senior research scientist emeritus at the University of Georgia, where he is a former adjunct professor of computer science at the Institute for Artificial Intelligence. I will have more to say about Lou Garner later, for I eventually followed him as the main columnist for *Popular Electronics* magazine.

A Travel-Aid Update

Shortly before Capt. Roger Mark returned to Massachusetts, we sat at a table in the Weapons Lab library to discuss questions I had about how to predict the detection range of the travel aid for the blind when it was pointed at various objects. Roger patiently explained how to derive an equation that would answer my question.

His equation was a mathematical model that predicted the distance over which the travel aid could detect an object if several variables were known. A key variable was the object's reflectance—the fraction of near-infrared the object would reflect toward the aid. Another was the sensitivity of the travel aid's infrared detector. A third was the amount of power in each pulse of invisible infrared. Roger's range equation added an important science element that had been missing from the travel aid project. I soon became determined to squeeze all the components of the handheld travel aid for the blind into a pair of slender tubes that could be mounted on eyeglass frames.

5. E. G. Louis, "'p-n-p'—'n-p-n' Oscillators," *Radio & Television News* 56, no. 1 (July 1956), 105–7.

Failure and Success

The ram-air rocket project seemed ridiculously trivial on Sunday afternoon, July 20, 1969, when *Apollo 11* touched down on the lunar surface. Six hours later, Minnie and I watched with amazement as Neil Armstrong became the first person to step onto the lunar surface only twelve years after the launch of the first *Sputnik*.

Three days later, when *Apollo 11* was returning to Earth, I resumed adding sketches about the ram-air project to my notebook. After Roger left, he was no longer available to witness the laboratory notebook William Holt of Varo had given me in December 1966. On August 3, Minnie witnessed the final page of the notebook, which described an idea for a rocket motor that incorporated a nozzle with recessed whorls that would cause a rocket to spin in flight. I then found an even better lab notebook at the University of New Mexico bookstore. This notebook came with a sheet of carbon paper that could be slipped between white and yellow pages having the same number to provide a carbon copy.

Page 3 included this status report on the rocket project: "I delivered the paper 'Ram Air as a Method of Rocket Control' [at the] research and development competition of the 1969 National Association of Rocketry meeting, NARAM-11., at U.S.A.F. Academy. This is the first public disclosure of the concept (15 Aug 69). One of the judges, Maj. R.F. Lopina took an interest in the concept. He's interested in having some advanced aero cadets do wind tunnel studies and flight tests of the concept. This will provide important data."

While my paper earned the first-place award for model rocketry research at NARAM-11, I was much more excited about Lopina's interest in the project. After Minnie and I returned to Albuquerque, Lopina and I quickly worked out an arrangement for his students to thoroughly research the ram-air control concept. During the next three weeks, I spent most evenings working on various ways to finally build a working ram air–controlled rocket, all of which were described in detail on some forty pages of the new notebook.

As before, the ram air–controlled rocket would include light sensors that would enable it to be guided toward the direction of the sun. The new design included a circuit that sent pulses to the motor instead of powering it with a continuous electrical current. This made it easier for the brake to stop the motor. By September 13, about half of the design had been built and tested. A notebook entry that day records that the brake immediately stopped the motor when the light detector was darkened. Six days later, the entire guidance and control system was finally completed.

Late on a Friday evening, September 19, 1969, the rocket's guidance section was suspended nose-down from a string directly over a flashlight pointing upward from the floor. When the position of the flashlight was adjusted so that all four of the light detectors were illuminated, I switched on the power and released my hold on the guidance section. The motor was nicely humming, and everything looked good. I then moved the flashlight slightly so that one of the light sensors was shadowed by the nose cone.

As expected, the pulsating power applied to the motor was immediately stopped and the brake was energized. Completely unexpected was what happened next, for the entire guidance section instantly jumped to one side and began rotating. The bottom of page 43 in my notebook closed with this sentence within an outlined border: "Precessional forces of [the] motor are significant when brake is applied—this may be an important problem."

The next day, September 20, 1969, was a Saturday. I sat down at our kitchen table and filled page 44 in my notebook with thoughts about the failed test the previous evening. I still recall the disappointment of that test of the carefully built guidance system twelve years after conceiving the idea about ram-air control in the seventh grade. Many test flights and wind tunnel tests had demonstrated that the ram-air principle worked. That was not the issue. The question was, how could I have overlooked what seemed so obvious? The energy in the rotating motor and the air-deflection cylinder had to go somewhere when they immediately stopped rotating, and that is why the entire guidance system had begun spinning.

My disappointment was tempered by a new development I noted at the bottom of page 44 in my notebook: "The tracking light article appeared locally today in *Model Rocketry*, September issue." Seeing my first magazine article in print was exciting, and I was especially surprised that George Flynn sent a check for $93.50 ($879 in 2023) for the article. I immediately thought about additional articles I could write for the magazine and told Minnie I wanted to become a freelance writer after leaving the Air Force. On Monday, I brought the magazine to the Laser Division at the Weapons Lab to show Ed Roberts.

5

MITS

H. Edward Roberts was one of those rare officers who began his military career as an enlisted man. After graduating from Oklahoma State University with a degree in electrical engineering, Ed was commissioned a second lieutenant in the Air Force and assigned to the Laser Division of the Weapons Lab. He arrived in the summer of 1968, four months after my arrival. Ed and I soon became friends, for we shared interests in history, photography, computers, and electronics.

Ed, an entrepreneur at heart, was a fan of Ayn Rand and her book *Atlas Shrugged*. He had four goals in life: become a millionaire, pilot his own airplane, become a medical doctor, and buy a farm in Georgia. These were not wishes; they were firm goals that he took seriously.

Before the Air Force sent him to college, Ed was assigned to Lackland Air Force Base, in San Antonio, where he established two one-man electronics companies in his spare time, Reliable Radio and TV and Reliance Engineering. His favorite and best-known part-time project was designing and building the hugely popular animated Christmas characters at Joske's, a large department store a block from the Alamo.

Ed and I read *Popular Electronics* and *Radio-Electronics* magazines, both of which featured projects available from the authors as reasonably priced kits. We often discussed how we could design projects for these magazines and sell kits. Both of us had experience building basic analog computers, and we seriously discussed developing an analog computer kit for hobbyists. But while Ed admired the simplicity of analog computers, he much preferred the idea of a do-it-yourself digital computer. He wanted to design a real digital computer that could be easily programmed to accomplish many diverse tasks.

Ed was reminded about this goal every day at work after our group invested $4,900 ($46,075 in 2023 dollars) in a revolutionary Hewlett-Packard 9100A desktop programmable calculator. HP ads proclaimed the 9100A as "Constantly available. At your fingertips whenever you need it. . . . *The new Hewlett-Packard 9100A personal computer.*"

Ed was assigned the task of acquiring the 9100A, and he spent hours punching transistor amplifier equations he devised into the keys of this powerful number-crunching machine. He became more determined than ever to develop one on his own.

Micro Instrumentation and Telemetry Systems (MITS)

On Monday, September 22, 1969, I brought to the Laser Lab the September issue of *Model Rocketry* magazine with my light-flasher article. The published version of that simple project was the breakthrough Ed and I had both talked about for more than a year. We quickly agreed that the light flasher and various other devices I had designed for model rockets could be sold to model-rocketry enthusiasts, and Ed suggested we meet at his house to discuss the details.

A week later, we met in his kitchen at 4809 Palo Duro Avenue NE. Also present were Bob Zaller, an Air Force officer and engineer who was teaching Ed to fly, and Stan Cagle, a civilian contractor at the Weapons Lab. After they reviewed the article about my light flasher and some of my model-rocket instruments, we agreed to form a company. Our second meeting was held in the smoke-filled front bedroom that served as Ed's office.

Our first major decision was naming the proposed company. Ed suggested Reliance Engineering, the name he used for his freelance engineering projects. I said we needed a name that would appeal to the model-rocketry audience. The Massachusetts Institute of Technology (MIT) was then home to a sophisticated group of amateur rocketeers, so I said having MIT in our name would be a boost. I first proposed MIT Systems and then suggested that the letters stand for *micro* and *telemetry*. Stan quickly responded with the counterproposal Micro Instrumentation and Telemetry Systems—"M-I-T-S."

Though I emphasized the MIT connection, Ed was unsure about the name and said, "Everyone will call us MITS," pronouncing it like *mitts*. I insisted that people would refer to us as M-I-T-S, just as MIT is referred to as M-I-T. Ed then pointed out that Reliance Engineering was an existing company with an established credit rating, so I proposed we designate MITS as a subsidiary of Ed's original company. This was acceptable to everyone.

Ed was right about the name, though. Within a few weeks, we quit calling ourselves "M-I-T-S" and simply said, "MITS." We even capitalized on the name by calling our products MITS KITS. Years later, Ed joked about the "hundreds" of times he had had to explain how MITS got its name.

Our first product was the light flasher described in my *Model Rocketry* article. Ed and Bob were surprised that the simplistic two-transistor flasher worked as well as it did. Stan then designed and made the etched circuit boards in his apartment. The TLF-1 Light Flasher was soon ready for production.

Within a month, Ed, Stan, and Bob had completed work on two transmitters plus a variety of modules while I was writing *The Booklet of Model Rocketry Telemetry*. Minnie typed each page on mimeograph stencils with our IBM Selectric typewriter, which I purchased for $400 ($3,761 in 2023) from a teen who had received the machine as a high school graduation gift from his grandmother. Minnie also typed the stencils for the data sheets and instructions I wrote for each of the MITS modules.

In October 1969, I wrote a press release that was published in *Model Rocketry*'s December 1969 issue. Meanwhile, Ed, Stan, and Bob assembled hundreds of MITS modules while Minnie and I finalized operating instructions, designed an order form, and published the mimeographed version of *The Booklet of Model Rocket Telemetry Systems*. I also installed MITS transmitters and modules in rockets I launched to test our products, all of which worked well.

Late in 1969, we incorporated MITS, with each of us receiving 950 shares of stock, with the remaining 200 shares going to our attorney. We agreed to each contribute $100 ($940 in 2023) to MITS, which provided crucial funds for buying parts and printing the instructions and the book.

The March 1970 issue of *Model Rocketry* carried our first ad, which proclaimed that MITS "manufactures the first and only model rocket telemetry modules and ground systems." Ordering information, a catalog, and *The Booklet of Model Rocket Telemetry Systems* were available for 50 cents in coin.

When orders arrived, we realized that the mimeographed booklet and instruction sheets gave Minnie a significant role in launching MITS. Ed's wife, Joan, was also crucial, for she received and shipped orders and made bank deposits. Most evenings, Ed and Bob were assembling circuits in Ed's garage while watching *M*A*S*H* and *Star Trek*. Stan was often locked in the darkened bathroom of his apartment making photographically exposed circuit boards for the light flasher and the telemetry products.

By May 1970, we had sold only a hundred or so transmitters and modules, and during one of our late-night meetings, I suggested we could cut our prices and increase sales by offering kit versions. We tested this approach by converting the TLF-1 Light Flasher into two kits, one with a fixed flash rate and the other with an adjustable flash rate. The first ad for MITS KITS appeared in the July 1970 issue of *Model Rocketry*. But sales did not significantly improve, and we began to realize that the market for our products was limited to precocious teenagers and a small number of engineers and university professors.

Leaving the Air Force

My superiors at the Laser Lab were aware of my model-rocketry research. Shortly before I left the Air Force, they sent me to MIT based on this authorization:

> *. . . to deliver the paper "Ram Air as a Method of Rocket Control" by F.M. Mims & Dr. R.F. Lopina. . . . The paper incorporates data from extensive wind tunnel and water table tests conducted by 4 cadets at the USAF Academy under the supervision of Maj. Lopina. [He will] chair a discussion group on "Model Rocket Guidance & Control." Air Force Benefit: Capt. Mims has promoted the official A.F. policy in support of model rocketry by his attendance at numerous technical meetings, several TV appearances, organization of one of the largest rocket clubs in the US, and regularly publishing technical papers in 'The Journal of Miniature Astronautics'. Attendance and active participation at the MIT meeting will continue this trend.*

On June 11, 1970, I completed four years of active duty and left the Air Force while continuing to work with MITS and hoping to become a freelance writer. A few months earlier, I had realized we would not be able to afford our apartment after I resigned. Minnie had saved $2,750 ($25,858 in 2023) while working for around $5,000 per year during her four years as a secretary at the Weapons Lab. Her goal was to buy a Ford Mustang, so I was more than impressed when she immediately agreed to my suggestion that we use her savings to purchase a 10-foot-by-56-foot (3 m by 17 m) used mobile home near the New Mexico State Fairgrounds. Without Minnie's generous donation of her savings, I would have had to stay in the Air Force. One of the tiny bedrooms in the mobile home became my office. A sheet of plywood atop a pair of two-drawer file cabinets served as a combined workbench and desk.

Immediately after leaving the Air Force, I began working as the night attendant at the parking lot of Albuquerque's airport, the Sunport. The plan was that a night job would provide plenty of time for writing. It did, but the salary of $1.60 ($15 in 2023) an hour was less than a fourth of what I had earned as an Air Force captain.

Worse were the pitiful looks on the faces of my former commanders and coworkers when they drove out of the airport parking lot after returning from their frequent trips. They must have thought I was crazy to trade a position in state-of-the-art laser research for a parking lot job.

In March 1970, I sold my first article to *Popular Electronics* magazine, a feature about light-emitting diodes. At one of our midnight MITS meetings, I suggested that we develop an LED voice-communication project for *Popular Electronics*. The article would give us free advertising for the kit version of the project, and the magazine would even pay us for the privilege of printing it. The magazine liked this idea and decided to hold my feature about LEDs and publish the two articles as a pair of cover stories. Ed and I then began designing what I called the Opticom.

In late July 1970, I received a call from Leslie Solomon, technical editor of *Popular Electronics.* Les was coming to Albuquerque with his wife and daughter and wanted to visit. For an aspiring freelance writer to be visited by a New York editor was a major development. Though I was proud to be writing a monthly column for *Model Rocketry*, this would be a chance to meet an editor from a magazine with a circulation of 400,000. Furthermore, this would provide the opportunity to introduce Les to Ed and Stan and discuss our kit ideas.

The Solomons arrived at our mobile home on Monday, August 3, 1970, and I proudly showed Les my tiny workshop. That evening, we met Ed and Joan for dinner at a steakhouse several blocks from Ed's house. A rare summer thunderstorm brought welcome relief from the afternoon heat. While it thundered outside, Les poked fun at the restaurant's decor while Ed and I silently wondered how we would pay the bill.

After the meal, our wives visited while Les, Ed, and I discussed MITS and our proposed Opticom project. Les was enthusiastic about the project and encouraged us to get it completed as soon as possible. When we asked how many kits we might sell, he said maybe a few hundred, maybe a thousand.

After explaining my plan to become a full-time freelancer, Les said it was next to impossible to become a successful writer outside of New York City. I like challenges, so that was OK with me. After all, a New York editor had come all the way to Albuquerque to meet one of his new authors.

Two weeks later, I flew to Santa Barbara, California, to attend the Guidance, Control, and Flight Mechanics Conference of the American Institute of Aeronautics and Astronautics, where Major Lopina and I presented "Ram Air as a Method of Rocket Control." That paper concluded six years of experimentation and twenty-one night flights of the light flasher that became the subject of my first magazine article and led to the founding of MITS.

When the first Opticom transmitter-and-receiver pair was ready, I field-tested the units by sending voice and music from a radio station over the invisible beam of light from the transmitter. I then wrote the construction article. *Popular Electronics* sent a check for $400 ($3,761 in 2023) for the article, which I deposited in the meager MITS checking account. Ed financed the Opticom kits by borrowing a few thousand dollars from an Air Force friend. We also moved our kit production to a small brick building formerly known as the Enchanted Sandwich Shop, which we rented for $100 ($940 in 2023) a month.

The LED feature and the Opticom article were on the cover of the November 1970 issue of *Popular Electronics*. When the magazine appeared in late October, we began receiving as many as a dozen orders a day. But within a few weeks, the surge slowed to a trickle. We eventually shipped a few more than a hundred Opticoms—far fewer than we had hoped. But we had established a relationship with *Popular Electronics* that would lead to far bigger developments for both MITS and my writing career.

The Split

Shortly before the Opticom article appeared, Ed and Stan had begun designing a desktop digital calculator. Ed wanted to move from the Opticom to a calculator kit before bigger companies became involved, but Stan and I held back. Stan wanted to use our remaining Opticom parts and lenses to continue our plan to develop an infrared intrusion alarm kit and a solid-state laser I had designed. We faced almost immediate competition after introducing our first model-rocketry modules, and Stan and I felt Ed's calculator plan was too risky.

Stan and I knew that Ed would move ahead with the calculator project with or without us, and Stan visited me at the parking lot one evening in early November to suggest that he and I buy Ed's stock in MITS. But on my parking lot salary and limited magazine sales, I could barely put food on the table. Soon Ed offered to buy our stock.

Stan, Ed, and I held our final meeting as MITS partners on November 10, 1970, in the parking attendant's booth at the Sunport. Ed offered to buy our stock for $300 ($2,821 in 2023) cash, $300 by the following March, and $350 in equipment. I took my equipment in the form of unsold model-rocket telemetry gear.

We met with Ed on friendly terms, for we had both agonized over our decision to leave MITS. Even though there was never enough money to receive a salary, leaving MITS was harder than resigning from the Air Force. Fortunately, the excitement of seeing my two articles featured as cover stories in *Popular Electronics* was still fresh in my mind. Leaving MITS would provide time to develop the writing career that I hoped would eventually lead to doing serious science.

Ed stayed with MITS and developed the first calculator kits. A few years later, he became famous when he introduced a product that changed the world, but that is getting ahead of the story.

6

THE SILICON SHACK

Before leaving the Air Force, I told Minnie that the $1.60 per hour parking lot job would be required for only a year, for by then I would be a full-time writer. But writing for *Model Rocketry* was not a full-time position. My first break occurred when Don Herrington, an editor at Howard W. Sams & Co., asked if I could edit a book about diode lasers by Ralph Campbell.

The book needed serious work, and I added so much new material that I was listed as coauthor of *Semiconductor Diode Lasers* (1972), my first book since *The Handbook of Model Rocket Telemetry*. The laser book was typed on my Selectric typewriter, which I brought to the parking attendant's booth each evening. I also brought adhesive tape and plenty of paper so I could suspend typed sheets from the windows in the little building while organizing the text.

Eleven months later, I had completed the laser book and several articles, but the only source of regular income was the parking lot checks. On May 22, 1971, I typed a one-page status report that began, "It is of value to periodically evaluate one's performance—particularly in my occupation. I have now been working at this parking lot job for almost a full year and still have not saved a year's salary (at a rate of $60/week) to quit and go full-time freelance writer. The frustrations are many. Lately, I have had quite a string of article rejections. In fact, rejections have far exceeded acceptances for this year. Problem is due in part to my attempt to write for general editorial magazines rather than just technical magazines . . . ."

It was time to make major career decisions, and the time came one month later when two boys were throwing handfuls of gravel at cars in the parking lot. The elderly security guard was also watching them, but he ignored me when I told him he should stop the boys. When I implored him a second time, he pulled his .45 pistol from its holster and pointed it at me! I ran to the back of the parking booth and hid there until he walked away. The next morning, the security guard was fired, and several days later, I quit.

While ending the all-night sessions at the parking lot in June 1971 provided much more time to write, we could barely afford to pay the rental space for our mobile home and buy groceries. Every time the money was nearly gone, we sat at our tiny kitchen table and prayed that a magazine would accept my latest submission. This worked, and a check arrived every time we were nearly out of money. The first major check was from *Popular Electronics* for an article about laser diodes for the October 1971 issue. The article included construction details for a laser kit I had developed for MITS.

Howard W. Sams and Co. was pleased with my work on the laser book, and I was assigned to write a series of books about electronics. One of the many circuits I built for those books was a voltage multiplier that provided high voltage for a device that converts invisible infrared images to visible images.

Back at the Laser Lab during my Air Force days, we had kept one hand in a pocket while working on high-voltage circuits to protect us from potentially dangerous shocks. I followed this protocol while working on the voltage multiplier, but one evening, the circuit slipped off the workbench and landed on my left leg. There was a loud pop as 600 volts stored in the multiplier burnt a hole through my jeans and the flesh below. The discharge flipped my left leg straight out, which threw me against the wall. I looked up and saw smoke rising from my jeans.

The Silicon Shack

Minnie and Eric needed more space, so I bought an 8-foot-by-10-foot (2.4 m by 3 m) metal building kit from Sears and assembled it atop concrete blocks adjacent to our mobile home. Winters in Albuquerque can be chilly, so I lined the inside of the building with insulation held in place by paneling. The two file cabinets and plywood work surface were moved from my mobile home office to one side of the new building. A second pair of two-drawer file cabinets and a plywood work surface were installed on the opposite side. The left side was for electronics and the right side was for writing.

From 1971 to 1976, I wrote more than a dozen books for Howard W. Sams and Co. in that tiny Silicon Shack. They included *Light Emitting Diodes* (1973), *LED Circuits and Projects* (1973), *Electronic Calculators* (1974, with Ed Roberts), *Optoelectronics* (1975), and *Light-Beam Communications* (1975). Sams's most important assignment by far was *Introduction to Electronics* (1972), the first of a series of twenty electronic project books for Radio Shack I wrote over the next five years. While Sams paid me a royalty fee on books under their imprint, I received a fixed fee for each of the Radio Shack books. Had I known that Radio Shack was selling hundreds of thousands of those books, I would have asked for royalty fees.

We were thankful that I was finally earning a livable income. We never applied for food stamps or other government assistance, and we saved enough money to buy a pair of bicycles and a bike seat for Eric. We could even afford to go to Taco Bell once a week for a special treat.

The Eyeglass-Mounted Travel Aid for the Blind

The travel aid for the blind was often on my mind, and on March 13, 1971, I described in Notebook No. 4 experiments in which a laser diode transmitter and a new receiver circuit could detect a wall from 10.5 feet (3.2 m) away. My model-rocketry friend Patrick Miller wrote a computer program based on Roger Mark's range equation that provided a neat graph of the device's range.

On March 15, I filled ten pages of the notebook with many tests of travel aid circuits, various versions of the range equations, and several charts showing experimental results. This intensive theme continued into Notebook No. 5 through the rest of March until April 27. I then took a two-week break to build a new prototype travel aid the size of a deck of playing cards that was considerably more advanced than the original travel aid from March 1966. The LED source was mounted inside a plastic tube that could be pushed up or down to alter the direction of the invisible beam of near-infrared. This provided a simple means to estimate the distance to an object. It detected unpainted plywood at 96 inches (2.4 m), unpainted concrete blocks at 60 inches (1.5 m), a shiny, outdoor light pole at 76 inches (1.9 m), and a white painted stucco wall at 100 inches (2.5 m).

In August 1971, I devised a much better receiver for the new travel aid. It consisted of a two-stage amplifier followed by a circuit called a monostable oscillator, like the one I had built for Roger Mark's monkey project. This circuit generated a pulse in an earphone each time an object was detected.

After completing several writing assignments, in December 1971 I began working long hours redesigning the travel aid so that it could be installed inside two small tubes affixed to eyeglasses. On December 23, I described on page 53 of the notebook important tests using new and more powerful LEDs that enabled detection of a tree trunk 20 feet (6 m) away. The page ended with this: "The past few pages (51–53) conclude that an eyeglass mobility aid for the blind is, at last, within reach. Thanks be."

I should have added "to God" but did not. Aren't engineers supposed to solve problems with the brain God gave them? Unsaid in the notebook was that I had begged God for assistance, for while I could design the circuits, I lacked the skills to miniaturize them. God heard my prayer, and the following day, Christmas Eve 1971, I mounted the circuitry on sunglasses and detected trees 24 feet (7 m) away. The total weight of the transmitter, receiver, and batteries was only 2.75 ounces (78 g).

The Blind Leading the Blind

Noel Runyan, a blind friend, was an electrical engineering student at the University of New Mexico. He had tested my earlier aids and provided important advice. He also taught me how to find the corners of a room while blindfolded, for facing a corner provides distinct sound clues.

The best advice Noel provided was how to walk across the campus at night while blindfolded. In a classic case of the blind leading the blind, he blindfolded me and handed me one of his canes. We then left his apartment and walked toward the campus. While walking along a sidewalk, I was able to detect cars parked along the adjacent street as we passed through their sound shadows. Those cars were blocking the sound of air conditioners and the breeze-blown branches of trees across the street.

Soon, we arrived at a street we needed to cross. Its presence was made obvious by changes in the sound created by our canes as we swept them back and forth in front of us. The canes told us when we arrived at the street, for their tips swept through the air instead of along the concrete sidewalk. As we crossed the street, Noel explained how to tap the end of the cane at each sweep and listen for the echo created by the curb on the opposite side. This worked! For the first time, I understood how my blind great-grandfather could count utility poles as we walked along a rural road.

Noel had always liked my idea of making an eyeglass-mounted travel aid, and he understood the technical difficulties involved. On New Year's Eve 1972, Minnie and I visited Noel to share with him the new eyeglass-mounted travel aid. Noel was eager to try out the device, and I filled four pages of Notebook No. 5 with his results and this summary: "Very useful in public places where cane is inconvenient. Potentially very useful when standing in line. Subject [Noel] presently taps person ahead with cane and this causes embarrassment. Useful for 'looking' at someone behind a counter, people during conversations and signs. Also potentially useful for avoiding tree limbs, signs, etc. Leaves and snow tend to weigh down branches and cause a hazard."

Overhead tree branches were a special concern for Noel, and he mentioned this several times during our experimental walk.

The Industrial Research Award

By February 1972, I had assembled the transmitter and receiver circuits on tiny, printed circuit boards and installed them inside a pair of 0.5-inch brass tubes. Each tube held a lens in one end and a miniature slide switch at the opposite end. The receiver tube included a thin plastic tube that slipped into the user's ear to provide beeps when objects were detected. The tubes were fastened to opposite sides of a pair of sunglasses. After the device received attention in technical magazines, I noticed an announcement in *Industrial Research* magazine about the tenth anniversary of the magazine's IR-100 Award, which was promoted as the equivalent of the Nobel Prize for applied physics research scientists and engineers. I entered the competition under the name Sensory Aids Research.

The application wanted to know how much money had been invested in the product being nominated. The previous year's winners had spent an average of $419,000 (about $34 million in 2023), so I did not think the $200 ($1,881 in 2023) I had spent on the travel-aid project would sound very impressive, so I added the income I might have earned had the time spent on the project been devoted to writing a book and produced a highly inflated total of $6,000 ($56,418 in 2023).

After submitting the award form, I built a simple instrument to measure how much near-infrared is reflected by various objects. This was important data for the equation that predicted how well the travel aid would detect objects. I installed the instrument in a camera case and bicycled and walked around Albuquerque, measuring the reflectance of fences, tree branches, walls, clothing, and the exposed skin of volunteers. The latter proved interesting, for the reflectance of dark skin was only 5 percent less than that of white skin.

This finding was reminiscent of earlier experiments when I noticed that people of various races looked the same shade of green when viewed through an infrared image converter. (I told teens in a Sunday school class I taught that God must have infrared eyes.) I also measured the reflectance of human hair by walking close behind women in a shopping center and holding the reflectance probe near their hair without their knowledge. While this protocol provided useful data, it was quickly abandoned when it attracted harsh stares from onlookers.

In July 1972, *Industrial Research* sent notice that Sensory Aids Research had received an IR-100 Award. Neil Ruzic, the magazine's founder and editor, wrote to ask if he could demonstrate the eyeglass travel aid on television during the awards ceremony. When I wrote that the device was not commercially available, and I did not want to give blind people false hopes, he replied that only a small number of the award projects would be on television. I finally agreed, but only if I were allowed to explain the device's limitations.

The IR-100 awards ceremony was scheduled for Chicago's Museum of Science and Industry, and 300,000 people were expected to view the winning projects. Ruzic was disappointed that I was unable to build or rent a display and said he would take care of me. And that he did. Famous explorer Thor Heyerdahl and the *Kon-Tiki*, his equally famous seagoing raft made of reeds, were stationed at the entrance to the exhibit hall. Ruzic arranged for a table to be placed next to Heyerdahl's raft where my travel aid and my small poster were exhibited.

The IR-100 award ceremony was a black-tie affair, but I did not own a suit, much less a tuxedo. I felt out of place in my plaid sports coat alongside the other ninety-nine winners in their sharp tuxedoes. They included executives from MIT and major companies, including Texas Instruments, General Electric, IBM, Bausch & Lomb, and Hewlett-Packard. The Air Force Weapons Lab was also a winner, and I was surprised to see Colonel Rowden, the new director of the lab. We became the first two IR-100 winners from New Mexico.

Many of the award winners visited my little table by Heyerdahl's raft. While they all expressed interest, none of them were sufficiently inspired to offer help in manufacturing the travel aid. I returned home with an engraved IR-100 Award plaque, Thor Heyerdahl's autograph, and a stack of letters from blind people and their relatives. The only company that responded to all the publicity the travel aid received was Ripley International Inc., the company behind the Ripley's Believe It or Not! entertainment franchise. They wanted to display the device, but I declined their request.

Several years later, I wrote an article for a popular magazine about my travel aids for the blind, and the editor sent it to a reviewer who held a doctorate in a technical field. The reviewer recommended my article not be published because, he claimed, it was impossible to make a travel aid small enough to fit on eyeglasses. That is what I thought before Christmas Eve 1971. When I sent the editor full details, he vetoed the reviewer and published my article.

Shortly after I developed the first travel aid in 1966, the American Foundation for the Blind gave me a grant to write a technical paper about the device for its *Research Bulletin*. A nonprofit organization also gave me a small sum, and the Veterans Administration had long provided advice. The Lions Clubs have long assisted the blind, and a wealthy philanthropist sent me to a meeting with some of their executives. While they were interested, I had only one eyeglass travel aid. I did not contact hearing aid manufacturers like I had back in 1966, for I assumed they would still be worried about liability. Instead, the project I had worked on for seven years was shelved.

While the travel-aid project did not help any blind subjects beyond test sessions in Texas, New Mexico, and Vietnam, the project does have a personal legacy. It provided a valuable introduction to electronic circuit design, miniaturization, optics, reflectance of natural and artificial materials, the range equations I learned from Roger Mark, and, with Patrick Miller's help, the computer simulation of the travel aid's performance. The first 1966 travel aid led to the light flasher for my rocket experiments and became the topic of my first magazine article, which directly led to the founding of MITS in 1969.

Today's technology is much more advanced than what I employed from 1966 to 1972, and I am glad to report that a variety of travel-aid projects are now under development. They include the expensive but sophisticated OrCam MyEye, which clips onto one side of an eyeglass frame and reads text, remembers faces, and identifies objects the user is looking toward.

The 1,000-Mile Bike Trip

During the winter of 1973, we drove to San Antonio to visit my family. Minnie was impressed by the greenery of Central Texas, and I began thinking about taking teenagers from our church on a bicycle trip from Albuquerque to Padre Island National Seashore. Previously, these teens had been taken by bus on church retreats. My goal was to teach them independence and the power of prayer under very difficult conditions they would have to manage without help from parents and traditional youth leaders. I had previously led a dozen of them on overnight bike trips into the mountains north and south of Albuquerque, and now it was time to plan the Texas trip with the help of Dennis and Karen Harbour, both of whom were experienced cyclists.

Fourteen teens survived the rigorous training rides for the trip, and we departed Albuquerque's First Baptist Church on June 10, 1974. Each rider carried all his or her food and supplies on their bike. During the trip, we raced pronghorns down hills and avoided dangerously close thunderstorms. We rode all night for six days to avoid the heat, and in West Texas, coyotes standing by the road howled at us as we rode by them in the dark. Those nights provided amazing views of meteors flashing overhead and the glows of distant towns and cities.

When checking out a possible campsite among a rare cluster of trees along the highway to Roswell, I encountered a large rattlesnake. The snake retreated into a hole when I kicked sand at it. Thinking the teens would not believe that this potential campsite was unsafe, I waited until only the snake's rattles were above the hole and broke them off. That was a bad move, for the snake instantly backed out of the hole, and I had to flee. When I showed the rattles to the teens, they were skeptical about my story and still wanted to camp under the trees. Instead, I required that we camp near the highway, a few hundred feet from the trees. The next morning was chilly, and we were surrounded by a dozen or more large tarantulas, one of which I employed to get some recalcitrant girls out of their warm sleeping bags.

This was an exceedingly difficult trip for me because the week before departure,

I had broken a collarbone in a bike accident and rebroken it while lifting my bike across a curb in Andrews, Texas, halfway to Padre Island. Shifting gears was a painful procedure.

When we arrived in Corpus Christi fifteen days after departure, the mayor presented us with a key to the city. The next day, we rode the final segment of the 1,000-mile (1,610 km) trip to Padre Island National Seashore, where the teens saw and swam in the Gulf of Mexico for the first time. We returned to Albuquerque on a bus provided by the church. When I wrote an article for *Event* magazine about the trip, I did not know how that article would impact my science career when I met with the editor of *Scientific American* magazine fifteen years later.

Back to Work

The Texas bike trip and cycling trips in the Jemez Mountains provided welcome breaks in what had become a very busy writing schedule. Cycling also inspired some new projects, including a pair of solar battery chargers, one of which was placed over my bike's panniers during the day to charge four AA cells used in a flashlight. Those chargers were made with surplus solar cells from NASA satellite programs, and they still work today.

I also built a miniature xenon strobe light that clipped on my belt during night rides and flashed a warning about my presence. An electrical engineer with a doctorate complained to *Popular Electronics* magazine that the strobe would not work. This reminded me of the PhD who had tried to veto my travel-aid article by claiming it would not work. Skepticism is an important part of both engineering and science, but skepticism fails when criticism is directed at a project that clearly works as claimed.

My friend Patrick Miller was at the University of California, Los Angeles, in 1974, and he invited me to participate in one of his workshops for gifted high school students. My job was to teach basic electronics and laser science, and one afternoon, Patrick and I decided to demonstrate the very narrow, brilliant red beam of my helium-neon laser from the roof of the building where the course was held. UCLA students were familiar with laser beam trickery, and someone called the campus police, who arrived and took an elevator to the roof while we were riding another one back down to the classroom.

We repeated this demonstration from a classroom overlooking the library at the University of New Mexico the following year. This time, a student volunteered to pretend he had been struck by a powerful laser beam as he entered the building. While I painted the beam across his back, he rolled onto the floor, gasping. This alarmed several students and a security guard, who called the police. When they arrived at our classroom several minutes later, I was calmly giving a blackboard lecture to the assembled students; the police said nothing and left.

The Radio Shack book series was interrupted during 1974 by an assignment from Howard W. Sams and Co. to update *Reference Data for Radio Engineers*, a massive, 2.5-inch-thick volume with a thirty-year history published by Sams for International Telephone and Telegraph Corporation (ITT). I spent several months updating nine of the book's forty-eight chapters.

As with the fixed-fee books for Radio Shack, I did not negotiate payment for this project. Sams editor Don Herrington asked what I wanted, and I told him to pay what was fair. This project was so major that I told Minnie that Sams should provide $2,500 ($23,508 in 2023) instead of the $1,000–$1,500 they paid for each of the Radio Shack books. After the revised and expanded chapters were sent to Sams, we sat in the tiny kitchen of our mobile home and prayed that Sams would be generous. Two weeks later, we were stunned to receive a check for $10,000 ($53,996 in 2023).

As 1976 approached, Minnie and I began making serious plans to move to Texas, and the check from Sams provided the money we would need. First, however, there was the unfinished business of another mountain bike trip with the church teens and a final assignment from MITS.

7

MITS MAKES COMPUTER HISTORY

Personal computers had not arrived when I changed my college major from physics to government and told Daddy I would continue my science on the side and hire a mathematician when necessary. I had no idea the "mathematician" would arrive as a series of computers loaded with programs that have become essential to the writing and science I have pursued since acquiring an early Radio Shack TRS-80 in 1977. Nor did I know that the model-rocket light flasher based on the travel-aid-for-the-blind project from my senior year at Texas A&M would play a role in initiating the arrival of the PC.

While I was busily pursuing a writing career after leaving the Air Force and MITS, Ed Roberts was working long hours at MITS in a rented set of offices near the New Mexico State Fairgrounds, five blocks from our mobile home. Ed's first calculator kit had appeared on the cover of *Popular Electronics*.

Ed and I had remained close friends, and he had hired me to write the assembly manual for his first kit calculator, the MITS 816, in return for the calculator I built while developing the manual. He later hired me to write the manual for the MITS 1440, a second, more sophisticated calculator kit.

Ed and I often discussed our goals. He still wanted to go to medical school and become a physician, as he had told me when we had worked together at the Weapons Lab. I was skeptical about that goal, and, despite the growing number of magazine articles I had written, he doubted I could become a successful freelance writer. More than once, he told me that he would believe I had arrived only if I wrote an article for *National Geographic* magazine. My friendship with *National Geographic* photographer Otis Imboden did not count; neither did my growing number of Radio Shack books.

Two exchanges with Ed especially stand out. One occurred when he was driving me south along San Mateo Street in Albuquerque, when I asked him to explain the operation of the analog-to-digital converter inside digital voltmeters. Ed was a great teacher, and his explanation was so simple that even I could understand.

Another memorable exchange occurred in 1974, when MITS introduced the mitScope, a device that displayed digital signals on an array of LEDs. This device resembled my homemade LED oscilloscope I had shown Ed in February. When I reminded him about this, he said he should pay me a royalty. I thanked him but noted that his design was digital and mine was analog. He had even mentioned that a digital version was possible when I showed him my prototype during a long discussion we had on February 19, 1974. Therefore, the mitScope was sufficiently different to stand on its own.

As Stan and I had predicted, the calculator business soon became highly competitive, and by 1974, MITS was on the verge of bankruptcy. That is when Ed decided to take the biggest step of his career. I did not know about this until the evening Ed called and asked me to stop by MITS to see something new. His office was only a few blocks away, so I hopped on my bike and rode over. The front door was unlocked, so I pushed my bike inside and walked to the main shop area. Ed and Bill Yates, a fellow Air Force vet who designed printed circuit boards for MITS, were standing in front of a blue metal box about the size of two small, stacked briefcases. Its front panel was lined with several long rows of switches and red LEDs. Behind the box, a giant printed circuit board outline decorated the wall.

"Well, what do you think?" Ed asked.

"What is it?" I replied.

Ed explained that the box on the table was a microcomputer that would be featured on the cover of '*Popular Electronics*'.

"How many do you think we'll sell?" Ed asked.

I was impressed by the computer. But Les Solomon of *Popular Electronics* had told us in 1970 that there is no way to predict kit sales. Our Opticom and laser projects in the magazine had sold less than a few hundred, so I simply said, "Maybe a few hundred." Ed looked devastated. I did not then know how that computer was his only hope for avoiding bankruptcy, and that meant sales must be a few thousand, not a few hundred.

The January 1975 issue of *Popular Electronics* was on magazine shelves and in mailboxes before Christmas 1974. The cover featured a photo of Ed's computer under a spectacular header:

PROJECT BREAKTHROUGH!
WORLD'S FIRST MINICOMPUTER KIT TO RIVAL COMMERCIAL MODELS . . .
"ALTAIR 8800" SAVE OVER $1000[1]

Below the headline was a photo of a blue box with a front panel lined with four rows of toggle switches and red LEDs.

Even Ed was not prepared for what happened next, for cars started arriving at MITS to buy Altair kits and assembled models a few months before there were any to sell. Ed called and asked me to write the operator's manual, but I knew nothing about the assembly language the Altair required, so Ed spent an hour giving me a short course on assembly language.

Altair's brain was an Intel 8080 microprocessor, and Ed gave me Intel's handbook about the 8080. I then spent the next two weeks at MITS typing the first operator's manual for the Altair. While I was working on the manual, two friends in Cambridge were working on something much more sophisticated: a version of the BASIC programming language that would allow users to easily program the Altair.

Paul Allen had seen the cover of the January 1975 issue of *Popular Electronics* at the Out of Town News newsstand in Harvard Square, and he described his reaction to the Altair 8800 article in his memoir, *Idea Man: A Memoir by the Cofounder of Microsoft* (2011):

> *I slapped down seventy-five cents and trotted the half-dozen slushy blocks to Bill's room in Harvard's Currier House. I burst in on him cramming for finals; it was that time of year. "You remember what you told me?" I said, feeling vindicated and a little breathless. "To let you know when somebody came out with a machine based on the 8080?"*
>
> *"Yeah, I remember."*
>
> *"Well, here it is," I said, holding out the magazine with a flourish. "Check it out!"*[2]

Allen and his high school friend Bill Gates immediately realized that the Altair 8800 had initiated the microcomputer revolution, and they called Ed to ask whether he was interested in a custom version of BASIC to run on the Altair. After Ed said he was, Allen and Gates began working nonstop to develop the language they promised. When they finished, Allen flew to Albuquerque to demonstrate their version of BASIC on an Altair. The demonstration was so successful that Ed agreed to acquire BASIC from Allen and Gates. Shortly after Allen returned to his job in

---

1. *Popular Electronics* 7, no. 1 (January 1975).
2. Paul Allen, *Idea Man: A Memoir by the Cofounder of Microsoft* (New York: Penguin, 2011), 7.

Massachusetts, Ed hired him as MITS's director of software development. Allen promptly moved to Albuquerque, and Gates joined him that summer. Their schedule soon became so full that Gates dropped out of Harvard and stayed at MITS. They named their partnership Micro-Soft, and you know the rest of that story.

As for MITS, the few hundred Altairs I expected them to sell became more than 5,000. The first retail computer stores were established to sell Altairs. Computer shows, user groups, newsletters, seminars, software exchanges, peripherals, software products, quality documentation, and cheap computers, all commonplace today, were pioneered for the personal computer market by the little company we had cofounded.

Ed even acquired a dual-engine aircraft for MITS and took Eric and me on a ride early in 1976. We took off at dusk and flew east directly over Central Avenue, the busy stretch of Route 66 that bisects Albuquerque. The dark streets below were outlined by porch lights and streetlamps that resembled a highly magnified view of an integrated circuit. The twinkling headlights of traffic mimicked moving electrons, and the traffic lights were the transistors that controlled their flow. Albuquerque at night resembled an enormous microprocessor.

Back in New York, Les Solomon was being criticized by nontechnical management at the Ziff-Davis Publishing Company for the racket his old-fashioned teletype made when he was using his Altair. He arranged to meet with Ed, Don Lancaster, and me in Ed's office at MITS. Lancaster was a well-known author of articles and books about hobby electronics. Could his famous TV typewriter evolve into a sophisticated, quiet video terminal for the Altair? The answer might have been yes, but they never got that far. Instead, the two electronics geniuses spent the meeting trying to outsmart one another when they could have planned a product that would have defeated the competition soon to come from Radio Shack and Apple.

Meanwhile, competition arrived from new companies that sold improved memory boards and other peripherals for the Altair. By 1977, MITS was selling sophisticated versions of the Altair, but so were competitors. Ed was growing weary of fighting the competition while managing MITS, and he agreed to sell MITS to Pertec, a major manufacturer of disk drives. During his brief stay at Pertec, he developed what would have been the first laptop computer, which Pertec declined to manufacture.

Ed soon left New Mexico and returned to Georgia, where he bought a 900-acre farm. By then, Ed had earned more than a million dollars and fulfilled all but one of the goals he had listed for me when we were working together at the Weapons Lab a decade earlier. His final goal was achieved when he graduated from medical school and became a country doctor in Cochran, Georgia. We kept up with each other during occasional telephone conversations.

StartUp

Shortly before we moved to Seguin, Dr. Uta Merzbach, the Smithsonian Institution's curator of mathematical instruments, visited our place in San Marcos to see my Altair 8800 computer and MITS-related materials she was collecting for the museum. In addition to donating what she had come to see, I gave her my high school–era language translator, which she described as an early example of a hobby computer. My Altair was displayed in the Information Age exhibit at the Smithsonian's National Museum of American History for fifteen years. The language translator has not been displayed in public, but it is shown on the Smithsonian's website along with a detailed explanation I provided.[3]

The museum exhibit most relevant to the development of the PC was *STARTUP: Albuquerque and the Personal Computer Revolution*, at the New Mexico Museum of Natural History and Science, which opened on November 18, 2006. Paul Allen conceived and sponsored *STARTUP*, which was dedicated to the history of personal computing and the roles played by MITS and Microsoft. The entrance to the gallery featured this inscription: "'The idea that you could have your own computer and do whatever you wanted to with it was fantastic.'—Ed Roberts, designer of the Altair 8800 microcomputer."

Before the opening of *STARTUP*, I called Ed to ask whether he was going, and he firmly said he was not. He still had hard feelings about a 1977 arbitration decision that cost MITS its exclusive license arrangement with Microsoft. That ruling allowed MITS to continue to use Altair BASIC while also permitting Microsoft to sell its version of BASIC to MITS's competitors. This decision was key to the stagnation of MITS and the explosive growth of Microsoft that quickly followed.

Fortunately, Ed changed his mind and arrived at the *STARTUP* opening with all five of his children. MITS cofounder Bob Zaller and his son were there, but we were unable to locate our fellow collaborator Stan Cagle. Minnie took a photo of Paul, Ed, Bob, and me.

The key exhibit at STARTUP was an Altair 8800. Nearby was a spool of paper tape with punched holes encoded with the Altair BASIC developed by Paul and Bill Gates. Also nearby was a photograph of Minnie typing the stencil for the MITS TX-1 telemetry transmitter instructions I wrote. The MITS test rocket I built in 1970 with a TX-1 transmitter installed in its transparent payload section was exhibited next to the photo. Perched nearby atop a wire was the light flasher I had built for *Model Rocketry* magazine.

---

3. "Mainframe and Output Section, Mims Analog Translator," National Museum of American History, https://americanhistory.si.edu/collections/search/object/nmah_1279676.

Various sessions were scheduled for the next day, but Ed had a better idea. He called and asked where he could take Minnie and me in the stretch limousine Paul had provided. I suggested Sandia Crest, and a few hours later, that *l-o-n-g* vehicle managed to negotiate the many curves on the road up the mountain, and we were enjoying the view of Albuquerque far below, just like Minnie and I had during our first date in 1968.

STARTUP closed on December 31, 2016. Some of my items that were displayed are now at Paul's Living Computer Museum in Seattle, which provides visitors with a rare opportunity to sit before the keyboards of many computers. As for the light flasher, it was carefully packed and shipped to Seattle in a climate-controlled truck and arrived at the Living Computer Museum on January 17, 2017. Hopefully, it will eventually be placed on display, for that simple circuit borrowed from the travel aid not only launched my writing career but also led to the founding of MITS and the subsequent arrival of Microsoft.

Remembering H. Edward Roberts

Ed's strong work ethic persisted when he became a country doctor in Georgia in 1986, for he often worked long hours seeing his many patients. In March 2010, Ed came down with pneumonia, yet he continued to work until he became a patient himself after a long day when he saw twenty-one patients.

Most of Ed's patients did not know that early computer users consider him the father of the personal computer. Bill Gates knew, and he surprised the staff at the hospital in Macon when he visited Ed shortly before he died on April 1, 2010. Ed's son David, also a doctor and a pilot like his father, was there when Gates spent time with his old friend recollecting MITS days.

Bill Gates and Paul Allen helped trigger worldwide publicity about Ed's achievements when they issued a joint statement after his death: ". . . Ed was willing to take a chance on us—two young guys interested in computers long before they were commonplace—and we have always been grateful to him. The day our first untested software worked on his Altair was the start of a lot of great things. We will always have many fond memories of working with Ed in Albuquerque, in the MITS office right on Route 66—where so many exciting things happened that none of us could have imagined back then."[4]

Minnie and I drove to Cochran for Ed's service, which was conducted by a Baptist pastor who had been one of his patients. Joan Roberts was there, along with

4. Dionne Walker, "Obituary: Dr. Henry E. Roberts Was Bill Gates' Mentor, a Tech Pioneer, a Doctor," *Seattle Times*, April 1, 2010.

their five children. Bob Zaller was also present, and Minnie and I drove Bob to the small church where Ed was buried. We enjoyed trading stories about early days at MITS when we drove back to Cochran.

Remembering Paul Allen

Paul Allen and Ed Roberts stayed in contact long after Ed sold MITS. After Ed died in 2010, Paul arranged a well-attended dedication of the *STARTUP* gallery to him. Paul had just completed his memoir, *Idea Man*, and he privately told me he was concerned that its candid revelations about some of the stormy times during his relationship with Bill Gates might offend his longtime friend and Microsoft cofounder.

Paul never forgot what he owed MITS, and he organized its 40th anniversary reunion in Seattle on the weekend of June 12–14, 2015. (It was the fortieth anniversary of Microsoft, not MITS, but it was MITS, its Altair 8800, and its employees that made Microsoft possible.) Paul provided transportation and hotel accommodations for several dozen of us and our spouses.

The reception was held at the Living Computer Museum, where Paul and his band, The Underthinkers, provided the entertainment for several hours. During a break in the music, I gave Paul a sealed glass vial of air from the Mauna Loa Observatory that I had marked with its carbon dioxide concentration. (I will discuss my time at MLO in a later chapter.) As he carefully examined the vial, I again invited him to visit MLO for a personal tour, but he candidly explained in detail how his health would not allow him to visit the high-altitude site.

Earlier, Paul had told us MITS veterans that he did not want to wait for the fiftieth anniversary of MITS to celebrate, because not all of us might be around. As the evening wore on, Paul looked exhausted, and he was having trouble playing his guitar. These revelations, and the fact that he had completed his memoir suggested that Paul's days were numbered. On October 15, 2018, the software specialist, researcher, sports team owner, investor, musician, billionaire, and philanthropist died from complications associated with non-Hodgkins lymphoma.

Idea Man's candid revelations about some major disagreements he had had with Bill Gates did not end their friendship. After Paul died, Gates wrote a poignant obituary: "I am heartbroken by the passing of one of my oldest and dearest friends, Paul Allen. From our early days together at Lakeside School, through our partnership in the creation of Microsoft, to some of our joint philanthropic projects over the years, Paul was a true partner and dear friend. Personal computing would not have existed without him. . . . Paul loved life and those around him, and we all cherished him in return. He deserved much more time, but his contributions to the world of technology and philanthropy will live on for generations to come. I will miss him tremendously."[5]

5. Jamie Reno, "What We Can Learn From Microsoft Co-Founder Paul Allen's 36-Year Battle with Cancer," Healthline, October 19, 2018, https://www.healthline.com/healthy/what-paul-allen-learned-from-his-36-year-battle-with-cancer.

8

THE BRIEFCASE OF A LASER SPY

There was a time when I was intrigued by the technology of spying, especially the incredible devices developed for Agent 007 by Q in the James Bond movies. While Q was explaining to Bond how to use his latest fictional gadget, I was trying to figure out how to make a working version. I did not want to be a traditional spy; I just wanted to design their instruments.

When I was growing up, I spent hours experimenting with invisible ink and devising secret codes. When I was a senior at Texas A&M, a CIA representative visiting the campus on a recruitment tour learned about my miniature radios and lightwave communicators and encouraged me to apply for an intelligence position after completing my four-year Air Force assignment. When I was attending the Armed Forces Air Intelligence Training Center after graduating from Texas A&M, a senior intelligence officer suggested I apply for a position with the Air Force's Human Source Intelligence program after service in Vietnam.

After a Pentagon intelligence officer read about my travel aid for the blind and off-duty rocket experiments in Vietnam, he requested that I apply for a position "within the collection area" and wrote, "There are a number of positions within the intelligence community that often go begging for people with both interests, i.e. intelligence and a scientific discipline." I submitted the biographical form he sent, but this potential career change ended when Col. David R. Jones arranged for me to be assigned to the Air Force Weapons Lab.

An interesting spying opportunity arose in early July 1975, shortly after I completed the Altair 8800 instruction manual, when a letter arrived from Alastair Gregor of the *National Enquirer* to Les Nelson, vice president of Howard W. Sams and Co., the book publisher with whom I had worked extensively. The letter began:

> *Further to our conversation last month, I am writing to ask you for your assistance in contacting one of your authors who are expert in the field of lasers.*
>
> *As I explained at the time this newspaper, which has an average weekly circulation of four million copies, is interested in carrying an article outlining the growing role of lasers in the everyday life of the people of America.*
>
> *You were kind enough . . . to suggest that Forrest Mims might be interested in doing this work for us. Perhaps you could ask him to contact me or my editor Mr. Bernard D. A. Scott (tel 305-XXXX) to discuss the matter further.*[1]

Don Herrington, my editor at Sams, had attached a note at the top of the letter: "Forrest, are you interested? Want to contact him? Don."

The *National Enquirer* had a terrible reputation, and I was concerned that writing for it could impact my reputation, especially because my twenty-one books were doing reasonably well and I had sold nearly a hundred articles to thirty magazines. But I was curious about what seemed to be a serious journalism project, so I called Bernard Scott and mentioned twelve topics related to lasers. Scott asked whether I was aware of reports that a laser could intercept conversations in a closed room by detecting the reflection of the laser beam from a window. Voices in the room would cause the window to vibrate in step with the sound, thereby imposing the voices as amplitude variations in the reflected laser beam.

Scott was surprised when I told him I could build a laser apparatus for this purpose. He then shifted the conversation to billionaire Howard Hughes, whose private life the *National Enquirer* was investigating to uncover what Scott described as suspicious business practices. Would I be willing to assist by building a laser bugging device? When I said maybe, he asked me to send him a proposal. He also strongly emphasized that our discussions should be kept secret.

The Laser-Eavesdropping Proposal

My main interest was to write the laser feature Alastair Gregor had asked about in his letter to Howard W. Sams and Co., so I quickly prepared an eight-page proposal that described twenty laser applications that could be covered in a major article to be titled "The Laser Comes of Age" or "There's a Laser in Your Life." I sent the proposal and a cover letter to Scott on August 21, 1975. The laser-bugging proposal Scott requested was also included. I titled it "Listening in on a Personality." The proposal listed three phases for the project:

1. Alastair Gregor, letter to Les Nelson, July 1975.
2. Forrest M. Mims, letter to Bernard Scott, August 21, 1975.

1. Feasibility Study—An experimental apparatus will be designed and assembled. The apparatus will be tested against a "typical" window and a tape recording produced of any sounds present (sound source will be a radio behind the window). The tape will be forwarded to the National Enquirer for evaluation.
2. Demonstration Experiment—The experimental apparatus will be demonstrated before representatives of the National Enquirer. If the demonstration is approved, a final apparatus will be assembled for the actual intercept.
3. Actual Experiment—The final apparatus will be taken to an on-site location and used in an attempt to intercept conversations.[3]

On September 5, Scott called. He ignored the laser-article proposal but was highly interested in the laser-eavesdropping proposal. He said the plan was for me to build the apparatus, fly it and me to Florida, and impress the publisher by using it to intercept conversations in the publisher's office. He provided a detailed description of the office, including its dimensions, the windows, and the location of the desk. Scott said that "this capability" would provide an important method for the *National Enquirer* to verify some of its major stories. He emphasized that the project must be considered top secret—that was a phrase I had not heard since working in the Air Force Weapons Lab.

On September 9, I called Scott to report tests in which the sound of a radio was detected by pointing the very narrow beam of a helium-neon laser at the window of the room in which the radio was playing. I repeated this experiment during the day by detecting sunlight reflected from the same window. Scott asked me to make a tape recording of these experiments. He then said that "the device" should be reliable and easy to use "when it's needed." That is when it became clear they wanted to keep whatever I built, and that is when I realized they were seriously planning to bug Howard Hughes. On September 15, I sent Scott a tape of a test session with the radio along with a letter explaining my concern: "I do not want any involvement with illegal operations but see an exceptionally unusual article in exposing the security weaknesses in various government offices. . . . In short, I see a good article on exposing the weaknesses of high-level offices and describing our little technique for the public."[4]

Scott responded positively to my suggestion about the vulnerability of government offices and said the *National Enquirer* had contacts who were close to Senator Barry Goldwater. He suggested we might get permission to spy on

3. Forrest M. Mims, letter to Bernard Scott, August 21, 1975.
4. Mims, letter to Scott, September 15, 1975.

Goldwater's office in the Senate Office Building to demonstrate the security problem posed by laser eavesdropping.

Scott's acquiescence to my concerns was welcome, so I continued to work on the laser apparatus. I was also interested in using sunlight to eavesdrop on a window, a method unknowingly devised by Alexander Graham Bell in 1880 when he invented lightwave communication with his famous photophone.

I had written about Bell's photophone in books and articles and expanded on the eavesdropping project by placing a cardboard box on a vacant lot across the street from our mobile home. I had fitted the box with a glass window from a picture frame and placed a radio inside. This arrangement worked reasonably well, but it nearly ended my new career as a high-tech spy.

It was necessary to stare at the reflected sunlight while adjusting the receiver lens, and even though I wore sunglasses, the brilliant beam of reflected sunlight produced afterimages called scotomas in both eyes. The effect was identical to that produced when a camera's flash is activated near one's face. However, while a flash-lamp scotoma disappears within a few minutes, the sunlight scotomas were still present the next day.

When I looked at a sheet of paper on my desk, there was a black hole. Additional tests with a card placed over one eye and then the other confirmed that both eyes were affected. My peripheral vision was normal, but the scotomas had done to both my eyes what Roger Mark and I had done to the eyes of the rhesus monkeys during the laser tests at the Laser Lab several years before.

I visited an ophthalmologist, who found no evidence of retinal damage in either of my eyes. After a week, the scotomas began to fade, and after ten days, my central vision was fully restored. So was my interest in the laser-eavesdropping project, which I called Scott to discuss on September 30.

He said that his superior remained interested and that I should send a note on the probability of success. He also seemed receptive to my suggestion about revealing Soviet spying. I then prepared a detailed progress report for Scott in which I reminded him about my concerns: "Assuming this new technique is reasonably successful, I feel the need to remind you that I assume its use will be constructive. For example, I see an excellent story in revealing the security weaknesses of our government and urging that they be strengthened. But the various espionage laws make the production of such a story fairly tricky unless prior arrangements have been made with an interested official who approves the venture."[5]

On October 18, I sent Scott a report about experiments conducted during

5. Mims, letter to Scott, September 30, 1975.

the recent World Hot Air Balloon Championships at Albuquerque the previous week. *National Geographic* photographer Otis Imboden had participated in an experiment I arranged in which he sent his voice over a beam of sunlight to balloonist Nikki Caplan. This may have been the first time voice had been sent from the ground to a balloon over a beam of sunlight, so I sent two photos to Scott.

Having never received anything in writing from Scott about our "top secret" project, I wrote to him, "I am reluctant to proceed without some written communications from you concerning both my fee and the projected use of the device. In looking through my National Enquirer file, I find not a single piece of correspondence on these two subjects. In fact, I have never received a letter from you!"[6]

The next day, Scott called, and sent the only letter I received from him. In the November 12 letter, he agreed that the *National Enquirer* would pay for the purchase of a telescope and precision micrometer for the project for up to $350 ($1,890 in 2023). He also agreed to pay a weekly rate for my time, but he did not mention the $500 ($2,700 in 2023) per week we had discussed by phone. I explained that while the red helium-neon laser beam worked well, the beam was so bright, it would be easily seen at night. Therefore, I needed to assemble a near-infrared laser illuminator that emitted an invisible beam.

The optical supplies arrived in late December, and on January 8, 1976, I wrote Scott that a battery-powered laser system with an invisible beam was completed. I then began work on a receiver system having much more sensitivity than the one used previously. On January 22, I called Scott to inform him that all the equipment was finally assembled.

Laser Bugging the *National Enquirer* Headquarters

On February 9, I arrived in West Palm Beach and assembled the laser apparatus in my briefcase at a Howard Johnson's hotel. The following morning, Scott and reporter Tony Brenna drove me to the *National Enquirer*'s main building.

Because of the paper's somewhat sleazy reputation, I was surprised by the impeccable landscaping and the neatly organized arrangement of desks and offices inside. Both Scott and Brenna sensed that I was concerned about the paper's reputation, for they explained in some detail how the paper carefully researched every story it published.

They were especially proud of the computer system that stored every article published by the paper, for this allowed them to quickly determine whether

---

6. Mims, letter to Scott, November 11, 1975.

a story had been previously published on a proposed topic. If so, an updated version would have to include significant new information or it would be rejected. They claimed that they were much more careful about fact-checking than other major newspapers. That was why they were so highly interested in the laser-eavesdropping apparatus, they explained, for they hoped it could be used to surreptitiously verify facts.

Before setting up the laser gear, I spent an hour at Scott's desk discussing the project. I especially wanted to know whether they had made any progress in contacting Senator Goldwater's staff about arranging a demonstration in Washington. But Scott had only one topic in mind: Could the laser equipment be used to spy on Howard Hughes? He produced photos of the thirteen-story Xanadu Princess Resort & Marina in the Bahamas and pointed out Hughes's bedroom on the right side of the top floor, the parapet that shielded the top floor from the one below, and the pyramidlike structure atop the roof that blocked helicopters from landing there.

I explained how the laser system could be used only against a normal window and not one set at an angle. This meant that a laser beam pointed at Hughes's thirteenth-floor bedroom window would be reflected toward the sky instead of the receiver on the ground. Scott then casually said that the *National Enquirer* would pay $100,000 (over $500,000 in 2023) for a full-face photo of Howard Hughes—without asking if I could obtain such a photo.

This was a good time to move the subject away from the laser project, so I suggested that the *National Enquirer* could fly a quiet, radio-controlled blimp equipped with a camera outside Hughes's window and take a photo when he looked outside. I was smiling, but Scott took me seriously and asked if I knew anyone with such a blimp. *American Modeler* magazine had published an article about such a craft several years before, so I suggested he contact the magazine. Scott immediately called an assistant to his desk, and within fifteen minutes, they were speaking with the blimp's owner by telephone.

I was curious why they were so interested in Hughes and asked whether he had any connections with organized crime. Instead, they told me that the *National Enquirer*'s owner might have such connections via his father. They said a quick call to the "right" person was occasionally needed to get the delivery trucks moving.

By midmorning, I had set up the laser apparatus on the carefully manicured yard outside the building, 100 or so feet (30 m) from the glass window of a conference room in which we had placed a radio tuned to a talk program. As I began a series of tests, the equipment and I began to attract a growing crowd of National Enquirer staff who came outside to watch what they were told was a top secret project.

A quick test of the system was to pass my hand back and forth across the laser beam while listening for clicks from the laser receiver. When a woman standing nearby assumed I was waving at her, she waved back and then walked over to ask what I was doing. While politely explaining that I was not allowed to explain the details, I noticed a man under a tree about 50 feet (15 m) away taking photos of us and the top secret equipment with a camera equipped with a telephoto lens.

When I looked at the photographer, he approached us while still taking pictures and introduced himself as the one-man show behind the *National Enquirer*'s in-house employee's newspaper. He was the junior paper's reporter, photographer, and editor, and he desperately wanted to scoop the senior paper with an article about the mysterious apparatus that had attracted the attention of most of the employees. He was as persistent in his questioning as I was in explaining why I could not answer them.

Scott and Brenna were nowhere nearby during this encounter, but occasionally I could see glances from a tall, distinguished-looking man when he stood up at his desk behind a window in the building. He was Generoso Pope, publisher of the *National Enquirer*. While he was fully aware of the project and looked my way from his office many times during two days of tests, he never approached for a closer view of the equipment or to introduce himself.

Back in Albuquerque, the laser system had worked reasonably well. But the *National Enquirer* demonstrations were temperamental. All the windows were made of extra-thick, hurricane-proof glass that refused to vibrate in response to sound as well as the window glass I had tested back home. I had brought a backup device, a short, 1-inch-diameter aluminum tube to which I had cemented an ultrathin glass mirror. When this gadget was mounted inside the conference room against the corner of a window with a bit of clay and the laser beam was aimed at the mirror, voices in the room were easily detected.

After I concealed this device inside the conference room, I asked Scott and Brenna to search the room for anything unusual. They found nothing and were suitably impressed by how well the tiny laser reflector worked, for it picked up every word, cough, and sound they made during their search. But they wanted a system that did not require inserting anything inside a room being monitored.

That night, we tried again and were successful. When the window was 112 feet (34 m) away from the laser and receiver, slight adjustments of the receiver lens within the reflected beam finally allowed conversations in the room to be intercepted. The next day, we had similar results using both the red helium-neon laser and the invisible-beam laser. All these tests required considerable time and careful adjustment of the position of the laser with respect to the window. This method might work if the laser system was installed in a van parked on a street opposite a window of interest. But that is not what the *National Enquirer* had in mind.

When I was preparing to leave, Scott gave me a *National Enquirer* statement and a "Story Payment" check for $2,020.85 ($10,912 in 2023). This included $261.47 for the flight to West Palm Beach, electronics expenses, and $500 a week for 3½ weeks of work. The following morning, I took a walk along the beach to reflect on the events of the previous two days and washed it all away with a swim in the Atlantic.

Laser-Eavesdropping Redux

On February 18, 1976, Scott called to discuss reports that the Soviet Union and other countries might be aiming microwaves into the United Nations building in New York. He asked me to send a proposal for conducting a microwave and optical survey of the area around the UN building. When I reminded him about the idea of demonstrating the laser-eavesdropping project in Washington, DC, with the cooperation of Senator Goldwater, he said he would propose this to his superiors. I quickly prepared a three-page proposal and sent it to him the same day.

A Senate committee staffer indicated interest in seeing a demonstration, but he never followed through. Years later, I wrote several articles about laser spying, including "Surreptitious Interception of Conversations with Lasers." This article was the cover story for *Optics & Photonics News*, a publication of the Optical Society of America. The cover photo showed a miniature laser source and the receiver built for the *National Enquirer* project in a miniature briefcase over the cover headline "The Case of a Laser Spy."[7]

Shortly before the *Optics & Photonics News* article appeared, John Horgan, a reporter for the Institute of Electrical and Electronics Engineers, published a full-page story about my *National Enquirer* adventures in the organization's tabloid, the *Institute*, after I gave a talk at an IEEE meeting.[8]

The articles Horgan and I wrote about laser eavesdropping attracted attention in the media. HBO flew Eric and me to New York to stage a demonstration of laser eavesdropping between two apartment windows for one of their shows. We used the helium-neon laser (which raised concerns in airport security due to its pipe bomb–like appearance) to listen in on a conversation in a nearby apartment occupied by HBO staff.

On April 29, 1988, a man who identified himself as an investigator for the

---

7. Forrest M. Mims III, "Surreptitious Interception of Conversations with Lasers," *Optics & Photonics News* 11, no. 11 (1985), 6–12.

8. John Horgan, "Inventor Seeks to Warn Government of Threat from Laser-Based Bug," *Institute*, October 1985, 8.

Internal Revenue Service called. He was intrigued by my laser-eavesdropping publications and said he would like me to assist with a criminal investigation of an organized-crime group in Chicago.

After the caller provided detailed information about the house to be targeted and the field surrounding it, I asked questions about how I would be protected if the project was discovered. I also said I would not do anything to violate eavesdropping laws. The caller assured me I would be kept safe and asked me to send technical details and the articles I had published. I asked him to send me a written request so I could verify that he was a federal government employee. On May 2, he sent the following under an Internal Revenue Service, Department of the Treasury, letterhead:

May 02, 1988

Dear Mr. Mims:
Pursuant to our conversation on Friday, April 29, 1988, I am requesting the articles and other material we discussed. Please send it to U.S. Treasury Criminal Investigation Division, One N. Wacker Drive, Room 819, Chicago, IL 60606.
Please forward any other material you feel would be of help.
Thank you, Forrest, for your help.

Sincerely,
[signed]
Forensic Investigator[9]

This letter was much more formal than the enthusiastic tone of the writer's voice when he called. It said nothing about my protection, eavesdropping laws, or a warrant to conduct the proposed operation. On May 18, 1988, I sent the requested materials and included a letter expressing serious concerns about the proposed eavesdropping project. He never replied.

Later, a man who said he was with the Federal Bureau of Investigation called and asked whether I could provide a confidential demonstration of my laser-eavesdropping equipment. I agreed only if the demonstration did not violate any laws, and he said that could be arranged. He never called back.

Finally, there was a computer convention in 1986 in Los Angeles, where I was exhibiting my Altair 8800 microcomputer while promoting my book *Siliconnections*

9. Internal Revenue Service, letter to Forrest M. Mims, May 2, 1988.

for McGraw-Hill. A woman who said she had heard me describing the *National Enquirer* laser-eavesdropping story during a radio interview called a publicist and asked for a private meeting.

After I was escorted to a curtained-off area, a remarkably beautiful woman dressed like a movie star arrived and earnestly pleaded for my help to prove that her ex-boyfriend was bugging her house with a laser. I explained the technical difficulties and how unlikely that was. We went back and forth for thirty minutes, but she was not convinced.

None of my projects have attracted such a diverse range of people as laser eavesdropping, and this especially includes John Horgan. Nontechnical reporters often made errors in their descriptions of the electronics devices I built, but not Horgan. He was a careful interviewer and first-class writer, which explains why he ended up as an editor at *Scientific American* magazine. As related in chapter 12, in 1989, I sent my proposal to Horgan to take over *Scientific American*'s column The Amateur Scientist. Had I not known John Horgan, my misadventure with that famous magazine might never have occurred, and I might never have begun doing serious science.

9
THE SILICON GARAGE

Thanks to the check for $10,000 ($53,996 in 2023) Howard W. Sams and Co. provided for the ITT reference book updates, during July 1976, Minnie, Eric, ten-month-old Vicki, and I left Albuquerque and moved to Live Oak, Texas, a town adjacent to San Antonio. We leased a small house for six months to provide time to look for a place we could buy. In January 1977, we moved to San Marcos, a college town in Central Texas, where we bought a home with space for a large garden.

Building the garden required removing a 10-inch layer of limestone rock with a steel wrecking bar and filling the open space with three truckloads of soil dumped in the front yard and hauled to the garden in a few hundred wheelbarrow trips. The leftover limestone rocks and slabs were used to build a border around the garden and a wall along the woods behind. The garden, which provided plenty of veggies, was an ideal teaching tool for Eric and Vicki.

My parents lived in San Antonio, and Daddy helped me convert half our garage into an office and electronics shop. In *Siliconnections*, I referred to Ed Roberts's garage in Albuquerque, where we had built the first MITS light flashers and telemetry transmitters, as the Silicon Garage. I now had my personal Silicon Garage.

I was then writing three monthly columns for *Popular Electronics*, including Experimenter's Corner, which followed Lou Garner's Solid State column. I did not then know that Garner had designed the code practice oscillator module I bought in 1966 that became the pulse generator for my travel aid for the blind and the model-rocket light flasher that led to the founding of MITS.

My first book after the move to Texas was *Radio Shack Introduces the World of Computing* (1977). This book accompanied Radio Shack's entry into personal computers, and it was followed by an assignment to write the much longer *Understanding Digital Computers: A Self-Learning Programmed Text That Will Teach You the Basics for the Microcomputer Revolution*. While I had written books about integrated circuits that counted and performed basic logic operations, I did not fully understand what happens inside the microprocessors that make personal computers possible. This must have been the case with the many computer authors who simplistically compared the operation of a microprocessor to how mail is sorted into letterboxes in a post office. That is not how they work.

Dave Gunzel, Radio Shack's technical editor, had assigned the book. During a status session in Fort Worth, I told Dave the book was complete except for the chapter on computer organization. I told him I was making progress, but I really was not until I read *Digital Computer Electronics*, by Albert Paul Malvino (McGraw-Hill, 1977). This remarkable textbook described the organization of a computer's processor circuitry so well that I was finally able to complete the book with a fifty-page chapter titled "Computer Organization." This chapter described in detail an imaginary programmable instruction processor (PIP-1) that I designed on paper and equipped with a set of microinstructions. The book sold 165,000 copies at Radio Shack, and even today, engineers still send thank-you notes to me for the book that assisted their entry into computer engineering.

After completing *Understanding Digital Computers*, I decided to transform the book's imaginary PIP-1 into a working processor. There was not space on my workbench to assemble the circuitry, so Minnie permitted me to do so on our kitchen table. Within a week, PIP-2 was a dozen integrated circuits interconnected by more than a hundred wires sprawled across the table.

Programs were loaded into PIP-2 by pulling a paper strip through a slot in a small wood block through which five parallel holes were bored. A row of five LEDs was installed in the holes in the upper side of the slot, and a row of five phototransistors was installed in the lower side. A series of twenty ¼-inch holes was punched along the center of the paper strip. Up to two holes were punched on either side of each center hole. The center holes allowed the center phototransistor to be briefly illuminated in sequence by its LED when the paper strip was pulled through the slot. Each flash of light received by the phototransistor created a clock pulse, a brief voltage spike that advanced the processor's memory address one step at a time. The holes punched on either side of the center holes represented instructions or data loaded into memory during each clock pulse.

PIP-2 and its six microinstructions were described in detail in "Microprocessor Microcourse," a five-part series I wrote for *Popular Electronics* (March–July 1978). I did not include the circuit diagram, because it was quite complicated and other assignments waited. Designing and building PIP-2 removed a huge gap in my electronics knowledge. Only after designing that simple processor did I finally understand the relationship between circuits and instructions that makes digital computers possible.

The Hand-Printed Radio Shack Books

Back in Albuquerque, I dictated most of the Radio Shack books into a small tape recorder during long afternoon bike rides. Minnie spent many hours at the Selectric typewriter transcribing the tapes. That is how most of *Understanding Digital Computers* was written.

While that method of writing books worked well, it was about to be abandoned. Dave Gunzel, Radio Shack's editor, was pleased with the success of *Understanding Digital Computers*, and during a meeting we discussed ideas for future books.

Dave was familiar with my lab notebooks, especially the entries about using LEDs as light detectors. During our meeting, Dave, the most creative editor of my career, next proposed an idea that had never occurred to me: Why don't we develop a book that I would print and illustrate by hand, just like my lab notebooks? Gary Burkhart, then Radio Shack's electronic-parts buyer, also liked the idea, for it would promote the many electronic parts for which he was responsible. We decided I would hand-print and illustrate a book on how to use each of the integrated circuits sold by Radio Shack.

I used an India-ink pen to print and illustrate the book on toothed Mylar sheets. Mistakes could not be corrected, so entire pages were redone when a mistake occurred. After a few months of building and checking the circuits for the book and transferring them to Mylar pages, the result was *Engineer's Notebook* (1979).

The first few introductory pages were typed on the Selectric, and the first hand-lettering was when I signed my name. By page 11, every page was printed and illustrated entirely by hand. When I drove to Fort Worth with the completed manuscript pages, I asked Dave if I could use an empty room to spray each page with a protective coating. While spraying the pages, Harry Helms, Dave's new editorial assistant, arrived to watch. That first meeting led to a long and productive friendship.

The Hand-Printed Books Become Bestsellers

Concerns about the fate of the new hand-lettered and illustrated *Engineer's Notebook* were alleviated when Radio Shack sold hundreds of thousands of them. Dave Wolf had become Radio Shack's parts buyer in December 1979, and he assigned me to update the book. *Engineer's Notebook II* was published in 1982. Total sales of the two editions exceeded 750,000 copies.

Dave Gunzel next proposed a major new hand-lettered book, which we called *Getting Started in Electronics*. This book would be a complete introduction to basic electronics followed by a hundred projects that readers could build, described in only 128 pages.

Using india ink on Mylar to produce the two editions of the *Engineer's Notebook* was hard on my right hand and even caused my index finger to bleed. Therefore, I requested to print and illustrate *Getting Started in Electronics* with a 0.7 mm mechanical pencil. Dave agreed, and this greatly sped up the production of the book, since mistakes were easily corrected.

Dave decided that Radio Shack and I, instead of Howard W. Sams and Co., should publish the new books. This arrangement would permit a lower cover price while maintaining a good profit margin for Radio Shack and a higher royalty for me. Dave took me to the Banta Corporation, introduced me to its staff, and showed me how huge rolls of paper are transformed into printed and bound books. While I was apprehensive about the new approach, it minimized the book's price while increasing my royalty.

After I made a storyboard summary of the book in 128 rectangles on several sheets of paper, Dave approved the organization of *Getting Started in Electronics.* All 128 pages were then hand-executed in fifty-six days, including building and testing all one hundred circuits four times each to make sure there were no errors.

Why four times? It is easy to build electronic circuits from memory, and that allows errors to slip through in the print version. Therefore, I rebuilt each circuit by strictly following my diagrams in the book and not from memory. *Getting Started in Electronics* was published in 1983 and quickly became Radio Shack's all-time bestseller. The book is still in print in several languages. It has sold 1.3 million copies and is pirated around the world on the internet and in print.

Between Radio Shack books, I worked on many other projects. In 1978, I wrote *The Programming Book,* a sixteen-page booklet for Hewlett-Packard that explained how its powerful programmable calculators are programmed. I also wrote a companion booklet of nineteen pages entitled *What to Look for Before You Buy an Advanced Calculator*. The publisher, the 13-30 Corporation, was under severe deadline pressure, which is not a positive environment for producing technical publications. I was flown to its offices in Knoxville, and an editor sometimes watched over my shoulder as I typed. Several times, he took pages out of the typewriter before I had a chance to review them. While this was a terrible environment for writing about program editing, subroutines, conditional transfers, flags, counters, and other technical topics, the finished booklets looked great.

In 1979, I wrote a hundred articles for Arête Publishing's *Academic American Encyclopedia.* This twenty-one-volume work was acquired by Grolier, which transformed it into the first electronic encyclopedia in 1985. This was an intriguing connection, for Grolier was publisher of *The Book of Knowledge,* the children's encyclopedia my brother, Milo, and I had regularly read when we were growing up.

A Practical Introduction to Lightwave Communications, based on a short course I developed for General Electric, was published by the Institute of Electrical and Electronics Engineers in 1982 along with a course I taught. *The Forrest Mims Circuit Scrapbook* was a compilation of many of my columns in *Popular Electronics* magazine that was acquired by Harry Helms after he moved from Radio Shack to McGraw-Hill in 1983.

In 1985, McGraw-Hill Osborne Media published *Forrest Mims's Computer Projects*, a collection of my magazine articles on how to use personal computers for science and technology applications well beyond writing and accounting. The book included programs I wrote that allowed personal computers to function as virtual instruments, including an oscilloscope, a light meter, an analog computer, an optoelectronic digitizer, and others.

Art Salsberg, my longtime editor at *Popular Electronics*, wrote the foreword. Even today, some skeptics and atheists unaware of my science pursuits claim that my objection to Darwinian evolution makes me an incompetent technical writer. Art opened his foreword with a different view: "Standing alone among electronics and computer writers, his written and diagrammed outpourings are largely the result of his imaginative pursuit of how electronic and computer technology really works beyond the traditional textbook. In the fashion of inventors of the past, out of reach of today's multimillion-dollar research laboratories, he continually probes the workings of electronic devices and how they can be utilized in ingenious circuits to accomplish new, more useful results, as well as doing the same with computer software programs to drive hardware. . . . Original work, though a rarity, is what Forrest Mims serves up on a regular basis."[1]

Around this time, I received a letter from an engineer at Panavision asking whether I would be willing to write the instruction manual for its newest camera. While this was a tempting offer, Panavision cameras are used to film some motion pictures that depict extreme violence or moral or ethical issues I cannot endorse, so I declined the offer.

Harry Helms

As noted earlier, I first met Harry Helms in 1979 at Radio Shack headquarters in Fort Worth. Harry was a ham radio enthusiast and avid shortwave radio listener who had worked at Texas Instruments. His background was a good match for Radio Shack editor Dave Gunzel, and the three of us became close friends. Harry was both

1. Art Salsberg, foreword to Forrest M. Mims, *Forrest Mims's Computer Projects*, by Forrest M. Mims (New York: McGraw-Hill Osborne Media, 1985), v-vi.

a writer and an editor, and a few years later, I recommended him to Hal Crawford, a senior editor at McGraw-Hill, and Hal hired Harry away from Radio Shack.

Harry had a wonderful sense of humor, and he could do a remarkable imitation of Steve Martin. He even did a few performances at a popular New York comedy venue. The performance side of Harry's personality was sometimes on display at McGraw-Hill editorial meetings, as I learned in 1983, after he assigned me to produce *The Forrest Mims Circuit Scrapbook*, a book (mentioned above) of many of my columns in *Popular Electronics* magazine. When the book was in production, an editor struck the concluding sentence of my "About the Author" section on the back cover, "A Baptist deacon, he and his wife, Minnie, and their children, Eric and Vicki, are active members of their church." Harry later told me that, infuriated, he went to her office, jumped on top of her desk, and told her to never censor an author's background from his book. The missing sentence was returned.

I assumed that Harry was exaggerating about jumping on the editor's desk, but another editor told me that Harry sometimes jumped atop a table to make his points during business meetings. A few years later, I visited McGraw-Hill to meet with Harry and his boss, Hal, to discuss a new book project.

While I was standing in front of Hal's desk, there was a loud sound at the door. Suddenly, the door swung open, and a man rolled sideways along the door and into the room while grunting and making loud, muffled sounds. It was Harry in a mock straitjacket made from wearing his coat backward. Hal stood up and seemed as unsurprised as I was surprised by the spectacle. My concern that Harry's performance might anger Hal was quickly ended by Hal's vigorous laughter. The professional version of Harry then joined our discussion about the new book project.

Harry eventually moved to Prentice Hall and then to Wiley. While he was at these publishers, he hired me to review technical books he was considering acquiring, and he eventually partnered with Carole Lewis to form HighText Publishing.

New York's Algonquin Hotel

Over the years, I met with a series of New York editors. When *True* magazine flew me to New York to discuss an article proposal, I was put up in the historic Algonquin Hotel, which I did not know was a well-known meeting place for famous literary figures. Maybe that is why three older men seated in chairs by the counter shifted their stares from the resident cat to me while I was checking in.

The next morning, *True*'s editor began our meeting by sternly warning me to avoid feeling famous just because I had stayed at the Algonquin and was now meeting with him. That strange introduction was on my mind a few years later during a second visit to *True*. This time, the former assistant editor, who was much more friendly, occupied the office of the former editor, who was seated at a lonely desk in the hall between offices and said nothing when I walked by.

Then there was my visit to *Popular Electronics*, when I heard a shout from behind as I pushed open the revolving door. I looked back, saw a man on a unicycle, and gave the door the extra push he ordered before he entered the door and unicycled to the elevator. These and many other experiences visiting New York editors were memorable, but nothing came close to straitjacketed Harry Helms's noisy entry into Hal Crawford's office.

The Evolution of a Debate

I occasionally wrote guest articles for the *San Marcos Daily Record* after we moved from Live Oak to San Marcos. After Dr. Tom Gray, an anthropology professor at Southwest Texas State University, now Texas State University, wrote a column advocating the teaching of evolution, I replied with an opposing view. We then exchanged a sequence of columns challenging one another's positions.

When he was a graduate student in 1974, Gray and Dr. Donald C. Johanson had jointly discovered Lucy, the fossil ape believed to be a human ancestor. When Gray invited me to visit his lab, I was surprised to see plaster replicas of Lucy's bones neatly laid across a table.

When I asked why a muscular student was standing nearby, staring at me, Gray smiled and explained that he was there to protect the fossil replica in case I displayed violent opposition to it. Another table in Gray's lab was covered with flint implements, and he handed me a quartz hammerstone. That ancient tool consisted of silicon dioxide from which pure silicon could be extracted. There was enough silicon in that hammerstone to make thousands of microprocessors, memory chips, and digital camera sensors.

During this and subsequent meetings, Gray and I candidly shared our philosophies, including his negative feelings about rivalries in anthropology teaching departments. During a meeting at his house, he served a delicious lunch of neatly engineered sandwiches. That was our last meeting, for he soon left the university and opened a sandwich shop in Austin.

Family Matters

The large lot and the Silicon Garage at our San Marcos home provided space for a variety of science experiments that could be shared with Eric and Vicki. For the annular solar eclipse of May 30, 1984, I mounted a cheap telescope on a section of two-by-four lumber, to which I attached a clipboard that held a sheet of paper. Vicki used this simple apparatus to trace the pattern of the solar disk as it was being obscured by the moon during the eclipse.

Eric and I launched model rockets from a hilltop field that is now home to offices and a hospital. I used kites and helium-filled trash bags to take aerial photographs over our place and nearby fields. It was easy to modify a Kodak camera to take photos when triggered by a signal from the radio-control transmitter I had used for rocket experiments in Vietnam.

During a visit to Corpus Christi, I flew the kite camera 75 feet (23 m) over the ground while the kids tossed food to seagulls. By adjusting our position relative to the wind, it was possible to fly the kite camera behind the birds to take pictures from their perspective of us down below. My kite camera was almost lost one afternoon when the kite crashed into a field. After I climbed the barbed-wire fence to retrieve the kite and camera, I had to run for my life when a curious bull gave chase.

I became very involved in optical-fiber communications and sent signals over a fiber to a helium-filled balloon tethered above. That and my transmission of infrared signals from a model rocket to the ground might have been technical firsts, but they were mere demonstrations with no practical value.

10
MIMS VS. BELL LABS

Since reading about Bell Laboratories in the July 1954 issue of *National Geographic* magazine when I was eleven years old, one of my ambitions had been to someday visit Bell Labs, the research arm of AT&T and one of the world's most productive scientific institutions. The transistor and the silicon solar cell were invented there, and the laser was first proposed by Bell Labs consultant Charles Townes and Bell Labs scientist Arthur Schawlow, his brother-in-law.

The Weapons Lab sent me to Bell Labs in 1970 to discuss its latest diode laser developments. In March 1979, I again visited Bell Labs, this time with Peter Purpura, curator of Explorers Hall at the National Geographic Society's headquarters in Washington, DC. The purpose of our visit was to discuss the Photophone Centennial, an event to commemorate the 1880 invention by Alexander Graham Bell of a method to send voice over a beam of light.

The Photophone Centennial

On February 19, 1880, Bell made history with the photophone, and he firmly believed it was the most significant of his inventions, even more so than the telephone. In 1978, I sent Melville Bell Grosvenor, Bell's grandson, a proposal to honor the centennial of Bell's invention of lightwave communications. Grosvenor, a former editor of *National Geographic* magazine, liked the proposal and invited me to visit him in Washington.

The walls of Grosvenor's office were covered with awards, plaques, and photographs of US presidents and famous explorers. When I showed him my homemade photophone, he listened intently as I explained its operation. The transmitter was a 1-inch-diameter aluminum tube with a thin glass mirror cemented to one end. (I did not mention its role in the *National Enquirer* adventure.) The receiver was a solar cell connected to an amplifier. Voice communication was established by reflecting sunlight from the transmitter mirror to the solar cell while speaking into the transmitter tube. The sound waves caused the mirror to vibrate in step with the voice, thereby imprinting the voice onto the beam of sunlight reflected by the mirror.

Grosvenor was fascinated by the simplicity of this version of his grandfather's invention ninety-eight years earlier. He excitedly insisted on a demonstration, but no sunlight was entering the room through the windows. That was no problem for the seventy-seven-year-old Grosvenor, who opened a window and scrambled onto a narrow ledge outside while holding my photophone transmitter. He then attempted to reflect a ray of sunlight to my receiver while speaking into the transmitter.

We next met with the National Geographic Society's board, which enthusiastically approved my proposal. Peter Purpura, curator of Explorers Hall, was assigned to implement the proposal, and our first mission was to visit Bell Labs to seek its cooperation.

Before arriving at Bell Labs with Purpura, I had arranged to meet William L. Keefauver, Bell Labs' vice president and general counsel, to discuss an invention suggestion I had submitted in 1973. While Purpura was meeting with Bell Labs' exhibit designers, I was escorted to the meeting with Keefauver, who headed a staff of nearly a hundred patent attorneys.

I did not know that then. Nor did I know that AT&T and its Western Electric and Bell Labs subsidiaries formed the world's largest corporation, with assets exceeding the combined worth of Exxon, Mobil, and General Motors. Among the 20,000 scientists, engineers, and support staff at Bell Labs were seven Nobel Laureates.

A Matter of Invention

As related earlier, in 1962, I discovered that photoresistors that detect light can also emit light. On March 14, 1966, during my senior year at Texas A&M, I had described in my notebook an experiment that showed that silicon solar cells that produce electricity from sunlight can also emit infrared light. In 1972, I discovered that light emitted by an LED can be detected by a similar LED, and I sent tones and voices from one LED to another LED several feet away and longer distances through an optical fiber. The roles of the LEDs were easily reversed, which meant information could be sent both ways through a single fiber. I have referred to this bidirectional use of LEDs as LED><LED.

I spent several days searching through scientific journals and books at the University of New Mexico library to determine whether these findings were novel. While I discovered that some solid-state physicists were aware of the basic principle at the quantum level, I found nothing about using LEDs or any other electronic components as both emitters and detectors of light. I could not afford a patent attorney, so I described my discovery in *Light Emitting Diodes*, a book I was writing.

This disclosure would provide a one-year grace period for applying for a patent.

On May 29, 1973, I entered in my lab notebook the latest experiments in which LEDs were used as both light emitters and detectors and then wrote a two-page letter to the Patent Department of Bell Labs. The letter included technical details and this specific application: "Fiber optic optical links employing a single LED at either end. This system would provide the potential for two-way communication over a single fiber optic strand."[1]

On June 5, 1973, Bell Labs' chief patent attorney, William L. Keefauver, wrote that ". . . we will consider your idea or suggestion" if I agreed to "Submission of Inventions or Suggestions," which stated, in part, "Should we want to make use of the suggestion, and if it is not already known to us and is not in the possession of the public and is original with the party submitting the suggestion or with a person whom he represents, we would expect to discuss the matter with such party in an effort to arrive at an agreement that is mutually satisfactory."[2]

I agreed in a letter to Keefauver dated June 11. Keefauver then asked Bell Labs staff scientists to evaluate my suggestion. Seven years later, I learned that one of them wrote, "A single . . . device usually cannot be designed to meet these conflicting requirements." His superior added, "I think it extremely unlikely that systems considerations would permit a single device to operate as both a source and detector. Certainly all our present thinking has been along the lines of separate fibers for transmitting and receiving."

The chief Bell Labs patent attorney summed up their thoughts: "The conclusion we have reached from our study of this suggestion is that it does not contain any novel features of sufficient interest to us at the present time to warrant our acquiring rights thereunder.3

While this response was disappointing, I continued many LED><LED experiments and learned that diode lasers, which are specialized LEDs, can both detect and emit light. I also built a working device that transmitted and received voice via an infrared beam. I showed this device to Ed Roberts and an engineering friend of his on September 12, 1973, and both witnessed its operation and wrote in my notebook. Ed wrote, "I have seen a demonstration of the device described on pages 8, 9, and 10 of this book. The LED receiver operates as described. H. Edward Roberts, 12 Sept '73."

1. Forrest M. Mims, letter to Patent Department, Bell Laboratories, May 29, 1973.
2. William L. Keefauver, letter to Forrest M. Mims, June 5, 1973.
3. William L. Keefauver, letter to Forrest M. Mims, August 2, 1973.

I described how to build a much neater version of my discovery for *Popular Electronics* in "Communicate Over Light Beams with the First Single-LED Transceiver.[4]

Dave Gunzel, my Radio Shack editor, was aware of my many LED experiments. In late November 1978, he told me that the December 4, 1978, edition of *Business Week* magazine reported that Bell Labs had announced a new kind of telephone system that incorporated the invention I had submitted in 1973. The new system ". . . would dramatically alter the basic nature of the phone network."[5]

In December 1978, *Optical Spectra* reported that the new phone employed a light detector that doubled as a light source, "thus permitting the sending and receiving of light signals over one fiber with a single device."[6]

In October 1979, *Electronics* magazine reported, ". . . the new phone could establish AT&T as the No. 1 provider of wideband services to home and industry. . . . The world's first phone powered by light carried on a fiber-optic cable could probably only have come from a place with the resources of Bell Laboratories."[7]

The idea that only Bell Labs could invent the new phone seemed rather silly when compared with my LED><LED experiments in the tiny Silicon Shack back in Albuquerque.

In February 1979, I visited the public search room of the US Patent and Trademark Office to look for patents related to my proposal to Bell Labs. I had no idea how to search for patents, so I was directed to a staff person seated behind a desk in an office lined with stacks of thousands of patent applications. While we looked at each other through a slot in the hundreds of patents stacked on his desk, the gentleman patiently explained the process. He then took me to a large room lined with rows of shelves filled with patents and then to one of the many narrow aisles extending from the room. At the very end of the aisle, he pointed toward a wall of shelves stacked with hundreds of patents and said, "If your idea has been patented, it will probably be somewhere in this section."

The task before me was so intimidating that I prayed for help before reaching for the first bundle of patents. Amazingly, that first bundle included a 1976 patent by Robert G. Hunsperger that described an LED that sent and received signals through an optical fiber. This precisely described a key aspect of the Bell Labs light telephone and my invention disclosure. While the patent preceded the Bell Labs

4. Forrest M. Mims, "Communicate Over Light Beams with the First Single-LED Transceiver," *Popular Electronics*, March 1974.

5. "A 'Light' Phone with Optical Fibers," *Business Week*, December 4, 1978.

6. "Down the Road: Light-Powered Telephones," *Optical Spectra*, December 1978, 33.

7. Harvey J. Hindin, "DeLoach Built the Fiber-Optic Phone," *Electronics*, October 1979, 231.

light phone, it was granted three years *after* my proposal for the same invention was sent to Bell Labs. After making a copy of the patent, I examined hundreds of additional patents and found none that were relevant.

Negotiating with Bell Labs

Bell Labs attorney William L. Keefauver agreed to meet with me when Peter Purpura and I visited Bell Labs to discuss the Photophone Centennial. After we met the public relations staff, I excused myself to find Keefauver's office. Although Keefauver was courteous, he insisted I had no claim. He then gave me two papers by IBM scientists that were published in an internal IBM bulletin. Both papers disclosed diodes that could both emit and detect light, but they did not mention optical-fiber coupling.

Keefauver also said they had applied for a patent on their light telephone, but it "never occurred" to them to include the fact the light source also functioned as a light detector. Keefauver then asked what I wanted to settle my claim. When I did not know what to say, Keefauver said, "It's going to cost us $100,000 ($432,430 in 2023) to fight you in court."

Back in Texas, I spent two months attempting to get Bell Labs to provide reasonable compensation and to promise they would cite my publications in their patent applications and publications about their light telephone. Keefauver offered $15,000 ($64,864 in 2023) for my volunteer work on the Photophone Centennial if I would construct a replica of Bell's photophone and drop all claims. He then offered $25,000 ($108,107 in 2023) if I would drop my claim and credit my next three books and two photophone papers to sponsorship by Bell Labs. I declined these offers.

Since the statute of limitations was approaching, I asked Keefauver whether I could return to Bell Labs to make my case with his boss, N. Bruce Hannay, vice president for research and patents of Bell Labs. Keefauver agreed, and I flew back to New Jersey. I arrived early and spent half an hour viewing exhibits in the reception area that proudly described Bell Labs as a national treasure.

Some claims in the exhibits were exaggerated or even wrong. For example, one exhibit falsely stated that Bell Labs had invented the semiconductor laser, a claim that would have surprised the scientists at General Electric, IBM, and MIT who almost simultaneously demonstrated the first diode lasers in 1962.

Keefauver then appeared and led me to what he said would be a ten-minute "audience" with Hannay. I had not flown from Texas to spend only ten minutes making my case in an "audience," and I fully intended to cover the major points of the dispute outlined on ten note cards. Twenty minutes into the meeting, the discussion became sidetracked, and I suggested that we return to the subject to avoid wasting time. Hannay curtly replied, "You are wasting my time!"

Hannay was offended when I said Bell Labs sometimes claimed inventions made by others. When he said nobody works harder than Bell Labs to verify its claims, I read the false claim about the invention of the diode laser I had copied on a notecard. Hannay was surprised and said that claim "might" need to be changed.

The ten-minute audience ended after sixty-six minutes without an agreement, which pleased Keefauver. He told me that when he retired, he planned to write a book about controversies involving priority in invention. I thought he was well qualified for the project.

Attorney Ted D. Lee

Shortly before meeting with Hannay, I met with patent attorney Ted D. Lee at his San Antonio office to discuss my options. Ted listened carefully as I related what had happened and showed him my notebook and correspondence. I was optimistic he would represent me when I noticed two of my books on the shelf behind him. One was *Light Emitting Diodes*, the 1973 book that disclosed the invention I had sent Bell Labs that year.

It might be unprecedented for a two-man law firm to agree to represent on a contingency basis a freelance inventor in a suit against the largest corporation in the world. But that is what occurred after I told Ted about the unsuccessful meeting with Hannay. He was well informed about the massive size of AT&T, and its huge legal staff. But none of this intimidated Ted, who in 1967 had worked as a patent agent for NASA's Apollo program, which had landed men on the moon. From 1970 to 1973, he had served as a judge advocate general in the US Marine Corps, where he had tried hundreds of cases before entering private practice. While Ted is a good-natured man with a sense of humor who strongly believes in settling disputes, he tenaciously pursues legal action against those who fail to satisfy his clients. As he told me, "I believe in the adage 'The best defense is a good offense.'"

Since the statute of limitations was fast approaching, Ted and his partner, Mark H. Miller, quickly prepared a lawsuit against Bell Labs, which they filed in the US District Court for the Western District of Texas in San Antonio. Both sides then requested the production of relevant documents.

In September 1979, Ted and I flew to New York, where we met with the Western Electric attorneys representing Bell Labs. They provided us with some 2,000 pages of documents in response to our suit. Nearly every page was stamped "Confidential," with the curious exception of the internal reviews of my 1973 suggestion quoted earlier. Those reviews clearly demonstrated that Bell Labs scientists were unfamiliar with the fact that LEDs can both emit and detect light, which was at the heart of my 1973 suggestion and their 1978 light telephone.

Those documents also included the names of the two scientists who had reviewed my 1973 suggestion. Their offices were within a few doors of one another and that of the scientist credited with inventing the light phone. It seemed to me that my 1973 suggestion might have influenced these scientists to conceive the bidirectional feature of the light phone several years later.

Because of hints that Bell Labs might want to settle the lawsuit and the time required to look through the discovery documents, both sides agreed to postpone the depositions until November 14, 1979. During five days of depositions in New York, I watched as former Marine Corps officer Ted D. Lee expertly questioned a series of scientists and a Bell Labs patent attorney about the case.

Ted had much more litigation experience than the typical patent attorney. Each time the team of Bell Labs and Western Electric attorneys aggressively objected to one of Ted's questions, he calmly revised the wording until he got the response he wanted. For example, when Ted asked Bell Labs attorney Michael J. Urbano whether his signature was on a document, Urbano said no. When Ted asked if the signature was a copy, he again said no. When Ted asked if the signature was a "conformed copy," Urbano finally said yes.

Western Electric attorney Patrick Leach was so combative that Western Electric replaced him with a backup attorney. Leach later returned with an older Western Electric attorney, who sat at the end of the table and said nothing. While there were now four attorneys facing Ted, the depositions proceeded with only a few clashes. They ended each day at precisely 5 p.m., when the attorneys immediately shoved their papers into briefcases and left. This annoyed the security guards, for Ted and I were left unescorted.

Because the 674 pages of depositions are confidential, their contents cannot be revealed. But I can say that a brief encounter with one of the Bell Labs scientists was more revealing than anything said during the five days of testimony. When he and I happened to be in the restroom during a break, I apologized for the inconvenience the lawsuit was causing him and the other scientists. "Don't worry," he politely responded. "It's not your fault."

While I did not believe that he or any of the others had overtly stolen my invention suggestion, I assumed that the essence of my suggestion might have been recalled several years after their formal review of the suggestion itself had been forgotten. Even today, I wonder why the scientist said the lawsuit was not my fault. Did he and his colleagues believe that Bell Labs should have paid a reasonable settlement, as suggested in their original letter of agreement? Did he know more about the review of my 1973 suggestion than Ted and I did?

On the morning of January 29, 1980, Western Electric attorney Edward Koziol, with Leach and Urbano, arrived at our home in San Marcos, Texas. Bell Labs had refused to disclose its famous light phone, so we decided to give its representatives full access to the Silicon Garage. They spent the entire morning searching through my file cabinets, desk, and bookshelves and selected some 1,800 pages of royalty statements, college transcripts, résumés, correspondence, and other personal documents not listed in their document request.

Urbano spent the day seated at my desk, inspecting every page of the eight lab notebooks I had kept since college. I objected, for Urbano was a Bell Labs patent attorney, and my notebooks described many potentially patentable inventions. These notebooks included dozens of pages about using LEDs as light detectors that I wanted them to see. But it also included many novel devices, including a nearly transparent microphone that I had proposed in Notebook No. 7 on September 6, 1977, and that was first built and described on January 10, 1979.

The transparent microphone consisted of a very thin mirror mounted at one end of a clear plastic tube. A bifurcated optical cable consisting of two clear fibers was inserted into the opposite end of the tube, and the mirror was illuminated by light from one of the fibers. Voice and other sounds caused the mirror to vibrate and alter the amount of light reflected into the second fiber, which illuminated a light sensor connected to an audio amplifier.

Urbano examined the light-microphone pages and every other page in all my notebooks. When I offered to show them my assembled LED><LED circuits, they declined and merely glanced at them. This response was likely due to their refusal to let us see the light phone.

While Urbano continued to review my notebooks late that afternoon, Leach and Koziol asked whether there were any other documents. I told them there were no documents in the rest of the house, and they were free to search if they did not believe me. I then remembered a box full of old book manuscripts stored in the attic. Leach demanded that I retrieve the box and promised that the box would be replaced. I offered them a flashlight and said they would have to retrieve and replace the box. I then placed a stepladder under the opening to the attic. Koziol climbed up to the attic in his three-piece suit and retrieved the heavy box, which he then jostled through the small opening and carried down a step at a time. They spent half an hour inspecting the old manuscripts, none of which they wanted.

When Mark Miller reminded them that they had promised to replace the box, Leach looked away and announced, "The box will not be replaced." Previously, all my communications with the lawyers had been through Mark. But I was unhappy with the way they had creased, folded, and generally mishandled my many files, so

I firmly told Leach, "The box will be replaced!" Leach replied, "The box will not be replaced!"

After a few more of these exchanges, I surprised all the attorneys, including Mark, by announcing, "None of the documents will be provided if the box is not replaced!" This tactic worked. Leach motioned toward Koziol and began shoving the box toward the ladder. As Koziol entered the attic, Leach removed his coat and climbed the ladder while holding the box. He then managed to insert the box through the attic opening. After they climbed back down, Leach replaced his coat and the three lawyers headed for the front door without a goodbye. Just as they reached their two rental cars, our living room clock chimed five o'clock.

The Photophone Centennial

Three weeks after Bell Labs searched my office, Minnie and I flew to Washington, DC, for the Photophone Centennial. An AT&T publicist had called to ask whether he could interview me for a film they were making about the event, but the interview was canceled. Fortunately, Bell Labs did not pull out of the Photophone Centennial itself, which featured one of Bell's original photophones, loaned by the Smithsonian Institution.

National Geographic photographer Bruce Dale was familiar with my ongoing battle with Bell Labs. During a meeting at the National Geographic Society where a Bell Labs official spoke about the centennial and thanked all but one of those involved, Bruce stated that they had neglected to mention the person who had initiated the event and mentioned me by name.

That was not the first time Bruce had defended me. A few years earlier, he had admired some photos I took during a mule trip in Big Bend Country organized by rancher Bill Saling for fifteen teens at our church, including my son, Eric. When I told Bruce that my article about the trip for *National Geographic World* had been rejected because the photos "lacked involvement with the young people," he strongly objected, stood up from his desk, and ordered, "Follow me!" He then led me to the editor of *World* and showed him the photos. After the editor admired the photos, Bruce asked why they had been rejected. The editor apologized, and my article and photos were soon published in *World*.

Thanks to Peter Purpura's planning and Bell Labs exhibit coordinator James Lowell, the photophone exhibit in Explorers Hall was well received. A bonus was Lowell's uncanny resemblance to Alexander Graham Bell.

The morning of February 19, 1980, Minnie and I were accompanied by representatives from *National Geographic*, the Smithsonian Institution, and Bell Labs as we walked to the site of Alexander Graham Bell's lab at 1325 L Street to reenact Bell's historic transmission of the human voice over a beam of sunlight exactly one hundred years before. *National Geographic* photographer and longtime friend Otis Imboden photographed our activities that morning, which I began by transmitting my voice over sunlight reflected from the same photophone transmitter I had showed Melville Bell Grosvenor two years earlier.

I had also made a pair of identical LED transceivers for the event. Each employed a red LED as a source and detector, and they were connected to opposite ends of a 100-foot (30 m) spool of optical fiber. I deployed the fiber, and we took turns speaking and listening to one another over a single fiber with a single LED at each end.

The irony of the centennial was that it occurred during a lightwave-communications lawsuit against the laboratory named after the inventor of lightwave communications. No one knows what Bell would have thought about this, but history records that he generously and openly acknowledged the scientists who had developed light-sensitive selenium detectors and his coworker Charles Sumner Tainter's contributions to the photophone project. He even listed Tainter as the coinventor of the photophone in the first patent.

On November 14, 1982, I was back in Washington, DC, and visited Otis, whom I had befriended back in Albuquerque. He had been assigned to photograph the first Washington parade for Vietnam veterans, and when I accompanied him to the parade, he insisted that I join the 15,000 vets while he walked beside me, taking photos. The cheers and smiling faces along the parade route were a very different reaction from what I experienced when I returned from Vietnam in 1968.

That was the second time that Otis provided a significant suggestion. The first was when he visited us when we lived in San Marcos. When I told him my ambition was to become a full-time *Geographic* photographer like him, he told me that the frequent long trips away from home could ruin my relationship with Minnie and our children. While I have since published many hundreds of photographs, most were taken from home to illustrate my columns and books, not during trips of many thousands of miles for the highly demanding *National Geographic*.

The Final Bell Labs Deposition

The Bell Labs suit took considerable time. During 1980, Bell Labs and Western Electric deposed at least ten people, including Dave Gunzel, my Radio Shack editor and close friend. They also deposed my editors at *Popular Electronics*. Bell Labs had failed to include my prior art in its patent application for a light phone. In September 1980, it abandoned the application.

The time had finally arrived for me to be deposed, and Bell Labs sent a Notice to Take Deposition demanding that I bring every relevant document produced during the ten months since their representatives had searched my office (". . . letters, memoranda, reports, notes, papers, books . . .").[8] I had developed a short course on lightwave communications for the Institute of Electrical and Electronics Engineers (IEEE), so I gathered every page of leftover drafts, carbon papers, and used typewriter ribbons. I even included a thousand or so quarter-inch paper disks punched from the IEEE manuscript so it could be placed in a three-ring binder. Magazine columns about LEDs and the lawsuit were also included.

All this was placed in a box. In a second box, I placed my camera in a briefcase and covered it with some of my lightwave circuitry and a science fair project my son, Eric, had made that used an LED. (Eric's science teacher had criticized the project, for it was based in part on a circuit published in one of my books. The teacher did not realize that many students across the country used my books to find circuits for their science fair projects.)

Bell Labs had hired Richard Page Keeton, a prominent Houston business attorney, to represent it, and the first day of the deposition was scheduled for November 10, 1980, at Keeton's office on the fifteenth floor of the City National Bank Building in Austin. I rolled a dolly with the two boxes into the elevator of the office tower that hosted Keeton's office and arrived before any of the lawyers. A secretary ushered me into Keeton's conference room, which featured a dramatic view of the Texas capital.

Before visiting Bell Labs, I had read several books on negotiations, all of which emphasized the importance of where to sit during a meeting. The best seat at the table faced the capital, so that is where I sat. Moments later, a man wearing a sports shirt arrived and introduced himself. He was Richard Keeton, and he said he enjoyed visiting his Austin office so he could look at the capital from where I was seated. Wouldn't I prefer to sit opposite his chair? I said, "No, thanks—this is just fine."

Ted Lee soon arrived, as did Leach, Urbano, and Koziol. Keeton then began asking about my life history. Keeton seemed intrigued by my background and the fact that my degree was in government and not engineering or science. He was a graduate of the University of Texas, Texas A&M's archrival, and this also seemed to be behind some of his curiosity. His questions were reminiscent of those asked by journalists and reporters who had interviewed me over the years. This brief exchange from the first morning illustrates Keeton's approach:

> KEETON. *Are you saying your goals had been to end up as a scientist or something? What goals were you telling them that led them to try to put you on a*

8. Bell Telephone Laboratories, Notice to Take Deposition of Forrest M. Mims III, October 2, 1980.

track toward the technical side?

MIMS. *When I was 11 years old I set forth a series of goals in my mind, what I wanted to accomplish in life. And I have been fortunate enough that all of those have been falling into place as the years go by. And I decided that—Shall I elaborate on this and go back to my childhood?*

KEETON. *Yeah. I am trying to get your background and maybe by elaborating on the goals I can understand it?*

MIMS. *I feel like I am in a psychiatrist's office, with all due respect. But I will be happy to do it.*

LEE. *Just stretch out on the couch and tell all.*

MIMS. *I will be happy to share with you my childhood aspirations.*[9]

At midmorning, Keeton decided it was time to inspect the new documents I had brought, which they spent an hour sorting and reviewing. I then opened the briefcase and placed its electronic contents on the table. When I reached down to retrieve the camera to photograph them inspecting Eric's science fair project and my lightwave gear, Leach immediately dove under the table and Urbano and Koziol ran from the room.

From under the table, Leach loudly ordered me to remove the camera. Keeton and Ted then engaged in a lively debate over whether I would be allowed to photograph the examination of the circuits on the table. Finally, they agreed not to inspect the circuits, and I reluctantly agreed to repack the camera. I was puzzled by the circus my camera produced until recalling that I had published photos of the Bell Labs legal team searching my office in one of my columns for *Popular Electronics*.

After two days in Austin, the depositions moved to Ted Lee's office in San Antonio. When I returned to the conference room after lunch on the third day, Keeton and the Bell Labs attorneys were seated in a corner of the room opposite where I was seated. Urbano was coaching Keeton about how best to ask a crucial question concerning my original 1973 suggestion.

They were whispering so loudly that they were easily overheard, and the court reporter and I exchanged knowing glances. An hour later, as the interrogation was concluding, Keeton asked the trick question that was supposed to derail my case. But I gave the correct response, and Urbano hurriedly wrote some follow-up questions for Keeton to ask. These questions were also easily answered.

On the fourth and final day of the deposition, Ted Lee arrived dressed in his white, western-style suit and cowboy boots. I had not seen this imposing outfit

9. Deposition of Forrest M. Mims, November 10, 1980, 125-126.

since the lawsuit had been initiated eighteen months before. The tone of Keeton's questions softened, and he seemed to be having doubts about Bell Labs' position. At the end of the day, when the Bell Labs and Western Electric attorneys were preparing to leave, Keeton startled them and us when he asked, "What are you turkeys going to do?"

The ordinarily aggressive opposition attorneys appeared stunned, but I wanted to cheer. Leach glumly mumbled that they were heading for the airport, and they quickly left. Keeton then asked me a few more innocuous questions. He then leaned back in his chair, propped his cowboy boots on Ted's table, smiled, and said it would be best for both sides to settle the lawsuit. Ted then leaned back in his chair and placed his cowboy boots on the table, and they quickly worked out a friendly agreement. Keeton even agreed to include a provision that would require Bell Labs to acknowledge my prior work in its future publications regarding light phones.

Thanks in part to Keeton's objective evaluation of the evidence against his client, former US Marine Ted D. Lee and his partner, Mark H. Miller, successfully defeated the world's largest corporation. Shortly before Christmas 1980, Ted and I met to sign the settlement agreement and pose for pictures while holding a sizable check from Bell Labs.

While Bell Labs abandoned its US patent application for the light phone, it later filed an application in another country. Thanks to Dutch patent law and the agreement arranged by Keeton, the application cited my 1974 article in *Popular Electronics*, which concluded that a single-fiber lightwave link using light emitters that doubled as light detectors "is a precursor of what telephone systems of the future are likely to resemble."[10] Perhaps because of the mobile phone revolution, that has never occurred. But several firms and engineers have developed optical Wi-Fi systems and a wide variety of other applications that use my discovery that LEDs can function as light sensors, which I designate as LED><LED.

Siliconnections: Coming of Age in the Silicon Era

After the settlement, Harry Helms said I should write a memoir that closed with what had happened at Bell Labs, so I prepared a proposal and sent it to two book agents, both of whom agreed to meet with me during a forthcoming trip to New York. By then, Harry had moved from McGraw-Hill to Prentice Hall, and I called him before the trip to ask what I should ask for an advance. He instantly said,

10. Forrest M. Mims, "Communicate Over Light Beams with the First Single LED Transceiver," *Popular Electronics*, March 1974, 70.

"$50,000" ($154,155 in 2023).

The first New York agent with whom I met was surprisingly friendly, and he said the proposal had merit. But he was highly skeptical of the $50,000 advance I wanted and declined to take the book. The reaction of the second agent was nearly identical. That afternoon, I headed for a meeting about the proposal that McGraw-Hill had arranged.

The McGraw-Hill meeting was chaired by a vice president and attended by half a dozen staff and editors. They were pleased by my proposal and asked about a delivery date. The vice president then asked how much advance I wanted, and I calmly replied, "$50,000."

Everyone around the table seemed surprised, and there was total silence as they turned from me to look at the vice president. I assumed he would suggest a lower figure, but he did not. Instead, he said my request was acceptable. There were smiles all around as everyone stood up to shake my hand and offer their congratulations.

After the meeting, I called Harry from a coin-operated telephone booth near the McGraw-Hill building to thank him for his advice and announce that McGraw-Hill had accepted the proposal. When Harry asked how much of an advance McGraw-Hill has offered, I said, "The $50,000 you suggested." Harry was stunned and exclaimed, "Forrest, I was joking! I was only joking!"

Nobel Laureate Arthur L. Schawlow

Siliconnections received several positive reviews, including one by noted technology writer Jeff Hecht in *New Scientist*. Hecht was particularly intrigued by the *National Enquirer* and Bell Labs stories, and he closed his review with this summary: "*Siliconnections* is not a finely polished masterpiece . . . but the homespun awkwardness lends to the atmosphere. Reading the book is like spending an evening swapping stories with other adventurers in the technological jungles—and Forrest Mims spins some."[11]

Arthur L. Schawlow, the Nobel Prize–winning coinventor of the laser and a fellow Christian, sent me his review of *Siliconnections*. Schawlow was a professor of physics at Stanford University, and I met him in November 1985 at a meeting of the International Congress on Applications of Lasers & Electro-Optics (ICALEO) in San Francisco. When I visited him at his office at Stanford, I was surprised to see some of my Radio Shack books in a nearby bookshelf. I sent Schawlow a copy of *Siliconnections*, and he responded with this unpublished review:

---

11. Jeff Hecht, "Yarns From the Technological Jungle," *New Scientist*, February 27, 1986, 51.

> *Siliconnections is a fascinating and highly readable account of one man's adventures in the exciting years of the birth of portable calculators, personal computers and other electronic marvels. Forrest Mims was there, and played an important role as an electronics designer and as a writer who taught the rest of the world what was happening, and how to join in. Before that, he learned his skills, not in college where he majored in government, but by constant experimenting, even under the most difficult conditions.*
>
> *"It took perhaps more courage than wisdom to test model rockets at night in wartime Saigon, but he did it. He worked on early military laser experiments, and later was enlisted by a newspaper to help search for the elusive Howard Hughes. It must have been great fun, although at times difficult, and reading about his account of those years is pure pleasure. Since he is still quite young, I can only wonder what he will do next, and look forward to another volume of his memoirs in the future.*[12]

Schawlow separately wrote, "As you can see, I enjoyed the book, and just had to finish it even though I really have less than zero time for reading these days. Your books make me want to reach for my soldering iron, but alas, there is no time. I am particularly tempted by diode laser communications. I have the December issue of *Modern Electronics* and am looking forward to the January issue for the rest of your articles on that subject . . . ."[13]

After I first learned about lasers, Arthur L. Schawlow became one of my heroes, and meeting him and receiving his review of *Siliconnections* were career highlights.

Though *Siliconnections* was popular with reviewers and was selected as runner-up in the Best General Non-Fiction Book Award category for 1987 from the Computer Press Association, it never paid back the advance. The young personal computer industry was growing much too fast, and McGraw-Hill and other publishers were sharply pulling back computer-related books. The same month *Siliconnections* was published, technology analyst John Gantz wrote about the "great crash of the mid-1980s" in the January 27, 1986, issue of *Infoworld.*[14]

Ironically, I had discussed this in the closing pages of *Siliconnections*, which itself became victim to the sharp decline in computer publishing. So few copies were printed that several of its key chapters have been condensed and retold

---

12. Arthur L. Schawlow, letter to Forrest M. Mims, December 4, 1985.
13. Schawlow.
14. John Gantz, "A Writer's 10 Fearless Predictions For 1986," *Infoworld*, January 27, 1986, 21.

herein. The computer-press crash also affected magazines, including two for which I wrote. My final column in *Computers & Electronics* described how to use a computer-driven pen plotter to make stencil characters. The two examples I selected were the words *MIMS* and *EXIT*.

11

THE SILICON FARMHOUSE

In 1985, we moved from San Marcos to a rural place of nearly 10 acres on Geronimo Creek, near Seguin, Texas. The new homestead fulfilled a dream of living on a wooded site next to a permanent creek. A century-old farmhouse adjacent to our new home provided more space to work than the Silicon Garage. It became the Silicon Farmhouse. An adjacent 1.5-acre field I call the Geronimo Creek Observatory (GCO) provided wildflowers and space for outdoor experiments. In 1986, daughter Sarah arrived. Eric and Vicki were both born in Albuquerque, and Sarah became our first Texas child.

The $50,000 advance from McGraw-Hill and our earnings from Radio Shack books allowed us to quickly pay off the mortgage. Not having to pay monthly mortgage fees removed a huge burden, for there are no career guarantees in the freelance-writing business. Daily walks through the woods to the creek and back were far more relaxing than the daily 20-mile bike rides back in San Marcos. There was more time to pursue science projects between developing new books and lab kits for Radio Shack. The Silicon Farmhouse became home base for what will soon be forty years of writing projects and science adventures.

Farewell to *Popular Electronics*

The first issue of *Popular Electronics* magazine was published by the Ziff-Davis Publishing Company in October 1954. The magazine's articles on do-it-yourself electronics projects quickly attracted many hobbyists, and *Popular Electronics* soon became the leading magazine in its field.

Popular Electronics editor Art Salsberg liked my articles, and, eventually, I was writing three columns for each monthly issue: Electronics Experimenter, Solid-State Developments, and Project of the Month. Hobby electronics changed dramatically after P*opular Electronics* featured the Altair 8800 microcomputer on the cover of its January 1975 issue. That single article began a major transformation of the hobby-electronics audience, and *Popular Electronics* was eventually transformed into *Computers & Electronics* in November 1982. I continued writing for the magazine, but all my articles were about computer projects.

When it closed, *Popular Electronics* had a circulation of around 401,000. When my article about the tenth anniversary of the Altair 8800 was published in the January 1985 issue of *Computers & Electronics*, the circulation was nearing 600,000. *Computers & Electronics* soon faced major competition from a new wave of computer-hobbyist magazines.

Popular Electronics editor Art Salsberg started *Modern Electronics*, and I moved my articles to the new magazine. Meanwhile, *Computers & Electronics* had not lost circulation, but the competition was soaking up its advertising income. The final issue of the magazine was published in April 1985, when we were preparing to move from San Marcos and the Silicon Garage to Seguin and the Silicon Farmhouse.

More Bestselling Radio Shack Books

The success of *Getting Started in Electronics* resulted in strategy sessions at Radio Shack to discuss another kind of hand-lettered and illustrated book. The result was a series of sixteen *Engineer's Mini-Notebook* volumes, each of which included forty-eight pages devoted to a single topic. Five volumes of the *Mini-Notebook* were primarily devoted to science projects. The electronic circuits for the *Mini-Notebook* volumes were designed and tested in the Silicon Farmhouse. I then hand-printed and illustrated the pages on our brightly illuminated kitchen bar with the two 0.7 mm mechanical pencils used to create *Getting Started in Electronics*.

The *Mini-Notebook* volumes were produced from 1984 to 1999. Before Radio Shack went bankrupt, they sold 1.3 million copies of *Getting Started in Electronics* and more than 7 million copies of all nineteen hand-lettered and illustrated books. All these books are still in print, though sales are not nearly as high as when Radio Shack was selling them. Sales might be better if the books were not pirated across the web.

Many engineers and scientists have sent messages about the influence the Radio Shack books had on their careers. One that arrived while I was writing this chapter was sent by Michael Mol, a site reliability engineer at Google who has allowed me to quote from his message:

> *I found the illustrations of the physics concepts in Getting Started in Electronics to be extremely accessible, even before I was out of kindergarten. I didn't get the math for a few years after that, but it was still enough to get me experimenting. Frankly, it would be amazing to have more books like yours that make such low-level concepts so accessible and comprehensible to people who can barely read. I've taken secondary-school and college classes that didn't do so good a job making such ideas stick, even when you were caught up or ahead on the*

underlying math. Let me put something simply: You have a talent for making these things accessible and comprehensible. I'm sure you've already been told that, but I'll just reaffirm it. :)[1]

I replied to Michael what I have replied to many others by explaining that I am not a degreed engineer. The books were successful because electronics has been my hobby since childhood, and I had to understand the topics at the level of fellow hobbyists before writing about them.

The Atari Punk Console

In the fall of 1973, I had used an Exar sound-synthesizer integrated circuit to create what I called the Marvelous Martian Music Machine, which I had hoped to sell as a kit. A decade later, I experimented with a chip that produced the whistling sound of a falling bomb. During a book tour to promote *Siliconnections*, I showed the device to the producer of Johnny Carson's popular TV talk show. The publicist politely told me I was not cool enough to survive the interview, but the sound-effect device did. Carson acknowledged me and staged a demonstration of the device during his show, but it also bombed. After Carson activated the machine's whistling sound, he and Ed McMahon waited for the sound of an explosion. When there was only silence, the puzzled, disappointed looks on their faces elicited considerable laughter from their audience.

In 1971, Hans R. Camenzind designed the 555 timer, which quickly became an extraordinarily popular integrated circuit. The 555 is easily used as a timer, light flasher, metronome, and tone generator, and I described these and other applications in my magazine columns. The 555 includes an input pin, and in Radio Shack's *Engineer's Notebook* (1980), I showed how to connect the output of one 555 tone generator to the input of another.

This arrangement causes the tone from the first circuit to modulate the tone from the second, and the result is a mishmash of undulating sounds ranging from cacophonous to melodious. By adjusting potentiometers that control the tone of each 555, the output sound is easily altered. The potentiometers can be replaced by light-sensitive photocells to control the circuit with a flashlight or hand movements.

Because I enjoy music but am unable to read it, much less play it on instruments, I had loads of fun playing with this circuit, which I called a sound synthesizer or a stepped-tone generator. It was my design, and it was much more fun than the Marvelous Martian Music Machine.

---

1. Michael Mol, message to Forrest M. Mims, August 10, 2019.

I did not realize how popular the circuit had become until years later, when I learned that Kaustic Machines had renamed it the Atari Punk Console, since some of its sounds resembled those produced by Atari digital consoles. Amateur electronic musicians even created a Wikipedia page for the Atari Punk Console, which concluded, "The circuit is a simple DIY noisemaker circuit that is relatively inexpensive and easy to build, easily adaptable and is configurable in many ways. It has been built into a wide variety of cases. Its flexibility has led to wide scale popularity among electronics enthusiasts. It is often suggested as a good circuit to build for beginners."[2]

Then *Make* magazine editor Mark Frauenfelder chaired the "Music Makers" panel for Moogfest 2014 in Asheville, North Carolina, a series of presentations and electronic music performances. Mark asked me to speak about the Atari Punk Console, and Drew Blanke (aka Dr. Blankenstein), who was introduced as "the mad scientist of analog synthesizers and Atari punks," demonstrated his own handheld version.

The back of the one he gave me bears an eye-catching label that reads:

ACTUALIZED BY
DR. BLANKENSTEIN
Designed, built & presented with thanks and gratitude
to the original creator of the ATARI PUNK CONSOLE
FORREST M. MIMS III

Drew's neatly crafted console is resting on my desk as I type this manuscript, and I look forward to another session with it. It provides a tangible reminder of the impact the Radio Shack books have had on two generations of hobbyists, budding engineers, and scientists—including, as Minnie has reminded me, mad scientists. You can see and hear many demonstrations of the Atari Punk Console online.

I am often asked why I do not patent some of the more creative circuits and projects in my articles and books, as I thought about doing with the travel aid for the blind. Don Lancaster is an electronics writer with a much stronger technical background than mine. He was among the first to popularize integrated circuits for hobbyists in his many articles for *Popular Electronics*. Don provided good advice about patents when he and his wife visited us in San Marcos after they had been on one of their caving expeditions. While still coated in mud and dirt, Don reminded me that patents are expensive, time consuming, and difficult to enforce. Openly publishing our projects brings instant income and motivates readers to come back for more.

2. The Wikipedia page for the Atari Punk Console no longer exists, but the article, under a Creative Commons license, appears on the Synth DIY Wiki at https://sdiy.info/wiki/Atari_Punk_Console.

Radio Shack Lab Kits

For years, Dave Gunzel, Radio Shack's editor, was hoping I could develop a new lab kit for the company. After I completed the eighteenth hand-lettered and illustrated *Mini-Notebook* in 1999, two of the buyers finally decided it was time for me to develop a lab kit I called the Electronics Learning Lab. This became a major project, for it was necessary to design the lab kit's front panel, develop more than a hundred projects, specify the parts, and write two camera-ready workbooks. The manufacturing side of the project began when my lab-kit plan was sent to a Chinese company for development and manufacturing.

The first stage was the preparation of a mockup lab kit based on my design. After the company and I were happy with the mockup, I began designing the circuits and testing them by installing components on the front panel of the mockup. After I was satisfied with a circuit's performance, I prepared a page in the workbook that listed all the parts and how they should be installed and tested.

The project was very time consuming, and I typically worked from around 3 p.m. to 5 a.m. the next morning, six days a week. This provided interruption-free working hours. It also allowed the Chinese company and me to exchange emails during its staff's working hours. Our goal was to solve problems during the night before Radio Shack checked in the next morning.

The Electronics Learning Lab appeared in the 2001 RadioShack catalog and began selling several thousand units per month. Schools bought many of them. I learned why when Radio Shack donated twenty kits to the University of the Nations for use during my science short course there. Students who had never seen a transistor or an integrated circuit thoroughly enjoyed building projects that flashed LEDs or produced siren sounds, sweeping tones, and beeps.

I developed two more lab kits for Radio Shack. The Electronic Sensors Lab was a neatly designed console with an array of sensors that detected light, sound, vibration, movement, and magnetic fields. It was an ideal training tool for connecting sensors to computers. But my favorite lab kit was the $29.95 ($50.56 in 2023) Sun & Sky Monitoring Station, a sun photometer and sunlight radiometer that measured direct and full-sky green, red, and near-infrared sunlight with four LEDs used as light sensors. I have used a Sun & Sky Monitoring Station regularly since May 19, 2003, at the Geronimo Creek Observatory, and I calibrated it annually at Hawai'i's Mauna Loa Observatory from 2003 to 2018. Its measurements of haze, total water vapor, and photosynthetic radiation are as good as those made by much more expensive instruments.

Mike Dziekan, who contributed articles to *The Citizen Scientist* (*TCS*), developed a comprehensive spreadsheet for processing data from the Sun & Sky Monitoring Station that was also published in *TCS*. Mike wrote "Using Light-Emitting Diodes as Sensors," which was published in *TCS*. He named the use of LEDs as light sensors the Mims effect.

My first two lab kits sold more than 20,000 units per year, the minimum required by Radio Shack, but the Sun & Sky Monitoring Station sold only 12,000 units. That is far more than any other sun photometer ever produced, but it was not enough to keep it in stock. The buyer and I suspected that the Sun & Sky Monitoring Station would not sell well enough to keep it in stock, but we both wanted to make it available to schools and amateur scientists before Radio Shack stopped selling lab kits.

Radio Shack was in trouble, and big changes were happening. The most obvious change for me was that Radio Shack's new attorney required me to sign a contract for the Sun & Sky Monitoring Station project. Since 1978, all my hand-lettered books and the two lab kits I had developed for Radio Shack were done with Texas-style handshake agreements and simple purchase orders. *The Wall Street Journal* even published a letter I wrote about this.

12
THE SCIENTIFIC AMERICAN AFFAIR

The Amateur Scientist was the longest-running and most popular column in *Scientific American*, America's oldest magazine. The publication featured a potpourri of beautifully illustrated projects in all areas of science, and for many years, I had faithfully read every installment. Becoming the author of The Amateur Scientist was a dream that came true, if only briefly.

While three of my columns were published in the venerable magazine, the editor so strongly disapproved of my doubts about evolution and abortion that he terminated my assignment. At the time, I thought losing the famous column was a major setback to my career as a freelance writer. But what happened next is a story worth telling, for I had no idea that this widely publicized misadventure would launch my career in atmospheric science.

Scientific American magazine began as a New York City weekly tabloid on August 28, 1845. The founder, Rufus Porter, was an artist and prolific inventor who was mesmerized by the technology of the day and who wrote in the premier issue that his publication would cover "New Inventions, Scientific Principles, and Curious Works."[1] Porter also advocated morality, traditional family values, political and sectarian independence, and what he called "Rational Religion":

> *First, then, let us, as rational creatures, be ever ready to acknowledge God as our Creator and Preserver; and that are each of us individually dependent on his special care and good will towards us, in supporting the wonderful action of nature which constitutes our existence. . . . Next to the worship of God by thanksgiving and prayer, we should repel and banish all feelings of anger and*

1. Mariette DiChristina, "Celebrating Science," *Scientific American*, August 1, 2010, https://www.scientificamerican.com/article/celebrating-science.

bitterness toward our fellow beings, and cherish love and kind feelings towards them. . . . It is also a rational duty to . . . trust in the goodness and benevolence of God for the present and future, and to feel willing to have it known amongst our associates, that we follow a rational course.[2]

Porter's tabloid eventually evolved into a popular magazine. In 1948, Gerard Piel, Dennis Flanagan, and Donald H. Miller Jr. acquired the rights to it. Flanagan served as editor until he and Piel retired in 1984, and Piel's son Jonathan became editor. But under Jonathan Piel, even Porter would have been precluded from writing for the magazine he founded.

C. L. Stong

C. L. Stong became legendary among amateur scientists during his time as the author of the column The Amateur Scientist. In December 1970, I called Stong to suggest possible topics for the column, and Stong, aware of my electronics books, said that someday I would replace him as writer of The Amateur Scientist.

The column for the March 1973 issue included a laser diode project that the contributor failed to inform Stong was based on "Solid-State Laser for the Experimenter," an article I had written for *Popular Electronics*.[3] On May 26, 1973, Stong wrote that he would forward my letter to the laser contributor and concluded, "I am glad that you are well launched in your career as a professional writer. Congratulations! To have cranked out six published books and more than 60 magazine articles in three years must have severely cut into the time that you would normally have spent in a local gin mill. I don't envy you all that labor."[4]

Stong and I corresponded several more times, and he concluded his final letter on July 16, 1973, with this sentiment: "It's great to learn that you plan to come east next spring. Please do let me know when you plan to be here. It would be fun to dip our beaks into a saucer of booze, mumbling amiably all the time."[5]

While I was unable to meet Stong before he died in 1975, I didn't forget the 1970 conversation when he told me I would someday replace him. In June 1977, I learned that *Scientific American* was seeking a replacement for Stong, and in a two-page letter dated June 15, I proposed taking over The Amateur Scientist. The letter included a two-page list of thirty-four possible topics, including sensory aids for the blind and deaf, a $12 ($61 in 2023) wind tunnel, pinhole photography, fossil

---

2. Rufus Porter, "Rational Religion," *Scientific American*, December 4, 1845.
3. Forrest M. Mims, "Solid-State Laser For the Experimenter", *Popular Electronics*, October 1971, 46–49.
4. C. L. Stong, letter to Forrest M. Mims, May 26, 1973.
5. Stong, letter to Mims, July 16, 1973.

identification, a solar battery charger, electronic music, how to collect spider silk, and remotely controlled cameras. The proposal went unanswered.

On January 18, 1984, I sent Dennis Flanagan, then editor of *Scientific American*, a letter pointing out the absence of fossil evidence for the transitional forms between wasps and ants that had been speculated in an article. If the only evidence for the so-called transitional form between wasps and ants was mere speculation, what about all the other speculated missing links between countless animals and plants? The letter concluded, "The credibility of evolutionary biology is not advanced by conjecture. Lacking fossil evidence, speculations on the evolutionary origins of Hymenoptera will remain just that."[6]

This attracted Flanagan's interest, and he replied on February 2, "Thank you for your letter of January 18. What you have to say is most interesting, and we may very well be able to publish it."[7] While the letter was not published, I was impressed that it was under consideration by the same editor who had commissioned articles by Albert Einstein, J. Robert Oppenheimer, Linus Pauling, and dozens of other world-famous scientists.

My first letter to the magazine after Jonathan Piel became editor was sent on October 23, 1986. The letter was about an article that erroneously reported that silicon is not satisfactory as a detector of light. This was the first time I had found such a glaring physics error in the magazine, and my letter was the first of a series about errors that were never acknowledged or published.

Scientific American's John Horgan Launches My Science Career

By 1988, my hand-lettered and illustrated Radio Shack *Mini-Notebook* was selling several hundred thousand copies per year. But writing those books was becoming tiresome. Radio Shack was not selling a sufficiently broad range of electronic parts to satisfy my interest in creating a broad array of more advanced projects involving rockets, helium-filled balloons, lasers, and telescopes.

On April 20, 1988, a letter arrived from Claus-Peter Oefler of Vancouver, Washington, proposing that I consider taking over The Amateur Scientist. I replied on April 22, "When Stong passed away, I submitted a proposal to take over the column. They never responded. Unfortunately, the present column is only a shadow of what it once was. Too often it's simple kitchen table experiments and demonstrations rather than solid projects that really do something. Often I think about submitting a new proposal to take over that column . . . ."[8]

6. Forrest M. Mims, letter to Dennis Flanagan, January 18, 1984.
7. Dennis Flanagan, letter to Forrest M. Mims, February 2, 1984.
8. Forrest M. Mims, letter to Claus-Peter Oefler, April 22, 1988.

These thoughts were on my mind two weeks later, when I was returning to the ancient farmhouse that serves as my office after the daily walk to Geronimo Creek. The moment I touched the latch on the backyard gate, I received a sudden inspiration to stop dreaming and develop a serious proposal to take over The Amateur Scientist.

Feeling as if ordered to follow through, I immediately called the magazine. John Horgan was the only member of the *Scientific American* editorial board I knew. As related earlier, he had interviewed me in 1985 for "Inventor Seeks to Warn Government of Threat from Laser-based Bug," an article published in the *Institute*, a tabloid of the Institute of Electrical and Electronics Engineers.

Horgan expressed interest in my goal to take over The Amateur Scientist and said that Jearl Walker, the columnist who had followed C. L. Stong, was thinking of retiring. I immediately went to work preparing a twenty-page proposal that included thirty topics for the column, including aerial photography from kites and balloons, measuring water quality, a simple seismometer, and an ultraviolet meter.

These were real projects, not proposed ones. For example, my radio-controlled camera could be flown up to 100 feet (30 m) by a cluster of helium-inflated trash bags or up to 500 feet (152 m) by a blimp 12 feet (4 m) long. The inflated blimp could be driven into the country while carefully tethered inside the bed of my pickup truck. Or I could attach it to my bicycle and pedal it a few miles to various flying sites (and trigger questions from surprised motorists who slowed and asked when the blimp, which could lift only a few pounds, would carry me skyward).

After working on the proposal for several days, I mailed it to Horgan on May 6, 1988, along with relevant background information and seven of my books in a package that weighed 8.2 pounds. Horgan, who received the package on May 13, liked what I sent and showed it to the magazine's editor, Jonathan Piel, who replied on July 8, "Dear Mr. Mims: Several of us have looked through the material you sent and read your compendium of topics, and find it all most impressive. . . . your proposal is still under active consideration, and we are most grateful to you for all the time and trouble you have put into it."[9]

I replied to Piel on July 1 and 14, and he responded on July 25, "Dear Mr. Mims: Thank you for the report on your UV measurements. The project, like the others you describe in your proposal for 'The Amateur Scientist,' is a fascinating one. I am sorry not to have any news for you yet; at the moment we are enmeshed in editing our annual single-topic issue. Your proposal is nonetheless very much on our minds."[10]

9. Jonathan Piel, letter to Forrest M. Mims, July 8, 1988
10. Piel, letter to Mims, July 25, 1988.

Having heard nothing by October, I sent letters with more column ideas to Piel, Horgan, and Timothy Appenzeller, an associate editor, and Armand Schwab Jr., the managing editor. On October 21, Piel wrote, "Dear Mr. Mims: Thank you for the delightful additions to your proposal for 'The Amateur Scientist.' We are finally getting around to giving all this the attention it deserves, and we should have news for you soon."[11]

While I was optimistic, two long months passed without hearing anything from the magazine, so, I called Piel shortly before Christmas to inform him about new column ideas. A letter dated December 20 arrived from Piel:

> *Dear Mr. Mims:*
> *The material that you have shared with us leaves no doubt that you would write a most engaging "Amateur Scientist" column. The hitch is simply that we are already publishing another engaging writer. Given the fact that Jearl Walker has a 50 percent readership there is no urgent reason to tinker with this part of our editorial engine. At some point in the next 12 months I will review the matter and be in touch with you. In the meantime I hope you will be patient. If, however, other opportunities emerge I could not in good conscience ask you to neglect them.*[12]

Piel and I exchanged several brief letters in January and March, but there was still no decision about the column. That summer my son, Eric, and I were planning a field trip to New Mexico, and on July 27, 1989, shortly before our departure, Piel called and said, "I've been remiss in not calling you sooner. How would you like to take over 'The Amateur Scientist'? The magazine needs a shot in the arm. . . . We think you can give us a shot in the arm." Piel then asked when I could fly to New York to discuss the column, and we agreed to meet on August 7.

"Red Alert! Red Alert!"

I arrived at *Scientific American*'s offices, at 415 Madison Avenue in New York City, at 10:15 a.m. on August 7, 1989. Piel was in a meeting, so he was not available until 10:55. What happened next is recorded in five pages of notes made the following day on the plane home and a detailed memo for the record begun two days later.

Following a brief discussion about *Science Probe!*, a science magazine I was planning with magazine publisher Larry Steckler, I showed Piel my unanswered proposal for taking over The Amateur Scientist sent to Dennis Flanagan on June 15,

11. Piel, letter to Mims, October 21, 1988.
12. Piel, letter to Mims, December 20, 1988.

1977. He read the letter and said, "I would have snapped you up! I don't understand why they didn't respond to you." He then gave the proposal to his secretary and asked her to make a copy. He also asked her to look for the proposal in their 1977 files. She quickly did but found nothing.

Piel said that Jearl Walker had done a fine job with The Amateur Scientist. He then said, "We just think you can do a better job." Piel then added, "We would sure like to have Forrest Mims on an exclusive basis."

I said that writing is how I earn an income, and that is all I have done for nineteen years. He then asked about the magazines for which I write. I listed several, including *Texas Parks & Wildlife, Popular Photography*, and *Bicycling*. I then said, "I've written for some Christian magazines."

Piel stopped me, leaned forward, and said, "You said Christian magazines. What have you written for Christian magazines?" I replied that I had written articles for two magazines on how to organize long-distance bicycle trips for church kids.

Piel's enthusiasm vanished, his face reddened, and his bow tie seemed to nervously bounce up and down. I may have imagined the bow tie's antics, but I did not imagine what happened next. Jonathan Piel looked straight into my eyes and firmly demanded, "Do you accept the Darwinian theory of evolution?"

I immediately realized that what I thought was a firm assignment to write The Amateur Scientist hung on my answer. Daddy and the Corps of Cadets at Texas A&M had trained me to always tell the truth, so I replied, "No, and neither does Stephen Jay Gould." I then mentioned Gould's ideas about punctuated equilibrium and Darwin's doubts about his own theory in *On the Origin of Species*.

Piel was not impressed, and was clearly troubled. I had entered sacred groupthink territory and changed the subject by informing Piel that I considered writing The Amateur Scientist to be the major goal of my career. I would give it the highest priority, and he would be free to reject any column that did not meet *Scientific American*'s standards. I then showed Piel some of the devices and chart recordings in my briefcase, and some of his former enthusiasm briefly returned as he examined these items. He was especially interested in the barometer watch, which Eric and I were planning to use during our delayed field trip.

It was nearly noon when Piel stood up and walked to my side of the table. He yawned and explained that I was not boring him but that he'd had only one hour of sleep the night before due to a family matter he related to me. He then told me to wait in his office while he collected the staff for lunch.

Tim Appenzeller, an associate editor, told me several years later that Piel then rushed into his office, exclaiming, "Red alert! Red alert! Mims is a creationist!" He then instructed Appenzeller to ask me an embarrassing question about evolution during lunch.

Five minutes later, Piel returned, and we left for lunch at a nearby Japanese restaurant accompanied by Appenzeller, Schwab, and Laurie Burnham, an associate editor. I took along some chart recordings of the ultraviolet-sunlight readings, several homemade devices, and some aerial photographs, all of which I showed them during lunch.

The staff was very interested and asked many questions about the homemade gadgets. They especially liked the aerial photographs made from my radio-controlled camera flown from kites and balloons and said the camera would be perfect for The Amateur Scientist.

Appenzeller said he had been a member of the Albuquerque Model Rocketry Club when he was eleven and a student at the Albuquerque Academy. He knew that I had organized the club in 1969 when I was assigned to the Air Force Weapons Laboratory. Piel, seated diagonally to me on my left, said very little during the show-and-tell session. I suspected that evolution was on his mind.

One of the devices I showed was described in one of my books. It featured a row of red LEDs that glowed in sequence to indicate changes in resistance. I asked everyone to hold hands around the table while two of us grasped the device's two probes. Piel sat back and did not want to participate, until Laurie Burnham said, "Come on, Jonathan." Everyone was impressed by how the LEDs glowed up and down in response to how tightly we squeezed our hands together.

Appenzeller later told me he tried to embarrass me by following Piel's instructions to ask an evolution question. I must have passed the test, for neither of us recalled what Appenzeller asked.

Back at the main office, I showed Piel and his staff more of my projects. Piel was particularly intrigued by my miniature LED sun photometer, which I am still using during my daily measurements of haze and the total amount of water vapor in a vertical column through the atmosphere. When I opened its case, Piel peered inside to examine the electronics and asked, "Did you actually build this?" He smiled broadly when I answered yes. He and the others were also interested in the radio-controlled kite camera, so I set it up and exposed a photo of them. When Piel and I returned to his office to resume our meeting, he gave me a copy of the *Scientific American Cumulative Index: 1978–1988*. When I admired the time required to produce such a comprehensive document, he said his retired father, Gerard Piel, sat upstairs and worked on things like that. I asked him to sign the index, and he wrote, "Best regards to a great Amateur Scientist. Jonathan Piel 8/7/89."

Piel then allowed me to ask some questions. He said I could keep book rights to the columns and that they would pay me "$20,000 to $30,000 per year" (about $48,500 to $72,500 in 2023). What happened next is best related by quoting from my memo for the record begun two days after the meeting:

> *Then occurred a most distressing and humiliating experience. Mr. Piel stated that now that I had asked some questions, he wanted to ask some questions. He then firmly stated I could do nothing whatsoever that might embarrass Scientific American. He said he didn't care what my personal views were. "You can be a member of a nudist camp . . . ," but under no circumstances could I write about creationism for any publication. Nor could I write anything that might be viewed as criticizing or ridiculing the findings of professional scientists. He then asked me if I will agree to this. I said that I would never use my position at Scientific American as a platform to espouse my views. As for an outright ban on writing, I said "I will have to mull it over." He asked me to get back to him with my decision.*
>
> *Mr. Piel mentioned his great concern about "embarrassment" several times. He said he would have to review my outside writings. I suggested I might use a pseudonym, such as "Ed Field," to write on topics that might trouble him. He said my identity would eventually become known, and he would have to review my outside writings under a pseudonym as well. [I don't believe in using pseudonyms and, looking back, wish I had not suggested this.] He twice used the phrase "Pass it by me first." As for a violation of this policy, Mr. Piel said there are two actions he could take. One would be a cut in pay. The other would be dismissal.*
>
> *I was profoundly and deeply troubled by the nature of Mr. Piel's remarks. Stunned might be a better word. All I could say was that I would do my best to deliver the highest quality product and that I would not miss my deadlines. And I would always give Scientific American first refusal on any science project articles.*
>
> *Mr. Piel closed the meeting by saying he would meet with the editors with whom we had lunch, make a decision, and get back to me by the time I return from the field trip on August 18.*

Appenzeller and Schwab had been listening to all this while leaning on the door to Piel's office, and they almost fell in when I opened the door to leave. I then met briefly with Horgan and Schwab, both of whom expressed their support for me writing the column. We were discussing my old 1977 proposal when the secretary arrived with a letter from the magazine to Jearl Walker from September 1977. The letter stated that

Walker was being sent an airplane ticket to come to New York to discuss taking over The Amateur Scientist from the late C. L. Stong. That explained why my 1977 proposal had gone unanswered, and reminded me to tell them that I paid my expenses for the trip Piel requested under the assumption that I was being hired.

I walked back to the hotel around 3:30 p.m. and called my wife, Minnie, to tell her what had happened. I then opened the Gideon Bible on the desk and read, "Whereas ye know not what shall be on the morrow. For what is your life? It is even a vapour, that appeareth for a little time, and then vanisheth away."[13]

That evening, I called Larry Steckler, an editor and publisher friend, who many years before had been rejected by a major magazine because he is Jewish. Larry shared my concerns. On arriving home the afternoon of August 8, I called my Radio Shack editor, Dave Gunzel, and fellow freelancer and editor Harry Helms, both of whom were amazed that Piel would demand to review all my submissions to other publications. The next day, I wrote, "As I conclude this memo at 11:22 am on August 9, I continue to feel personally violated by the condition Mr. Piel has placed on me. Writing 'The Amateur Scientist' is the highest career goal I can imagine. I'll be glad to work with Piel or anyone else to write the column. But I cannot in good conscience agree never to publish anything that might happen to discuss my personal views on creationism."

Back home, I began researching the history of *Scientific American* and the Christian views expressed by founder Rufus Porter in the first installment in 1845. I then found the November 1925 cover story on amateur astronomy by Albert G. Ingalls, which was titled, in large type, "The Heavens Declare the Glory of God."

This quotation from the Bible's book of Psalms was painted on the eaves of a structure called Stellafane by the Springfield Telescope Makers, in Springfield, Vermont. That article led to more articles in *Scientific American* on how to make telescopes, and by May 1928, Ingalls, an amateur astronomer himself, was writing a regular column called The Back Yard Astronomer.

Ingalls wrote the column under various titles for many years and in April 1952 expanded it to include all of science under the title The Amateur Scientist. After Ingalls retired in May 1955, the column was taken over by Stong and, later, Walker. My goal was to follow them with columns of my own creation instead of those contributed by readers. But my agreement with the faith of the magazine's founder and the motto of the amateur astronomy organization responsible for the origin of The Amateur Scientist stood in the way.

---

13. The Gideon Bible, James 4:14.

The New Mexico Field Trip

The week after returning home from New York, Eric and I left on a field trip to New Mexico along with the new instruments I had been building for *Scientific American*. We made many sunlight measurements and used a Geiger counter to measure the elevated background radiation at Soda Dam, where hot springs are sprinkled along the bottom of the chilly Jemez River and at surface sites nearby. We also measured higher-than-normal radiation at one of the gates of the Los Alamos National Laboratory, home of the first three atomic bombs.

When we were camping at Coronado State Monument, I noticed some harvester ant mounds. Reasoning that the ants might excavate artifacts from deep underground, I looked carefully at one mound and found a turquoise bead, which we gave to the attendant at the museum.

That evening, I needed to buy some ethyl alcohol to clean the UV filters on my instruments, so I walked to the drive-up window of a nearby liquor store and asked the clerk if I could buy their smallest bottle of Everclear, which is ethyl alcohol. When I explained that the liquid was needed to clean some optics, the woman smirked and said, "Yeah, sure."

At the Carlsbad Caverns gift shop, I noticed some souvenir wood postcards and bought one printed with the image of a skunk. I wrote a brief note on it about our field trip and mailed it to Jonathan Piel. He had said he would give me a decision by August 18, and that day, I stopped at a pay phone in Seminole, Texas, to call Piel. He apologized for not having decided and said he would call back the following Monday. I said I would "eagerly await your call," and he apologized a second time and assured me he would "definitely call." On Monday, I remained by the phone all day, but Piel failed to call.

Schwab Invokes the Deity

During September, I had long telephone conversations with Horgan, Appenzeller, and Schwab, the magazine's elder leader. He had worked there since 1961 and was displeased by Piel's mismanagement of my situation. He, Horgan, and Appenzeller all told me multiple times that I should have been hired. After a conversation with Schwab on August 28, I wrote in the memo for the record: "I said that Jonathan had twice by phone offered me the position before I went to New York. I said that, 'Jonathan specifically asked if I would take on "The Amateur Scientist."' Armand responded, 'Oh, Lord!' a response he [also] used when I told him the trouble it was to try to call Jonathan from West Texas."

On August 30, Schwab asked me to send an invoice for the expenses for the New York visit, which totaled $494.72 ($1,198 in 2023). I then recorded this in my memo:

"On a very positive note, Armand then said they wanted me to prepare and submit a trial column. He said that given the many pounds of books and proposals I had already sent, 'It's something we should have [had] you do long ago.' He said I would be paid $2,000 [$4,845 in 2023] for the column, and that there is a 'likelihood we could use it, even if only a trial.'"

We discussed sunspots and aerial photography as possible topics, and Schwab liked both. As for the various kinds of projects I can handle, he said, "God knows we all agree with that." I then said I might do both with the understanding that they want only one. On September 7, I told Appenzeller I was also working on a third column, on measuring the sun's UV. He did not object.

Writing the first two columns took three busy weeks. On Saturday morning, September 23, I drove to San Antonio and shipped the columns to Piel, Schwab, and Appenzeller from Federal Express. The following Wednesday, Schwab said he, Piel, and Appenzeller liked the columns and that they "are very promising." He said he did not know when Piel would decide, and that he might mention the matter at an editorial retreat in October. Schwab then said that *retreat* is a misuse of the term, because "*retreat* is almost a religious term." In a letter postmarked September 27, Piel wrote, "Dear Forrest: Thank you for the trial installment of columns. Please do not write another one until I have had time to read the ones that you have sent and decide what course to take with respect to the department."[14]

"Your Religious Views"

Piel did not reply to two Federal Express letters about the status of the two columns I had submitted, so on October 3, 1989, I phoned Laurie Burnham, the associate editor at *Scientific American* who had been present during the lunch meeting with Piel and his staff. Burnham was a biologist, and during the forty-five-minute call, I suggested that my columns could cover such topics as eyeshine, spider webs, bacteria-contaminated water, reflectance of leaves, and so forth. I also told her about my collection of insects encapsulated in amber.

As noted in the memo for the record, she had seen the columns I sent and said, "Those are great." She added that Piel had shown her the photos from the aerial-photography column and said that they would look good in the magazine. She specifically said, "[Jonathan] loved the columns." She also said he was "excited" about the columns, and that he had added that they stimulated one to want to go out and try such a project.

---

14. Jonathan Piel, letter to Forrest M. Mims, September 27, 1989.

After expressing deep concerns about my religious views, however, Laurie introduced a new issue: "We would be worried that you might write about the sanctity of life." I asked what she meant by "the sanctity of life." She said she was referring to the view that life begins with conception. I said I considered life to be sacred, and that I was glad my mother bore me and that I assumed she felt the same about her mother. "Aren't you glad your mother did?" I asked.

Laurie then said, "Our concern is that you might speak out on creationism or speak out at pro-life rallies." I asked what any of this had to do with writing an effective amateur-scientist column. She said they were convinced I would write a "great" column. But she added that they remained "concerned about your religious [views]." Laurie closed the conversation by saying, "Your honesty speaks highly of you" and asked why I had not called Piel. I explained how I had tried to call him, and that he had not replied to my two letters. Laurie said she would ask Piel to call me.

Piel Calls

After the lengthy discussion with Laurie, I telephoned Ted D. Lee, my patent-attorney friend who had defeated Bell Labs. Ted said I needed evidence that Piel was discriminating against me, and this justified recording Piel should he call. Ted said that Texas law permits the recording of telephone calls so long as one party is aware that a recording is being made. I then learned how to record a call using my telephone answering machine.

The next day, Minnie and I were preparing to go to a midweek function at our church when my cordless phone rang. I suspected it might be Piel, so I told Minnie to leave without me as I switched on the phone. I was right. I hurriedly walked to the office, picked up the regular phone, and switched on the recording function of the office answering machine, which I had never used.

The call was lengthy, and it seems best to defer to Paul Tough, a writer at *Harper's Magazine*, to describe what happened next. Tough called in November 1990 to request a transcript of the recording. After discussing his request with Ted D. Lee, I made a twenty-seven-page transcript and sent it and a copy of the recording to Tough. *Harper's Magazine* published Tough's lightly edited portion of the transcript:

Telephone Transcript
"SCIENCE'S LITMUS TEST"

JONATHAN PIEL: *The problem lies, as nearly as I can make out, in the following area: namely, that you've got certain attitudes and beliefs that are in conflict with editorial positions and trends and traditions in this magazine. Okay, so what? Well, the what is this. I'm sure that you're a man of honor and integrity. That's very clear and obvious. And I'm sure that you would never try to use your column to express views that were contrary to our editorial policy and traditions. And you've also assured me that you would not express such views in a context where you would be identified as a contributor to this magazine. That may be well and good. But the problem still exists that if you were The Amateur Scientist columnist for Scientific American and some third party saw this contrast, it would be impossible for us to stop our name from being used in a way that could be damaging to causes and policies which we and the vast majority of our contributors believe in. And, you know, that's a problem I'm really wrestling with.*

FORREST M. MIMS III: *In other words, you are suggesting that if I were talking to somebody in the media, for example—are you suggesting that maybe they would expose this or something?*

PIEL: *Yeah, sure, that would be one possible way this kind of thing could happen. Another possible way could be that creationists, advocates of adopting a textbook in Texas or somewhere, would say, "Why, even Scientific American is in favor of this position. Their columnist is on record as believing that creationism deserves equal attention in the schoolroom." You know, someone could put words in your mouth.*

MIMS: *Well, you know, that is so utterly hypothetical.*

PIEL: *That's the kind of hypothetical I've got to watch out for as editor of the magazine.*

MIMS: *Listen, I sympathize with where you're coming from. But I totally disagree. For all both of us know, there is a Democrat or a Republican or a Marxist on board the magazine and somebody could do the same thing. Prior to the visit to your offices, there was never even a hint that religion would become an issue.*

PIEL: *Forrest, come on, that's why I had the meeting with you. You know, I didn't know you were just a voice on a telephone and a name on a piece of paper. That's why that kind of meeting is had, so two people can get to know each other. As a result of that meeting, I heard you say things that awoke serious reservations that were not there before.*

MIMS: *Well, you asked me if I accepted Darwin's theory of evolution, and I said no.*

PIEL: *Stephen Jay Gould, as you point out in your letter, in some sense doesn't accept it either, but I think that anyone who really views—*

MIMS: *He would have answered, "No, but." I answered, "No, and here's why." I have read many of his papers, and he makes a brilliant case for punctuated equilibrium.*

PIEL: *But, you know, Darwin is not libeled [in Gould's interpretation]. Darwin is nice. There's a real difference between a Gouldian take-off on Darwinism and a flat rejection of evolution.*

MIMS: *Look, I really need, you know—*

PIEL: *There's no question that on their own merits the columns are fabulous. If you don't do them for us you ought to do them for somebody, because they're great.*

MIMS: *If I'm not good enough to write for Scientific American—*

PIEL: *Forrest, that's not—come on, I just made it very clear to you what the issue is.*

MIMS: *Well, I'm sorry, if my belief system does not qualify me to write for Scientific American, then my entire career is totally fractured. Let me put it this way: My wife is Hispanic, and when we go to restaurants, people make fun or make snide remarks. Can you relate to how we might feel about that?*

PIEL: *Very deeply.*

MIMS: *We feel very deeply about that. Being discriminated against is not fun.*

PIEL: *As far as the issue of discrimination, you know, there is no discrimination here.*

MIMS: *It's absolute discrimination.*

PIEL: *Forrest, a different analogy would be whether the Democratic National Committee is right in refusing to print an essay by Ronald Reagan.*

MIMS: *And a religious magazine doesn't have to accept a contribution from an atheist. That is totally allowed. But nonreligious institutions and nonpolitical institutions are simply prohibited from discrimination based on age, race, gender, and, of course, religious beliefs.*

PIEL: *Oh no, I think the former analogy is more accurate. I wish you wouldn't bring that implication in here, because I totally respect your right to express your opinions and hold them. But I also have a right as an editor to control what opinions are expressed in my magazine—*

MIMS: *Absolutely. And you could fire me the minute I did it.*

PIEL: *or that become associated with my magazine.*

MIMS: *I really think your concern is highly exaggerated. I've been in this business for nineteen years, and these things just don't come up.*

PIEL: *I've been in this business a few years myself, and I've seen the shrapnel fly from unexpected directions. What you've written is first-rate. That's just not an issue. It's the public relations nightmare that is keeping me awake.*

MIMS: *Well, you know, I look at it from the other side. I see a public relations nightmare if magazines discriminate.*

PIEL: *I refuse to accept that characterization. If you're going to keep saying that, this conversation is over. I'm not discriminating. I'm exercising my constitutional right, sir, to protect an institution that I love and the values for which it stands. And I may add that the behavior of some of your co-believers is a profound violation of First Amendment principles.*

> MIMS: *What co-believers? I don't know who you're talking about. I know what my views are, but—*
> PIEL: *You're telling me that I'm discriminating; I'm telling you that I am not. Why don't we just leave it at that and try to conduct the rest of this conversation as gentlemen.*
> MIMS: *All right, we'll leave off the discrimination. I just want you to know that I've been very troubled about the whole matter of religion being brought up.*
> PIEL: *I have been wrestling with this question. I've been very honest with you about laying it out. You have been equally honest with me about what you would do, and I accept that. But there is the irreducible danger that third parties could make use of your political positions and your firmly held beliefs to misuse the good name of this magazine. And, you know, that's a risk that I have to weigh very carefully.*
> MIMS: *Well, I—*
> PIEL: *That's all there is to it.*[15]

The full transcript included other troubling points. For example, during the meeting with Piel in New York, he had said that I would have to pass all my outside writing on to him for his approval. While I did not agree to allow Piel to censor my outside writing, I told him I would not embarrass the magazine. This was alluded to in the unabridged recording:

> PIEL: *You know, you've assured me that, that these views that you're surely not going to try to express in this magazine, and that they are views that you would certainly not express them in any context where your identity as our columnist would be a question.*

"Chew on That!"

The day after the conversation with Piel, I called him to suggest that he postpone a final decision until after my two columns had been published. According to my memo for the record for October 5, 1989, he replied, and I made these notes about the conversation:

> *"We haven't discussed the money, but we can get on to that." He seemed to ramble for a moment about these details. Then he became rather strident in his tone of voice and said he had thought about our discussion yesterday and was very*

15. "Science's Litmus Test," *Harper's Magazine*, March 1991, 28–32.

troubled over my comments about "discrimination." He said he was offended by my use of the term. He said that before we could proceed any further, "Perhaps you should chew on that!" Without allowing me to respond, he then [loudly] said, and I quote as best as I can remember, "If you really believe I've discriminated against you and yet you are willing to compromise yourself so that your columns can be published, then that suggests a serious lack of integrity on your part." Before I could respond, he [loudly slammed down] the phone. Jonathan did not tell me to stop work on the third column, so I will proceed.

The next afternoon, a Federal Express truck rumbled down our long rural driveway. The driver handed me a large package from *Scientific American* that contained the two columns, the proposal and books sent in 1988, and a letter copied to Schwab, Appenzeller, Burnham, and Horgan:

Dear Forrest:
In the course of our telephone conversation yesterday, you referred to my concern about an inadvertent linking of your beliefs with the good name of this magazine as discriminatory. I take profound exception to this characterization. Indeed, I see no basis for a working relationship with anyone who would so describe my motives or actions. I, therefore, think it best to terminate all further discussion of the possibility that you might contribute to the pages of this magazine.

Sincerely,
Jonathan Piel, Editor[16]

Piel had previously told me he would buy and publish three of my columns, so I ignored his dismissal language and sent him a Federal Express letter to inform him I was working on the third column. Piel did not respond, and the column about a do-it-yourself solar UV monitor was sent to him on October 23.

Piel was surprised to receive the third column. In a letter dated October 27, which was accompanied by a $4,000 ($9,690 in 2023) check for the first two columns, Piel denied what he told me during the October 4, 1989, telephone conversation I recorded. My memo about the letter reads, in part, "'Neither I nor anyone else . . . agreed to "buy and publish" any of the installments that you prepared for "The Amateur Scientist,"' Piel wrote. 'I had made no decision with respect to publishing your articles.'"

16. Jonathan Piel, letter to Forrest M. Mims, October 5, 1989.

Yet the unabridged transcript of our October 4, 1989, telephone conversation I recorded included Piel's agreement to buy and publish three of my columns:

PIEL: *There's no question that on their own merits the columns are fabulous. If you don't do them for us you ought to do them for somebody, because they're great.*
MIMS: *There's nowhere I can market that material. I had the opportunity to—*
PIEL: *I'll, I'll, I'll, you know, I'll, I'll, give me three of them and I'll run 'em and give Jearl [Walker] a vacation.*
MIMS: *Well, I, I very much appreciate that, uh—*
PIEL: *Okay. I'll, I'll buy 'em from you, uh—*

"What Can We Do to Make You Happy?"

It was time to appeal to Piel's boss, so I wrote Claus-G. Firchow, the president of Scientific American Inc. In 1986, the magazine had been acquired by Holtzbrinck Publishing Holdings Limited Partnership, a German company. The Berlin Wall was coming down and was very much in the news, so, I wrote Firchow that while walls against freedom are falling around the world, *Scientific American* had erected a wall of religious discrimination and prejudice.

I knew nothing about the enthusiastic support of Holtzbrinck's founder for the Nazi Party during World War II. Could this have influenced what happened several days later, when Jeffrey Rich, an attorney for *Scientific American*, called?

"What can we do to make you happy?" Rich asked.

"Publish my columns and pay for them," I replied.

In only minutes, we reached an agreement that the magazine would fulfill my request if I would agree that they had no further obligations to me. Rich seemed pleased that we had resolved the matter so quickly and amicably. Within a few weeks, Russell Ruthen, a young and very competent editor, was assigned to edit the three columns.

Learning to Measure the Ozone Layer

While working with Ruthen to prepare the columns for publication, I continued developing several new projects for the magazine in the hope they would also be published. I had kept Tim Appenzeller informed about progress on these projects, the first of which was a handheld instrument half the size of a paperback book that, when pointed at the sun, measured the total amount of ozone in a vertical column through the atmosphere. I named it TOPS (Total Ozone Portable Spectrometer).

Converting data from TOPS into the total ozone amount required a complex equation beyond my understanding. Therefore, I devised a simple method that compared TOPS data with the ozone amount over my site measured by NASA's Total Ozone Mapping Spectrometer (TOMS) aboard the *Nimbus-7* satellite. The plan was to do this for five months and then plot the satellite ozone against the TOPS data. A computer program (TableCurve) would then find the formula that best fits the TOPS data with the actual ozone amount and provide a simple way to convert TOPS data into the ozone amount. The five-month comparison, which began on August 28, 1989, would have been impossible without the advice and assistance of Arlin Krueger, a brilliant ozone scientist at NASA's Goddard Space Flight Center.

Ozone is measured in Dobson units (DU), and 100 DU is equivalent to a layer of pure ozone just 1 millimeter thin. On February 6, 1990, the first day when the new ozone equation was used, TOPS measured 276.2 DU and the satellite gave 271.3 DU, which meant that ozone measured by TOPS was within 2 percent of what the satellite measured. I was optimistic that Jonathan Piel would publish how to assemble TOPS in *Scientific American.* I also hoped he would publish an article about my LED sun photometer, which had impressed him during my visit to his office.

The Columns Are Published

The first of my three columns was published in the June 1990 issue of *Scientific American*, seven months after the publication's attorney asked what would make me happy. It was titled "Sunspots and How to Observe Them Safely," and it featured the methods and data used by my young daughter, Vicki, in one of her science fair projects. When the magazine arrived in the mail, I could not believe that the column was really in *Scientific American.*

Had they sent me a fake issue? Only when I saw that the column was in a newsstand issue of the magazine did I believe that the lengthy battle was finally over and I was finally writing the column that, two decades earlier, C. L. Stong had told me I would someday write.

"How to Monitor Ultraviolet Radiation From the Sun," the second column, appeared in the August 1990 issue and attracted a good deal of reader mail. The two UV monitors described in this column had led to TOPS, which would soon find an error in NASA's ozone satellite and jump-start my science career. The final column, "A Remote-Control Camera That Catches the Wind and Captures the Landscape," was published in the October 1990 issue.

So far, my goal of writing The Amateur Scientist had succeeded. Would TOPS and the sun photometer impress Piel enough for him to reconsider and reassign me to write the column?

13

THE PUBLIC RELATIONS NIGHTMARE

Because of the very positive mail that arrived when my three columns were published in *Scientific American*, I was optimistic that Jonathan Piel would allow me to continue writing The Amateur Scientist. But Piel rejected six new proposals I sent that included measuring the length of lightning bolts and detecting killer bees by the frequency of their hum. I had also made major progress with the sun-photometer project and TOPS, the handheld instrument that measured the ozone layer. Letters sent to Claus-G. Firchow, president of the magazine's parent company, went unanswered. It was time for Jonathan Piel's public relations nightmare, but I had no idea how best to trigger it.

The *Houston Chronicle*

Joe Abernathy was a computer writer at the *Houston Chronicle* I knew from his time at a San Antonio paper. Maybe he could provide some tips on how best to publicize the *Scientific American* situation, so I called him.

"This is a national news story!" Abernathy exclaimed. He then asked whether he could tell his editor about it, and I said yes. His editor happened to be standing nearby, and I overheard a brief conversation in which the editor also echoed Abernathy's comment. Abernathy then interviewed me about every detail of what had happened. The *Chronicle* sent a photographer, and Abernathy's story and a photo appeared at the top of the page in the paper's City & State section on October 8, 1990.

What happened next was summarized by Mark D. Hartwig in "Defending Darwinism: How Far Is Too Far?": "On October 8, 1990, Jonathan Piel's public relations nightmare became horribly real when the *Houston Chronicle* broke Mims' story to the public. The reality, though, was much worse than Piel could have ever dreamt. After two weeks, *The Wall Street Journal* picked up the story, followed by *The New York Times*, *The Washington Post*, and countless other newspapers across the country. *Scientific American* was now fully in the public eye, caught in a most unflattering light."[1]

Minnie and I were as surprised as *Scientific American* by the media coverage sparked by Abernathy's article. While I sent no press releases and did not hire a publicist, we had to order a new telephone line to receive and answer the many faxes arriving from the media.

The *Wall Street Journal*

A week after the *Chronicle* article appeared, Bob Davis of the *Wall Street Journal* called to begin a daylong series of interviews and fact-checks. He was rushing, for he knew that the *New York Times* had already started its story. If the *Times* story ran first, he told me, his story would not run at all.

Davis had already interviewed Piel, who, Davis said, "clearly implied that you were lying." This was Davis's conclusion about Piel's reaction when he read to Piel some quotes I attributed to him. Davis said Piel asked, "Why do you believe him?" Davis told me he responded that he knew me to be a credible writer and had no reason to doubt me.

That is when I remembered the recording I made of the telephone call with Piel on October 4, 1989. Despite attorney Ted D. Lee's advice about the importance of recording Piel, after *Scientific American* began publishing my columns, I felt guilty about the recording and had decided to erase the tape with a magnetic degausser in my shop.

Perhaps a phone call distracted me, for I left the tape by the degausser for weeks before noticing it and deciding to keep it in case it was needed. Now it was needed, so I called Ted, who asked whether I had any other proof of Piel's discrimination. I said no. Since Piel was denying to a reporter my account of what was recorded on the tape, Ted said I had every right to play the tape to the reporter. I called Davis and played the entire tape for him over the phone.

Davis was finishing his story at home the following Sunday when he called that afternoon to clarify the date of *Scientific American* associate editor Laurie Burnham's sanctity-of-life question. He told me he had interviewed Burnham's colleague Tim Appenzeller, former colleague Dennis Flanagan, the American Civil Liberties Union, and noted attorney Laurence Tribe.

Davis also told me about Piel's reaction to the recorded conversation. He said he had called Piel again to report that "Forrest" has more information and that Piel responded,

---

1. Mark D. Hartwig, "Defending Darwinism: How Far Is Too Far?" Access Research Network, November 21, 1990, https://www.arn.org/docs/orpages/or131/mimsrpt3.htm.

"Oh, 'Forrest'—you're on a first-name basis." Davis and I had never met, and Davis said he explained that we had simply communicated a few times over the years. Davis added that he then read some of the statements on the tape to Piel, whose reaction was one of "shock." Piel declined to respond, saying he had not heard the tape.

My reaction to this was to tell Minnie and Daddy that without the tape, Piel's version of the story would have prevailed. After all, he was the distinguished editor of the world's best-known and longest-published science magazine, and I was a mere freelance writer who rejected one of science's key paradigms.

Bob Davis's story, "Scientific American Drops Plans to Hire Columnist Who Believes in Creationism," was published in the *Wall Street Journal* on October 22, 1990. Davis wrote (in part):

> *The dispute raises explosive constitutional issues. On one side is an individual's right to practice his religion; on the other is a publication's right to decide whom it will employ and what articles it will publish. "It's certainly an ugly business," said Arthur Schawlow, a physicist and Nobel laureate [for the invention of the laser] who is a fan of Mr. Mims's work. . . . Mr. Mims on Friday played for a reporter a tape he made of a conversation late last year with 'Scientific American' editor Jonathan Piel. "What you've written is first rate," Mr. Piel said. "It's the public relations nightmare that's keeping me awake." . . .*
>
> *In an interview, Mr. Piel said that "'Scientific American' has never discriminated against anyone for religious reasons," but he wouldn't comment further on the controversy. On the tape recording, however, he told Mr. Mims several times that he worried how creationists somehow might exploit Mr. Mims in their efforts to push creationist textbooks to schools . . .*
>
> *Laurence Tribe, a Harvard University law professor, agreed that the magazine appeared to have discriminated against Mr. Mims. The magazine's rationale was "distressing," he said. "A company could say we don't want to be seen identifying with a point of view, so we don't want to hire Jews."*[2]

The *Houston Chronicle* Redux

The day after the *Wall Street Journal* story appeared, Joe Abernathy published a follow-up piece that led with Jonathan Piel's denial of religious discrimination and his unwillingness to explain his decision. He then quoted two well-known technical writers:

2. Bob Davis, "*Scientific American* Drops Plans to Hire Columnist Who Believes in Creationism," *Wall Street Journal*, October 22, 1990.

"Did Einstein throw the theory of relativity into questioning because he had certain beliefs about God and peace and morality?" asked Alcestis Oberg, a Houston technical writer who is internationally recognized. "No, they're separate domains completely."

"The nation was founded on the belief that nobody should be discriminated against on the basis of their race, creed, or color. That should be applied in journalism as in life," added Oberg, a self-described agnostic whose views on creationism are the opposite of those held by Mims.

Harry Helms, who has served as a senior book editor at McGraw-Hill, Prentice-Hall and Harcourt Brace Jovanovich, said Mims' personal beliefs are irrelevant to his abilities as a writer.

"It's like asking is Forrest black, or is Forrest a Jew," said Helms, who has served as editor on a number of science books written by Mims.[3]

The article concluded:

[Oberg] found the magazine's treatment of Mims chilling.

"I've been in science journalism for 12 years; I wrote about medical technology, space technology . . . ecology, the environment, you name it," she said. "And never in all that time were my beliefs—religious, social or otherwise—raised with any of my editors at any time.

"And that's proper; that's as it should be."[4]

The *New York Times*

On October 16, Maria Moss of the *New York Times* called to request an interview. I wanted to be open and told her that the *Wall Street Journal* was also planning a story. "That's OK," she said. "A story isn't news until it's printed in the *Times.*" Moss spent three hours in my office conducting a detailed interview about every aspect of the *Scientific American* affair. Her story, "Hire a Creationist? A Nonbeliever in Darwin? Not at a Proud Science Journal," became news when it was published at the top of page 2 of the *Times*'s US section on October 24, 1990, two days after Davis's article ran in the *Wall Street Journal.* The article featured two highlighted text boxes: "It is not what the columns say, but what people might think" and "I even told them I could be their token Christian."[5]

---

3. Joe Abernathy, "Journal's Firing of Creationist Draws Fire," *Houston Chronicle*, October 23, 1990.
4. Abernathy.
5. Maria Moss, "Hire a Creationist? A Nonbeliever in Darwin? Not at a Proud Science Journal," *New York Times*, October 24, 1990; https://www.nytimes.com/1990/10/24/us/hire-a-creationist-a-nonbeliever-in-darwin-not-at-a-proud-science-journal.html.

Like Davis, Moss interviewed the principals and wrote:

> *By the summer of 1989, according to Armand Schwab, who was* Scientific American*'s managing editor at the time, Mr. Mims was a strong candidate. "The trial columns he sent were quite good," said Mr. Schwab, who retired in March after nearly three decades at the magazine, "and his proposal outlining future projects was impressive."...*
>
> *Mr. Piel denied in a recent telephone interview that Mr. Mims had been the victim of religious discrimination. "*Scientific American *has never discriminated against anyone on the basis of their religious beliefs and it never will," he said. He declined to elaborate or answer other questions.*
>
> *But others who worked at the magazine at the time said there had been considerable debate over what to do with Mr. Mims. "His work in amateur science was perfectly good," said Tim Appenzeller, who was associate editor at the magazine then. "He is involved in projects of testing and measurement, and if they don't work, readers know it," Mr. Appenzeller said, emphasizing that there had been no question about Mr. Mims's qualifications to write the column. "I was among those who felt we should have hired him," said Mr. Appenzeller, now senior editor at* The Sciences *magazine. If the creationism issue had not come up, he added, "I'm sure we would have."*[6]

The American Association for the Advancement of Science

In mid-October, I contacted attorney Mark Frankel of the American Association for the Advancement of Science to ask whether the AAAS could intervene with *Scientific American*. I provided the details he requested, and the media was as surprised as I was when the organization's Committee on Scientific Freedom and Responsibility sent the following letter to me and copied it to Jonathan Piel:

> *29 October 1990*
>
> *Dear Mr. Mims:*
>
> *The Committee on Scientific Freedom and Responsibility of the American Association for the Advancement of Science has received the materials you submitted in connection with your complaint regarding SCIENTIFIC AMERICAN. The legal questions that may be involved in this matter are beyond the purview of the Committee. However, the Committee does wish to affirm its commitment*

6. Moss.

to the principle that articles submitted for publication in journals devoted to science, technology and medicine should be judged exclusively on their scientific merit. A person's private behavior or religious or political beliefs or affiliations should not serve as criteria in the evaluation of articles submitted for publication.

We emphasize, in particular, the consensus of the Committee that even if a person holds religiously-derived beliefs that conflict with views commonly held in the scientific community, those beliefs should not influence decisions about publication of scientific articles unless the beliefs are reflected in the articles.

We wish to stress that, in expressing this opinion, the Committee is not taking any position on the particulars of your dispute with SCIENTIFIC AMERICAN.

Sincerely,
[signed]
Sheldon Krimsky, Ph.D., Chair
Committee on Scientific Freedom and Responsibility
cc: Jonathan Piel[7]

I told Minnie this letter was the most important outcome of the controversy, and someday it might even be viewed as historic. After making some copies, I slipped it into a protective sleeve and locked it away. After fellow writer Harry Helms read the letter, he replied:

Seriously! Send out press releases quoting this to NY Times, WS Journal, Wash. Post, etc. and pour it on! How long can Piel hang on now that he's proven a bigot (per AAAS) and liar (per ferrous) [magnetic tape recording]?

Harry[8]

The *Houston Chronicle* Again

After the AAAS letter and an unsolicited, enthusiastic defense of my position by Lamar Hankins, acting director of the Texas branch of the American Civil Liberties Union, Joe Abernathy wrote a third installment about the controversy for the *Houston Chronicle.* In his article, published on October 30, Abernathy reviewed the matter and the magazine's worries and wrote that the AAAS "released a statement

7. Dr. Sheldon Krimsky, letter to Forrest M. Mims, October 29, 1990.
8. Harry Helms, letter to Forrest M. Mims, October 29, 1990.

Monday defending Mims."[9] He then quoted from the ACLU letter:

> *"I'm urging* Scientific American *to reconsider their position, because it strikes me as being similar to the type of blacklisting that went on in the McCarthy era," said Lamar Hankins, acting director of the Texas office of the ACLU . . . .*
>
> *"It went on in the McCarthy era because of people's political beliefs or associations, real or imagined. In this case, we're dealing with someone's religious beliefs, but it's still blacklisting," he said.*[10]

The *Washington Post*

On October 31, I flew to Dallas to participate in a debate about the controversy on CNN's *Crossfire*. After exiting the plane at the Dallas–Fort Worth International Airport, two men with cameras were shouting my name. They were photographers on assignment for the *Washington Post*, and they asked if they could take my picture. While deplaning passengers watched, they took several photos of me and then raced away to meet their deadline.

I was surprised when the photo of me ran in the *Post* the following day alongside the headline "Big Bang Over Belief at *Scientific American*" over an article by reporter Charles Trueheart, who had interviewed me several days earlier. Trueheart wrote about both the ACLU letter and the AAAS committee's decision. He also interviewed Dr. Sheldon Krimsky, who had signed the AAAS letter, and wrote, "Krimsky, the Tufts University professor who chaired the AAAS committee, said in an interview, 'The criteria to be used in evaluating someone's work should be what's in his work—not on the man and his complex life and thoughts, both personal and scientific. . . . We have a long history of scientists who have divided their personal beliefs and their scientific beliefs. Galileo was one of them. So was Newton.'"[11]

Houston Chronicle Editorial

The editors at the *Houston Chronicle* were watching the increasingly widespread media coverage of the *Scientific American* controversy that followed their own Joe Abernathy's three stories. On November 2, their lead editorial at the top of the opinions page was titled, in large letters, "Blacklist Specter." The editorial began:

---

9. Joe Abernathy, "Scientists, ACLU Take Up Creationist's Cause, *Houston Chronicle*, October 30, 1990.
10. Abernathy.
11. Charles Trueheart, "Big Bang Over Belief at *Scientific American*," *Washington Post*, November 1, 1990, https://www.washingtonpost.com/archive/lifestyle/1990/11/01/big-bang-over-belief-at-scientific-american/33a72112-eddb-46a7-8116-f8d44fc272d2.

Scientific American magazine has the unquestioned right to publish or not publish whatever authors and for whatever reasons it wishes.

The corollary is that the scientific community, the public and other publications have a right to question the wisdom of Scientific American's judgments, if that seems appropriate.

It does seem appropriate in the case of Forrest Mims III of Seguin, a well-known, well-regarded and widely published science writer. It is fairly clear from reports by the Chronicle's Joe Abernathy that Mims lost a columnist position at Scientific American because of a personal belief in creationism.[12]

The editorial continued with the magazine's concern about becoming associated with my personal beliefs and concluded, "That may be worrying to Scientific American, but it is nowhere near as worrying as what the magazine has done. It has raised the specter of blacklisting an author because of personal religious beliefs. That does Scientific American no credit."[13]

Media Saturation

For several months, the media, especially radio stations, called and faxed. Sometimes the telephone and the fax machine were ringing at the same time. When media people could not get through, they would send a fax and ask me to call back. Or they would call Minnie on our house line and give her the message. For the first month, I was rarely able to eat lunch. Minnie had to squeeze supper between afternoon and evening radio interviews, and she relayed messages to me when I was taking calls. This was not simple, because we were dealing with four time zones.

Many newspapers and magazines printed opinion pieces and letters that chastised *Scientific American*. C. H. "Max" Freedman, a columnist for the *New York City Tribune*, titled his column for October 26, 1990, "Evolutionist Thought-Police Jeopardize America's Prestige." In it, he thoroughly chastised the magazine and observed, "Evolutionists are desperate people. Each new finding confounds their already discredited theory makes them act more hostile and irrational."[14]

Freedman closed by coupling an optimistic look at the future of the collapsing Soviet Union with the *Scientific American* affair: "... cataclysmic reforms are

12. "Blacklist Specter," *Houston Chronicle*, November 2, 1990.

13. "Blacklist Specter"

14. C. H. "Max" Freedman, "Evolutionist Thought-Police Jeopardize America's Prestige," *New York City Tribune*, October 26, 1990.

taking place in the Soviet Union. For all we know, they may presently be enacting legislation making it a crime to discriminate on the basis of an individual's inner beliefs—or even to ask about them as a condition of employment. Just the thought of such monumental embarrassment to this nation should make every concerned American drop whatever he or she is doing and rally to the cause of Forrest M. Mims 3rd."[15]

In a top-of-the-page commentary in the *Los Angeles Times*, Dr. Phillip E. Johnson, a University of California, Berkeley, law professor, wrote a critique of Darwinism on November 3, 1990, that concluded, "The Mims episode shows us that science is beset by religious fundamentalism—of two kinds. One group of fundamentalists—the Biblical creation scientists—has been banished from mainstream science and education and has no significant influence. Another group has enormous clout in science and science education and is prepared to use it to exclude people they consider unbelievers. The influential fundamentalists are called Darwinists."[16]

Articles about the controversy appeared in numerous newspapers and magazines, including *Human Events, Modern Electronics, The New Republic, Science, The Scientist*, and *Texas Monthly*, as well as *Acts and Facts, Christianity Today*, and an assortment of other Christian publications.

Piel's Nightmare Continues: "All the Lines Were Lit Up"

After each major newspaper and magazine article appeared in print, a fresh burst of calls arrived from the media. My phone was often so busy that faxes would arrive asking me to call when I could. Then there was the one-hour talk show on radio station KFI, a clear-channel station in Los Angeles, on election eve, November 5, 1990. Before the program, the producer told me there might not be much interest, because several controversial ballot initiatives and a close governor's race would be decided the next day. As I wrote Harry Helms, "After the first hour, the producer said all the lines were lit up, and they asked me to stay a second hour. So I did. The program went very well, with most callers in my camp. Those few who were not came across as very narrow-minded theophobics."[17]

Some of the radio interviews were broadcast on many stations. A one-hour interview for Moody Radio was heard on 83 stations. An interview on *Issues in*

15. Freedman.
16. Phillip E. Johnson, "Unbelievers Unwelcome in the Science Lab," *Los Angeles Times*, November 3, 1990.
17. Forrest M. Mims, letter to Harry Helms, November 6, 1990.

Education was broadcast on 41 stations. Accuracy in Media did two separate *Media Watch* broadcasts on 300 stations. An interview with evangelist and Christian media figure Dr. D. James Kennedy was also heard on 300 stations. An interview on Armed Forces Radio was heard around the world. In addition to the many interviews, the controversy was covered by many news and talk shows, including Rush Limbaugh's radio program.

Preparation for the dozens of radio program interviews was simple: Enthusiastic quotations by Jonathan Piel about my columns and his quote about a potential public relations nightmare were placed on the left side of my desk. Charles Darwin's *On the Origin of Species* (1859) and *Voyage of the Beagle* (1839, originally titled *Journal and Remarks* and reissued later as *Journal of Researches*) were placed on the right side. My Bible was left on its shelf. Why resort to a creation story that critics disbelieve and ridicule when Darwin himself provided strong criticisms of his own theory?

When radio talk-show hosts wanted to know what *Scientific American* thought about my columns, I simply read from Piel's correspondence ("They're great!"). Nearly all the callers were on my side. When an occasional call arrived from a skeptic, I read Darwin's own words from *Voyage of the Beagle* about a curious phenomenon he observed during his visit to Australia. He noticed a familiar insect larva slinging jets of sand toward an ant that had fallen into its conical pitfall and wrote: "There can be no doubt that this predacious larva belongs to the same genus with the European kind, though to a different species. Now what would the skeptic say to this? Would any two workmen have hit upon so beautiful, so simple, and yet so artificial a contrivance? It cannot be thought so: one Hand has surely worked through the universe."[18] Darwin also praised the civilizing influence of Christian missionaries in his intriguing and entertaining travelogue.

In addition, I read brief passages from Darwin's *On the Origin of Species* that questioned his famous theory—for example: "There is another and allied difficulty, which is much more serious. I allude to the manner in which species belonging to several of the main divisions of the animal kingdom suddenly appear in the lowest known fossiliferous rocks."[19]

Darwin, referring to fossils of advanced creatures such as trilobites and many others that then and now have no known predecessors in the fossil record, then wrote, "To the question why we do not find rich fossiliferous deposits belonging to these assumed earliest periods . . . I can give no satisfactory answer. . . . The case

18. Charles Darwin, *Voyage of the Beagle* (New York: Penguin, 1989), 325.
19. Charles Darwin, *On the Origin of Species*, (New York: New American Library, 1958), 308.

at present must remain inexplicable; and may be truly urged as a valid argument against the views here entertained."[20]

If Darwin could doubt his theory, why couldn't I? Simply reading Darwin's doubts quickly ended inquisitions by skeptical callers. Darwin was invoking the skepticism scientists are supposed to apply to their theories, a fundamental tenet of science that the National Academies (which includes the National Academy of Sciences) inexplicably removed from the latest edition of its landmark book *On Being a Scientist: A Guide to Responsible Conduct in Research*. Darwin's skepticism was tempered by his optimism that fossils older than the oldest known in his time would eventually be found. Earlier fossils of a wide range of complex animals were eventually found in Canada and China, but they, too, suddenly appeared.

While the Darwinists had little or nothing to say after I read Darwin's doubts during the radio interviews, they might have ridiculed me had I read any of the biblical passages that influenced me to study and then abandon traditional Darwinian evolution. One of these is found in Job, believed to be one of the oldest books in the Bible: "But ask now the beasts, and they shall teach thee; and the fowls of the air, and they shall tell thee: Or speak to the earth, and it shall teach thee: and the fishes of the sea shall declare unto thee. Who knoweth not in all these that the hand of the Lord hath wrought this? In whose hand is the soul of every living thing, and the breath of all mankind."[21]

While the theory of evolution by natural selection is widely proclaimed as the foundation of modern biology, a growing number of skeptical scholars are taking a fresh look at Darwin's "valid argument." Former geophysicist and college professor Dr. Stephen C. Meyer, who directs the Center for Science and Culture at the Discovery Institute, has written a *New York Times* bestseller devoted to Darwin's reservations about his theory: *Darwin's Doubt: The Explosive Origin of Animal Life and the Case for Intelligent Design*. Meyer digs much deeper than the most ancient fossil beds to explain how the origin of the astonishing complexity of biology at the molecular level is a far greater mystery than Darwin's missing fossils.

While Darwin's doubts remain today, they are rarely taught in public schools, and I have met very few biologists who are aware of them. That is why I responded to an editorial in *Nature*, which advocated the teaching of Darwin to school students:

---

20. Darwin, *Origin*, 309.
21. The Bible (King James Version), Job 12:7–10.

Compulsory Read?

SIR—Nature is tempted to involve "all of Darwin" to defend evolution from California creationists (Nature 364, 746; 1993). To begin with, I suggest that Darwin's The Origin of Species and Journal of Researches be made compulsory reading for the students of California. The risk, of course, is that the former may put many students to sleep while the latter, which was Darwin's favourite, is much more lively and better written. Moreover, it contains inspirational passages about Christian principles and an essay about the hollow conical pitfalls of the lion-ants of England and Australia, which reads, in part:

There can be no doubt that this predacious larva belongs to the same genus with the European kind, though to a different species. Now what would the sceptic say to this? Would any two workmen ever have hit upon so beautiful, so simple, and yet so artificial a contrivance? It cannot be thought so: one Hand has surely worked throughout the universe. (Voyage of the Beagle, 325; Penguin 1989).

Yes, let "all of Darwin" be used to defend evolution from California creationists.

Forrest M. Mims III[22]

The skeptics might have ridiculed me had I read any of the biblical passages that influenced me to abandon traditional Darwinian evolution. But few of them had much to say after their prophet's words were read. Even more remarkable was that the highly prestigious *Nature* published without change what I would have never written, much less been allowed to write, in the *Scientific American* column.

In the *Crossfire*

In late October 1990, a producer for CNN's *Crossfire* show called to ask whether I would be willing to appear on the program to discuss the *Scientific American* affair. I agreed and asked if I could bring a prop, but she said it would not be possible to display anything, because only heads and shoulders would appear onscreen. That is all I needed to know, so I did not ask permission to place a trilobite fossil in my coat pocket.

My advocate was cohost Cal Thomas. The opponents were cohost Michael Kinsley, and Dr. Eugenie Scott of the National Center for Science Education.

As mentioned above, Mark Hartwig, whom I would soon meet, wrote about the debate for the Access Research Network in a report titled "Defending Darwinism: How Far Is Too Far?" These excerpts from Hartwig's report sum up Scott's position during the debate:

22. Forrest M. Mims, "Compulsory Read?" *Nature* 366 (November 11, 1993), 104.

Another disturbing development is the way in which some Darwinists have chosen to justify the magazine's actions, resorting to libelous ad hominems, and displaying distressing attitudes toward freedom of conscience. One of the more egregious examples of this occurred on October 31, when Dr. Eugenie Scott, Executive Director of the National Center for Science Education, appeared with Forrest Mims on the CNN's Crossfire. During the broadcast, Dr. Scott made some astonishing comments.

Early in the broadcast, co-host Cal Thomas asked Eugenie Scott: "Dr. Scott in San Francisco, let me jump in here and ask you a question. There are an awful lot of Americans, not only religious Americans, who believe that you evolutionists are trying to censor and silence people who don't agree with you. Isn't the essence of scientific inquiry free and open access and debate?"

Eugenie Scott replied, "It is indeed, but I think what we have to look at is what are we—what are we giving—what are we calling equally valid ideas? We're not dealing with political speech, we're not dealing with opinions on art. We're dealing with what science is . . . ."

In essence, Dr. Scott seemed to be saying that freedom of conscience doesn't extend to science, because science deals in matters of truth—as opposed to matters of opinion. Such a viewpoint indicates a grave misunderstanding of both the nature of science and the meaning of freedom of conscience.

. . . In addition to her disturbing remarks on science and freedom of speech, Dr. Scott also made a libelous attack on Forrest Mims.

. . . Scott: I am not an employee of Scientific American, so, you know, I am not defending them. They can defend themselves but what I would consider if I were in this—in their position is whether you would be—whether they would be limiting the scope of this column by hiring somebody who is so far out of the scientific mainstream. This man would not be able to write about a wide variety of scientific topics because of his views which are basically religious.

The point that Dr. Scott tried to make is one that has surfaced repeatedly in the creation/evolution debate: evolution is so integral to science and so well established that disbelievers cannot possibly be competent as scientists—or science writers.[23]

The producer who had told me not to bring any props was happy when I held the trilobite fossil next to my face and stated that this ancient creature has no

23. Mark D. Hartwig, "Defending Darwinism: How Far Is Too Far?" Access Research Network, November 21, 1990, https://www.arn.org/docs/orpages/or131/mimsrpt3.htm.

predecessor in the fossil record. During a commercial break, she said through my earphone that everything was going great, and "Go after her!" I wondered whether she was giving the same message to Scott.

As for Scott, she was the first person with a scientific background to openly defend *Scientific American*. I wondered about her background during the debate and several years later sent her a request for her vita. She replied in a much friendlier tone than she had exhibited during the debate. She even reversed the position she took on *Crossfire* and wrote, "You have a reputation for high-quality work."[24]

While the number of Scott's writings critical of creation was impressive, her peer-reviewed scientific publications were limited to rather basic papers on human body size, nutrition, and dentition, none of which make a significant contribution to supporting the evolutionary hypothesis she advocates.

Scott's critics blame her atheism for her promotion of evolution, but her letter closed with a clue at least as important, for she wrote that a pencil sharpener was the limit of her mechanical aptitude.[25]

Designing electronic circuits and writing computer programs require precise attention to detail for the result to properly perform. Mobile phones did not evolve from engineers randomly soldering components into an electronic circuit or blindly punching keys while writing a program. Nor did random mutations over millions of years lead to the creation of Eugenie Scott.

But if Scott's mechanical aptitude does not extend beyond a pencil sharpener, how can she understand the obvious design of the complex chemistry of vitamin D synthesis in her skin, the leucocytes and their kin that patrol her body, searching for pathogens to devour, and the astonishing kinesin molecules that carry huge burdens while walking along spontaneously assembled microtubules inside her cells? Perhaps Scott's critics should offer to teach her some basic mechanics, chemistry, and electronics. If molecular biologists were required to take courses in these fields, they would better understand the complexity of life and might even be open to the idea of design.

The Meeting of Intelligent-Design Scholars

On the evening of November 5, 1990, the kitchen phone rang, and the caller identified himself as Mark Hartwig. He did not tell me he held a doctorate in educational psychology, that he specialized in statistics, and that he was involved in the early phases of the intelligent design movement. Instead, he apologized for

---

24. Dr. Eugenie Scott, letter to Forrest M. Mims, October 15, 1996.
25. Scott, letter to Mims.

imposing, informed me about a special meeting of an ad hoc committee that would discuss the *Scientific American* controversy, and asked whether I could attend. He explained that the committee consisted of prominent attorneys, scholars, and writers who questioned Darwinian evolution. When he said that the Stewardship Foundation would provide airplane tickets and a hotel room, I agreed to attend.

The committee meeting was held at a hotel in Tacoma, Washington, November 23–25. There, I learned that the Stewardship Foundation was headed by Christian philanthropist C. Davis Weyerhaeuser, who took a serious interest in the budding intelligent design movement, which I had not heard about before.

UC Berkeley law professor Phillip E. Johnson was among the key men at the meeting. His book *Darwin on Trial* was soon to be published, and he provided an update on its status. He also read some hilarious email exchanges he had with several prominent Darwinists completely unprepared to debate a prominent law professor with a deep knowledge of both Charles Darwin and evolutionary theory.

My role at the meeting was to speak about the *Scientific American* affair and the recent *Crossfire* debate. I had given many such accounts on radio and television, but never before a group of prominent scholars. They found my presentation as amusing as Johnson's emails, loudly proclaimed their support, and expressed surprise at the extent of media coverage. They then moved on to the next topic.

This was not what I had expected, for I was hoping the committee, or at least some of its attorneys, would back my campaign to regain The Amateur Scientist. Only after I had a conversation with one of the attorneys did I learn that their goals were far broader than mine. The media attention given my case had shown the public the closed mind of the Darwinists' paradigm, and that is why they wanted to hear from me. My campaign to get back the column would be a distraction.

The following morning at breakfast, the hotel restaurant was packed, and the twenty-five or so committee attendees sat at several tables scattered around the room. Though all our plates were loaded with food, everyone at my table was talking and no one was eating. Suddenly, I heard a loud *bing-bing-bing* and looked up to see Johnson standing at a table across the room, striking a glass with a spoon.

"Gentlemen," he proclaimed, "let's thank the Lord for our breakfast."

With every head in the room bowed, including those who were not part of our group, Johnson led a blessing that answered any question I might have had about the group's Christianity.

Arthur Caplan Defends *Scientific American*

Arthur Caplan, then director of the Center for Biomedical Ethics at the University of Minnesota, wrote a guest editorial in the *St. Paul Pioneer Press* on December

10, 1990, titled "Victim of Religious Discrimination Deserved to Be Fired." Caplan wrote, "Forrest M. Mims III recently became unemployed . . . . He got canned because of his religious beliefs." After mocking me and the AAAS letter, Caplan continued, "I believe Mims is not qualified to write a regular column about science for the general public." After he continued this diatribe, he wrote that science magazines have a right to discriminate.[26] The newspaper was surprised by what happened next, which it covered in an editorial on December 29:

> ***Science Columnist's Rejection: Wrong Test Applied***
> *With their letters to the editor, nearly 50 of our readers have told us that they believe Arthur Caplan misfired in his Pioneer Press column . . . .*
> *For us to receive that many letters in response to a single story or column in our newspaper is highly unusual. Five letters critical of Mr. Caplan appear today on this page, along with a reply to the Caplan column from Mr. Mims himself. (Six other critical letters were published Dec. 22.) . . .*
> *Clearly, the magazine has a right not to hire someone whom editors believe would impair the publication's integrity. But if the publication concluded Mr. Mims' religious beliefs were incompatible on their face with responsibilities to write a column about amateur science, Scientific American exercised bad judgment.*
> *The appropriate test to apply was whether Mr. Mims could be objective in spite of his personal beliefs. Belief in creationism, by itself, should no more disqualify someone from writing about science than belief in evolution should disqualify someone from becoming a minister.*[27]

The letters printed below this editorial sternly chastised Caplan for what the writers described as his persecution, hypocrisy, and intolerance. My response was also published by the paper:

> *'Scientific American' is the oldest and most distinguished magazine in the United States. Its founding editor, Rufus Porter, advocated belief in a "Creator God" in the magazine's premier issue. The current editor, Jonathan Piel, has withdrawn his assignment for me to write the prestigious column known as "The Amateur Scientist" solely because I believe in a Creator God and do not believe in Darwinian evolution or abortion. . . .*

---

26. Arthur Caplan, "Victim of Religious Discrimination Deserved to Be Fired," *St. Paul Pioneer Press*, December 10, 1990.
27. "Science Columnist's Rejection: Wrong Test Applied," *St. Paul Pioneer Press*, December 29, 1990.

Caplan wrote, "I believe Mims is not qualified to write a regular column about science for the general public."

I was so surprised by this statement that I called Caplan and asked what he knew about my background. He was totally unaware that for 20 years I have earned a living by writing more than 50 books and many hundreds of articles and papers for some 75 magazines, journals, and newspapers.

He didn't know that I have designed, assembled and operated virtually every day for more than 18 months the only ground-based total ozone monitoring station between California and Florida at my latitude.

He had no knowledge of my development of infrared travel aids for the blind, my photographic study of the life cycle of the black widow spider, my two-year program to measure the sun's ultraviolet every day, the instruments I have designed to measure the reflectance of leaves and other objects, my study to identify killer bees by the frequency of their wing beat, and the hundreds of other investigations and projects I have conducted.

'Scientific American' knew about my qualifications and never once questioned my record as a writer about science and engineering. That's why they assigned and published my columns about observing sunspots, monitoring solar ultraviolet radiation and aerial photography from helium-filled balloons.

Even though the senior editorial staff at 'Scientific American' endorsed me to the end, I was dismissed from "The Amateur Scientist" and that venerable column was killed after 38 years of publication solely because the editor was afraid he might be embarrassed if my personal belief in a Creator God somehow became known.

Had he accepted my written agreement that I would never use the column as a forum to espouse my beliefs, there would be no problem. Instead, for two months he has been embroiled in the public relations nightmare he wanted to avoid. . . .

Caplan and the newspaper editors who ran his column without comment should carefully consider the implications for themselves if 'Scientific American' goes unchallenged. For next it may be they whose personal beliefs are questioned, they whose works are censored, and they who lose their First Amendment rights. It's happened before, and it can happen again.[28]

When the *Birmingham News*, in Birmingham, Alabama, published Caplan's column, so many protests were received that the editor called me to apologize. He

28. Forrest M. Mims, "Victim of Religious Discrimination Speaks Out," *St. Paul Pioneer Press*, December 18, 1990.

then assigned Mark Hartwig to write an eloquent rebuttal. Meanwhile, Knight-Ridder Newspapers, which had distributed the Caplan column, distributed the response I sent to the *Pioneer Press*. It was printed at the top of the Viewpoint page in the *San Antonio Light*, in San Antonio, Texas.

The Mims-Caplan Debates

On November 29, 1990, Tom Ewing, executive editor of *The Scientist*, a popular tabloid distributed to some 85,000 scientists, called to ask whether I would provide a 1,500-word essay about the *Scientific American* affair for the February 18, 1991, issue, to be distributed to some 5,000 members of the American Association for the Advancement of Science during its annual meeting in Washington, DC. When I agreed, Ewing suggested that his newspaper sponsor a debate between me and a supporter of *Scientific American*. On January 2, the opinion page editor of the *Scientist* wrote that Arthur Caplan had agreed to write a counterpoint essay that would appear in the same issue.

Our essays appeared on the opinion page of the *Scientist*. In "Intolerance Threatens Every Scientist—Amateur or Not," I reviewed the widespread criticism directed at *Scientific American*, described my science projects, and closed with, "In retrospect, it's apparent that Jonathan Piel had good reason to be worried about the Arthur Caplan's of the scientific community when he canceled my assignment to 'The Amateur Scientist.' Now that I've been subjected to some of their attacks, I can almost sympathize with Piel's plight. Even though he anticipated their reaction and dismissed me, to his credit he never once stooped to disparage my qualifications."[29]

Caplan's rebuttal was titled "Creationist Belief Precludes Credibility on Science Issues." My call to him after his essay defending Scientific American softened his criticism, for he began:

> *It is tough defending the position that Scientific American was right to fire Forrest Mims as the author of "The Amateur Scientist" column. Mims meets one of the central requirements of the job—he is a competent amateur scientist. He is also an excellent writer, as anyone who has had the pleasure of reading any of his numerous popular science writings can attest.*
>
> *Not only does Mims have many of the requisite credentials for the job, it is no fun defending the conduct of those at Scientific American who gave him the boot.*

29. Forrest M. Mims, "Intolerance Threatens Every Scientist—Amateur or Not," *The Scientist*, February 17, 1991, https://www.the-scientist.com/opinion-old/intolerance-threatens-every-scientist--amateur-or-not-60771.

> *Having discovered that Mims is a creationist, the magazine's staff went on to ask him some obviously inappropriate questions about his views concerning abortion, subsequently and inexcusably killed off "The Amateur Scientist" column and, judging from media reports that I've read, continue to manifest the alacrity of an ostrich under predatory attack in defending the decision to let Mims go.*
>
> *Forrest Mims does not need me to toss encomiums in his path. Ever since he was fired, he has received an outpouring of unqualified support. Everyone from the American Civil Liberties Union to the American Association for the Advancement of Science apparently belies that dropping Mims as a columnist is nothing short of invidious religious discrimination.*[30]

With an opening like this, I thought Caplan had switched sides. Unfortunately, he devoted the remainder of his essay to advocating the discrimination to which I was subjected to by *Scientific American* and closed with this: "Forrest Mims is a competent writer and amateur scientist. But his personal beliefs about creation limit what he can and cannot tell his readers about all the nooks and crannies of science. They also distort the picture he conveys regarding what scientific methodology is all about. It is a hard line to draw, but Forrest Mims and others who espouse a belief in creation and reject the scientific standing of evolution are on the wrong side of the line."[31]

Although I did not know at the time, the same Caplan who wanted me blacklisted over my rejection of Darwinian evolution had previously raised questions of his own about contemporary evolutionary theory and its true believers. Phillip Johnson and Mark Hartwig were aware of this when they read the *Scientist* debate, and they provided the vital key to upending Caplan during an upcoming debate between us on the Voice of America.

Caplan's Fail on the Voice of America

For several weeks, Victor Morales of the Voice of America (VOA) had been in touch about a possible interview on his daily show. This dialogue led to a lively debate between Arthur Caplan and me on March 21, 1991. The evening before the debate, I called Phillip Johnson to inform him about what was about to happen, and he told me to call Mark Hartwig. All I knew about Caplan was that he had claimed that because I did not accept Darwinian evolution, I should never write about

30. Arthur Caplan, "Creationist Belief Precludes Credibility on Science Issues," *The Scientist*, February 17, 1991.
31. Caplan.

science for the public. What I did not know was that Caplan also had issues with evolutionary theory. When I called Mark, he briefed me on this and then sent a thirty-six-page fax to the hotel.

I was tired from travel that evening and was not sure I would be able to read all the pages that Mark sent. But after scanning the fax, I discovered why he was so anxious for me to read several key passages.

The fax included two of Caplan's publications that raised fundamental questions about the status of evolutionary theory and its adherents. In one seventeen-page paper, Caplan neatly stripped away the philosophically simplistic veneer of Darwinism in an insider's criticism aimed mainly at the practitioners of Darwinism, not the notion that life has evolved. But if Caplan could question Darwinism, why couldn't I? And that was Mark's key strategic point when I called him back. I then stayed up late filling thirteen pages of a yellow-lined notepad with quotes from Caplan and these two pages of points to make during the debate:

1. *What about my 20 years of writing about science?*
2. *What do I do with the thousands upon thousands of ozone and solar ultraviolet radiation measurements I've made for the past 3 years?*
3. *Tell me one thing I've written that fails the scientific method.*
4. *[Caplan wrote that I] 'deserved to be fired' & [I was] 'not qualified' before knowing who I am or what I do!*
5. *Should Orthodox Jews be banned from science magazines? Should the many scientists & philosophers who question & criticize evolution be banned?*
6. *EVEN YOU! (apply same criteria)*

The final point was Mark's strategic trap, and I followed it by copying on the notepad this remarkable Caplan quotation from the fax Mark sent: "Evolutionary theory has been charged time and again with the sins of excessive malleability, ad hocness and unfalsifiability. It is precisely these concerns that led Karl Popper and various other philosophers to relegate Darwinism to the dustbin of metaphysics. The charge lingers on to the present day, particularly in paleontological and taxonomic circles. Biologists in these fields have shown an astounding capacity for self-flagellation as they bemoan the theoretical docility of many of their peers who are unable to see the modern synthetic theory of evolution rests upon sandy mythological foundations."[32]

32. Arthur Caplan, "Say It Just Ain't So: Adaptational Stories and Sociobiological Explanations of Social Behavior," Philosophical Forum 13:3 (Spring 1982), 144–60.

I arrived at the Voice of America studios the following day wide awake and ready for the thirty-minute debate. Victor Morales, a cordial gentleman, explained how the debate would be conducted and informed me that Caplan would be speaking to us by telephone.

I had made most of the points on the yellow note pages before the one-minute break halfway through the debate when the clock on the wall indicated there was time to spring Mark's trap using Caplan's own words. After asking whether scientists and philosophers who raise questions about evolution should be banned, I asked, why not Caplan himself? I then read his own words from the quotation given previously. Before Caplan could respond, Morales announced the mid-debate break, during which he said how pleased he was with what had just transpired.

During the first half of the debate, Caplan addressed me as Mr. Mims, and I called him Dr. Caplan. After the break, Caplan's radio personality was transformed. Thanks to Mark's fax, Caplan's aloofness was gone. He was now my friend, calling me Forrest and avoiding his own words that I had quoted. Before the VOA debate, I appeared on some fifty talk shows broadcast on many hundreds of stations around the country. The VOA debate was in a different league, for it was taped for a Palm Sunday broadcast to a potential audience of 140 million listeners, all of whom could hear for themselves the double standard of the before-and-after Caplan.

Some skeptics were upset by the VOA's interest in the *Scientific American* affair. Perhaps they were unfamiliar with the VOA's charter, which includes this statement: "VOA will represent America, not any single segment of American society, and will present a balanced, comprehensive projection of significant American thought and institutions."[33]

The German Embassy Joins My Campaign

My unconventional campaign to win back the *Scientific American* column took a new twist after I sent a complaint to the German embassy in Washington. The magazine had been acquired by Holtzbrinck Publishing Holdings Limited Partnership, a German company, and on February 8, 1991, an embassy official called to express his concern about possible embarrassment to Germany. He had already spoken with John Moeling Jr., the magazine's publisher, and he asked me to provide more information. I did, and he later sent a copy of his letter to the magazine's owner. The letter was in German, and I found a woman of German descent who translated it into English. The letter began with a brief introduction to the controversy and continued with a recommendation:

33. "The VOA Charter: Telling America's Story," Voice of America, July 12, 2016, https://editorials.voa.gov/a/the-voa-charter-telling-americas-story/3414626.html.

Mr. Dieter von Holtzbrinck:
... Please allow me to recommend to the Publishing House to try to remove and resolve this matter. I realize that Mr. Mims with his problems goes beyond the usual means with extreme intensity of solving this matter.

He manages to generate the interest of the American Media which is especially interested in those cases of religious discrimination, which I suspect is not in your best interest. I would be thankful if you would inform the Embassy of the necessary steps to take to solve this conflict. I allow myself to send a copy of this letter to Mr. Mims.

Friendly greetings,
Count Lambsdorff[34]

The National Press Club

In March 1991, prolific author and technologist John Ash McCormick called to ask whether I would like to give a Morning Newsmaker talk at the National Press Club in Washington, DC, on March 28. I thanked McCormick for his generous offer and agreed. The timing was good, for the previous morning, I was scheduled to debate Arthur Caplan on Voice of America.

After Gershon Fishbein, a longtime journalist and publisher of the *Genetic Engineering Letter*, gave the introduction, I began with bad news about threats against me by *Scientific American*'s lawyer and good news about her reminder that I had a legal responsibility to tell the whole truth. I looked into the C-SPAN camera and said that that is what I intended to do, and that is what I did for the next thirty-seven minutes.

I began the talk by holding up the 1845 premiere issue of *Scientific American* and reading a few lines from Rufus Porter about the Creator God. I said that today, Porter would be unable to work at the magazine he founded. During the talk, I showed one of the TOPS ozone monitors that I was to have written about in *Scientific American*. I also showed my homemade total-water-vapor instrument, a fossil trilobite, and some fossilized insects encapsulated in ancient amber. I ended the talk by again holding up the 1845 issue of *Scientific American* and reading these words: "In conducting this publication, we shall . . . exercise a full share of independence, in the occasional exposure of ignorance and knavery, especially when we find them sheltered by ignorance and aristocracy."[35]

---

34. Otto Friedrich Wilhelm Freiherr von der Wenge, Graf Lambsdorff, letter copied to Forrest M. Mims (translated), February 15, 1991.
35. Rufus Porter, "To the American Public," *Scientific American*, August 28, 1845.

I then asked, "Where are those words found? Rufus Porter, founding editor of *Scientific American*, August 28, 1845."

Only half a dozen or so reporters showed up for the National Press Club talk, which was on the Thursday before Easter weekend. But that was OK, for C-SPAN was there and the reporters asked good questions. The news must have been slow that weekend, for C-SPAN telecast the entire talk six times on March 28, 29 and 30. John Ash McCormick later wrote, "The club told me at the time your newsmaker appearance/story got the most interest of any to that date." It's still getting some attention on the internet.[36]

A Battle with the *Journal of Molecular Evolution*

One of the few publications that supported *Scientific American* was the *Journal of Molecular Evolution*, the leading science journal about the latest discoveries in biology at the molecular level. Pages 1–2 of its July 1991 issue were devoted to ridiculing me in an editorial by Thomas H. Jukes, one of the journal's founders.

Jukes's piece was sprinkled with erroneous and even bizarre statements about both me and the controversy. He ignored *Scientific American*'s infractions detailed in its own public statements and the recording of Jonathan Piel. He closed with a mock inquisition of almost-juvenile questions.[37]

I wrote Emile Zuckerkandl, the journal's founding editor, to request equal space to respond to Jukes. He responded that I should reply in a letter to the editor. Following Mark Hartwig's advice about being tough, I wrote that Jukes took nearly two pages to libel my reputation, and I would settle for no less than two pages to defend myself.

After a few more exchanges, the editor caved, and my two-page defense appeared on pages 1–2 of the January 1992 issue of the journal in the same Random Walking feature that had featured Jukes's diatribe.[38] Thus, a skeptical view of Darwinian evolution and some of its advocates was published by one of the world's leading journals devoted to evolution! Ironically, Jukes was well known for his provocative 1969 paper *Non-Darwinian Evolution* with Jack Lester King in the journal *Science*.

Jukes passed on in 1999, and I will do so someday. The *Journal of Molecular Evolution* and its essay by an intelligent design advocate will be preserved so long as there are libraries and the internet.

---

36. "A Scientist Who Believes in Creationism," C-SPAN, March 28, 1991, www.c-span.org/video/?17314-1/scientist-believes-creationism%20and%20www.press.org/news-multimedia/videos/cspan/17314-1.
37. Thomas H. Jukes, "Creation vs. *Scientific American*," *Journal of Molecular Evolution* 33, no. 1 (July 1991), 1–2.
38. Forrest M. Mims, "Response to Jukes," *Journal of Molecular Evolution* 34, no. 1 (January 1992), 1–2.

Meanwhile, I continued to fax messages to Claus-G. Firchow and Jonathan Piel that I was ready to resume work on The Amateur Scientist. For example, on October 19, 1991, I sent the following message to Firchow: ". . . I close by reaffirming your right and mine to choose whether or not to believe in a Creator God and the sanctity of life. While I have not written about these topics, I will never permit the publication of my works to be determined by my personal religious beliefs. Some of my ancestors came to this country to avoid religious discrimination. So did some of the German immigrants who farmed the land visible in the aerial photograph in the final "The Amateur Scientist." One of their descendants built my house and fenced its land."[39]

The Moody Church

Several churches asked me to give talks about the *Scientific American* affair. The most significant invitation arrived from Rev. Erwin W. Lutzer, of Chicago's Moody Church. Lutzer arranged for me to speak to a large adult class and to make a brief statement before his Sunday morning congregation on January 13, 1991. He also arranged for me to speak at the Moody Business Network's bimonthly luncheon. Prominent educator Richard P. Aulie was in attendance, and he followed up with a carefully composed ten-page letter agreeing with most of my points and disagreeing with a few. He also sent a copy of a letter of support he had sent to Jonathan Piel and the chairs or directors of eight major corporations.

During the Chicago visit, CNN asked to do a television interview with me about the *Scientific American* affair. After I arrived at the tiny two-room studio, I was seated at a chair facing a TV camera a few minutes before the interview was to begin. A monitor tuned to CNN was near the camera. The technician said he needed to place a phone call and told me to start talking when the interviewer asked me a question. He then lay on the floor of the adjacent room to place a bet with his bookie. He was whispering into his phone during the first half of the interview.

Piel's Nightmare Ends

Sometimes, Jonathan Piel's public relations nightmare sent dark clouds my way, and I became concerned about how negative statements by *Scientific American* and its supporters might harm my reputation. I was also concerned about the impact of the controversy on Minnie, who still recalls how overwhelmed she felt. Radio Shack editor Dave Gunzel, fellow writers Harry Helms and Mark Hartwig, and my

39. Forrest M. Mims, letter to Claus-G. Firchow, October 19, 1991.

father were among those from whom I sought advice. In a November 7, 1990, letter I thanked Harry: "Thank you again for your advice and counsel. This may seem hard to believe, but the loneliest times during all this have been when the phone rings off the hook all day and all evening and I wish I could get some feedback from a friend about how best to respond. Thanks for supplying some of that feedback."[40]

On the positive side, virtually all the radio and television interviews were friendly. Many newspaper editorials and letters to the editor lambasted *Scientific American*. Even some of the reporters who wrote objectively about the controversy privately wished me well.

Many letters of support arrived from a surprisingly diverse collection of attorneys, professors, Orthodox Jews, homemakers, pastors, engineers, scientists, and editorial writers. Letters of support also arrived from such diverse organizations as the ACLU, the National Right to Life Committee, the American Family Association,Center for Law & Policy, Jews for Morality, the Institute for Creation Research, the Western Center for Law and Religious Freedom, the Christian Legal Society, and various churches.

After nearly six months, the incessant ringing of the phone and the fax machine trickled to an end. Jonathan Piel's public relations nightmare was finally over. The first day no calls arrived from the media was a letdown, until I realized they had thoroughly covered the story. They needed to move on, and so did I. It was time to resume writing books for Radio Shack and designing more instruments to monitor the atmosphere and to continue the daily atmospheric measurements I had begun on February 4, 1990, to prove I could do real science. All the records of the *Scientific American* affair were placed in folders and stashed in a file cabinet.

Meanwhile, the *Scientific American* controversy continued to pop up in emails and online discussion groups and interviews, and those who opposed my position did so adamantly. To prepare myself for worst-case criticisms during discussions of the controversy, I considered the writings of Dr. Richard Dawkins, who is famously known as an atheist who ridicules those who question evolution. For example, in his review of *Blueprints: Solving the Mystery of Evolution*, by Maitland A. Edey and Donald C. Johanson in the *New York Times* (April 9, 1989), Dawkins wrote, "It is absolutely safe to say that if you meet somebody who claims not to believe in evolution, that person is ignorant, stupid, or insane (or wicked, but I'd rather not consider that)."[41]

40. Forrest M. Mims, letter to Harry Helms, November 7, 1990.
41. Dr. Richard Dawkins, "In Short: Nonfiction," *New York Times*, April 9, 1989.

I found Dawkins's email address and wrote him on February 2, 1997:

Dear Mr. Dawkins:
In 1990 SCIENTIFIC AMERICAN ended my assignment to write "The Amateur Scientist" column when the editor learned I reject Darwinian evolution and abortion. In view of your well-known advocacy of Darwinism, I respectfully wish to know if you feel that a person who rejects Darwinian evolution and holds faith in God can do credible scientific research capable of being published in peer-reviewed, scholarly, mainstream journals.
Your response shall be most appreciated.

Forrest M. Mims III[42]

This was sent nearly two years after Dawkins had ceased his laboratory research to become Oxford University's first Simonyi Professor for the Public Understanding of Science. He did not permit me to reprint his extreme ridicule of people like me who do not believe that we are descended from apes and their supposed ancestors. But he permitted me to quote one sentence from his emails: "I warmly congratulate the Editor of *Scientific American* for firing Forrest Mims."[43]

Looking back, I agree with Dawkins. While losing The Amateur Scientist was depressing, had Jonathan Piel's concern about my religious faith not intervened, today I might be working full-time producing a challenging column each month instead of doing science and writing this memoir. Instead, my childhood dream of becoming a free-thinking, self-taught amateur scientist was fulfilled with long-term projects that went far beyond those I published in The Amateur Scientist. My life as a part-time science fair project developer and writer of electronics books was about to expand in unexpected directions.

42. Forrest M. Mims, email to Richard Dawkins, February 2, 1997.
43. Richard Dawkins, email to Forrest M. Mims, February 16, 1997.

14

TRANSITIONING INTO SCIENCE

My tenure as writer of The Amateur Scientist column in *Scientific American* was brief. Only three of my columns were published, and Wikipedia explains what happened next: "Although the [Mims] incident did not diminish *Scientific American*'s commitment to the column, it did make the editors reluctant to offer the column to another amateur scientist. The magazine invited a number of potential columnists to submit articles, some of which it published. But *Scientific American* was unable to find anyone with both professional credentials and the breadth of scientific interests necessary to recapture the popularity the column enjoyed under Stong and Ingalls. Without a regular columnist, the department languished, appearing only sporadically between 1990 and 1995."[1]

Meanwhile, *Science Probe!*, the magazine I had cofounded with Larry Steckler and edited, reached a circulation of 60,000. We even published an iconic article by Dennis Flanagan, who had edited *Scientific American* for many years before Jonathan Piel took over.

Scientific American readers wanted The Amateur Scientist to return, and the editors finally found a solid candidate for the position in the fall of 1995. That is when I received a telephone call from Dr. Shawn Carlson, the new candidate, who calmly explained he had been offered the position but would not accept it without my permission.

I had met Shawn at various science education meetings and was impressed that his doctorate in physics did not come with an aloof view of amateur scientists like me. His grandfather had been an active amateur scientist, and Shawn had begun the Society for Amateur Scientists. I congratulated Shawn, thanked him for requesting my endorsement, and encouraged him to accept the assignment.

1. "The Amateur Scientist," Wikipedia, https://en.wikipedia.org/wiki/The_Amateur_Scientist.

Shawn's version of The Amateur Scientist appeared from November 1995 to March 2001, when the magazine dropped all its columns. Shawn nearly lost the assignment when he devoted the May 1997 column to an LED sun photometer based on the one I had developed for the column in 1989. (My piece about the photometer was never published.) The column was titled "When Hazy Skies Are Rising," and it opened with:

> *Haze is a vital indicator of our atmosphere's health . . . surprisingly little is known about how the amount of haze is changing globally because no one is coordinating haze observations over widely dispersed areas.*
>
> *That may change with the latest design from Forrest M. Mims III. (Mims may be familiar to readers from his columns in this section in 1990.) He has invented an atmospheric haze sensor that costs less than $20 [$38 in 2023] and is so simple to construct that even the most hardened technophobe can put it together in under an hour. Mims's instrument could revolutionize this important area of study by opening the field to all comers, that is, to amateur scientists.*[2]

The column nicely summarized my work, mentioned me by name several times, and included my circuit diagram and a chart with several years of my haze data. It closed with, "I gratefully acknowledge informative conversations with Forrest Mims."

Because *Scientific American* had been widely chastised following my dismissal from The Amateur Scientist, some high-level staff at the magazine strongly objected to Shawn's use of my name and material. But Shawn refused to alter the column, and it was published. I was glad to be back in The Amateur Scientist column, and I have never forgotten Shawn's loyalty.

Robert Tinker

My daily measurements of ultraviolet sunlight, begun on September 23, 1989, from the field next to the Silicon Farmhouse, continued during *Scientific American*'s public relations nightmare, described in chapter 13. I also continued near-daily measurements of the ozone layer, haze, UV-B (the ultraviolet responsible for sunburn), and the total amount of water vapor and oxygen overhead begun on February 4, 1990. After the *Scientific American* affair was finally over, I told Minnie I was taking a year off to prove that a person without a science degree who rejects Darwinian evolution can do serious science, make discoveries, and be published in journals of science.

---

2. Shawn Carlson, "When Hazy Skies Are Rising," *Scientific American*, 276, no. 5 (May 1997), 106–7.

Haze and total water vapor were measured with a companion to the LED sun photometer I had shown Jonathan Piel in his office at *Scientific American*. This instrument, which I still use every day the sun shines, is novel, for it replaces traditional silicon photodiodes and optical filters with LEDs that function as light sensors that respond to a narrow band of wavelengths.

In operation, the sun photometer is pointed directly at the sun. One LED detects near-infrared (near-IR) at 880 nm, and a second LED detects near-IR at 940 nm. Water vapor absorbs much more near-IR at 940 nm than at 880 nm, and the ratio of the two signals is proportional to water vapor between the instrument and the upper atmosphere. The 880 nm LED is much less sensitive to water vapor, so it is also used to measure the presence of dust, smoke, and air pollution in a narrow column through the atmosphere.

In the fall of 1991, I received an invitation to speak about my research at the Second International Conference on Education and Global Ecology at Pocono Manor, Pennsylvania. The meeting was sponsored by the Technical Education Research Center, a prominent science education consulting firm. The meeting was planned by Dr. Robert Tinker, who had a doctorate in physics from MIT and was TERC's chief science officer.

The Pocono conference was among the first of many such meetings I attended after losing the *Scientific American* column. I was somewhat apprehensive about how a disgraced former science columnist would be received, for other speakers included Dr. John King, professor of physics at MIT; Brian Rosborough, chair of Earthwatch; Dr. Boris Berenfeld of the Academy of Sciences of the Soviet Union; Monica Bradsher, director of the National Geographic Kids Network; and other well-known science educators. But not one person at the meeting said anything negative to me about the widely publicized *Scientific American* affair. Instead, I was surprised when several privately told me they supported my stand. After the meeting, Bob Tinker's wife, Barbara, sent a reassuring note: "You and John King were the most spectacular duo I have heard! Thank you so much for bringing your energy and inventiveness to our gathering. Everyone loved meeting you, and we look forward to further correspondence and meetings as the years go on."

Bob Tinker was an early advocate for incorporating computers and sensing devices into the high school science curriculum, and he was extremely interested in my miniature atmospheric monitoring instruments, especially TOPS, my instrument that measured the ozone layer, and the LED sun photometer. He and I became friends, and I eventually worked with TERC to develop an LED sun photometer for students. But when I told him I was working on a formal paper about the LED instrument, he cautioned me to avoid writing scholarly papers. He said they take too much time and are too often challenged by unfriendly reviewers.

I was puzzled by Bob's advice, for I had long assumed that the only way I could enter the world of science without an academic degree was through the publication of scientific papers. I had already published several scholarly papers and many electronics articles. Therefore, I continued working on the LED sun photometer paper, and "Sun Photometer with Light-Emitting Diodes as Spectrally Selective Detectors" was published by *Applied Optics* on November 20, 1992.

Bob was right about the considerable time required to write scholarly papers and respond to peer reviewers. But I am glad that I persevered, for this paper, which has been cited in 170 scientific publications, stimulated the development of low-cost LED sun photometers, including the GLOBE sun photometer, by scientists and educators around the world. This paper also helped establish my credibility and led to several more papers. As this is written, the original instrument is in its thirty-third year of measuring both total column water vapor and haze.

After completing the LED sun photometer paper, I built a second TOPS (TOPS-2), which was placed in use on June 27, 1990. The ultraviolet filters in TOPS deteriorate over time, and TOPS-2 was added to track TOPS-1. The ozone indicated by both TOPS devices agreed so well that a difference would indicate that one, or even both, had developed a problem.

As the data accumulated month after month, it was possible to see seasonal trends—ozone, water vapor, and haze decreased during winter and rose during spring—so it seemed best to continue making the measurements, which I had planned to continue for only a year. I had no idea at the time, however, that those measurements would eventually lead to breaking Jonathan Piel's promise that I would never be back in the pages of *Scientific American*.

Science Probe! and the Solar Eclipse of 1991

As *Scientific American*'s public relations nightmare wound down, Larry Steckler was pushing ahead with *Science Probe!*, which Harry Helms named and which Larry hired me to edit. For each issue, I wrote the editorial, news stories, at least one major feature, and Science Notebook, a two-page project feature modeled after my hand-illustrated Radio Shack notebooks. The most I contributed to a single issue was the editorial and two major feature articles (which covered twenty-nine pages) for the January 1992 issue cover story about the total solar eclipse of July 11, 1991. That eclipse was historic, for the duration of totality was six minutes and fifty-three seconds, the longest totality of any eclipse until 2132. In addition to my articles, the eclipse issue of *Science Probe!* included reports and eclipse photographs from others.

The eclipse issue of *Science Probe!* required considerable planning and instrument preparation. The eclipse would peak over the Gulf of California, and Minnie and I made reservations for the *Viking Serenade*, a Royal Caribbean ship chartered for a cruise to the eclipse by Ed Love of Love to Travel. The ship would depart from Los Angeles. Because of my many instruments, we decided to drive to Los Angeles rather than shipping the gear or risk loss or damage by checking it on a flight. I used a Geiger counter to measure the background radiation along the entire drive.

While I made measurements of barometric pressure, temperature, sunlight, and background radiation during the eclipse, my main goal was to measure the ozone layer with TOPS-1. Soviet scientists had measured waves in the ozone layer during a previous solar eclipse, and my goal was to determine whether I could, too.

Two sets of waves in the ozone layer could be caused by an eclipse. One set would follow the shadow of the moon and be detected directly along the eclipse centerline. The second set of waves would spread outward from the eclipse path like waves from the bow of a boat. These waves should be detectable far away from the centerline. Accordingly, Eric's assignment was to measure the ozone layer using TOPS-2 at our Texas home while Minnie and I were at the centerline. Amazingly, both Eric and I detected waves in the ozone layer, and we submitted "Fluctuations in Column Ozone During the Total Solar Eclipse of July 11, 1991" to *Geophysical Research Letters*. The paper was published on March 5, 1993.

There were many other highlights on the eclipse trip, including the opportunity to spend quality time with Dr. Harrison Schmitt, the only Apollo astronaut who had walked on the moon who possessed a science degree. Schmitt is a trained geologist, and his major lunar achievement was finding the oldest moon rock of the Apollo lunar landings. Schmitt followed his career as an astronaut by serving two terms as a US senator for New Mexico, Minnie's home state.

The ship had a small gambling area through which passengers had to walk while headed for various destinations. While Minnie and I never saw anyone gambling, all of us really were doing so, for the art of eclipse forecasting deals only with the precisely known positions of the sun and the moon and not the weather. The scientific literature includes many sad stories of costly expeditions to distant places to observe a solar eclipse that turned out to be blocked by clouds. Our view was almost blocked by clouds, but the cruise's meteorologist knew exactly where the ship should head. We were in a perfect location when the moon approached the sun.

As the sky darkened, lights around the ship began to switch on, and moths could be seen flitting about the upper decks. Before and after totality, I was busily engaged making as many ozone measurements with TOPS as possible. During totality, it was impossible to make ozone measurements, so I was able to join

Minnie and stare at the solar corona in the dark sky and feel the same emotions as everyone else during those magical seven minutes.

After totality, the sky brightened as the sun returned, and loud splashes could be heard near the ship's bow. A large school of dolphins was passing by, and many of the animals were leaping from the water and doing backward flips. They seemed to be enjoying themselves as much as we were, and one could not help but wonder if they were signaling their appreciation of the eclipse.

After the eclipse, we gathered in the ship's auditorium to share our reactions. A few observers described technical details about the eclipse, but most described its emotional impact. As I reported in *Science Probe!*, "Some read hastily composed poems. A college professor said he was 'totally awed.' A mathematician said: 'I've had a real religious experience here.' A University professor said: 'The most sophisticated science is not incompatible with wonder.' Others described the eclipse as 'a wonderful cosmic gift' and 'like being in the cathedral of the cosmos.'"[3]

The Green Flash

That evening at supper, I told Minnie and our tablemates that we might be able to see a rare green flash as the sun dropped below the horizon. I kept an eye on the sun, and as it was about to disappear behind the sea, I told the others. As all six of us watched, the final sliver of the sun was suddenly replaced by a brilliant green flash. Our spontaneous cheer attracted questions from those at nearby tables who had missed the phenomenon.

That was my first and best green flash. I have seen many paler versions since, which I call green glows. The phenomenon we observed off Baja was a distinct, strobelike flash.

While driving back to Texas, Minnie and I visited Petrified Forest National Park in Arizona. It was near sunset, and as I pulled off the road and waited, the thin sliver of the setting sun briefly became bright green. The road continued uphill, and I reasoned we might soon see a second green sunset, and that is what occurred. We saw a third at the top of the hill we were climbing. While those green sunsets were much more subtle than the flash we observed from the ship, they were definitely green.

When I began spending nights at Hawai'i's Mauna Loa Observatory, I often saw and photographed green flashes. Twice, I observed the even rarer blue flash. None of these events matched the brilliant green flash off Baja, but they were still spectacular.

---

3. Forrest M. Mims, "The Great Eclipse of 1991," *Science Probe!*, January 1992, 35–55.

Farewell, *Science Probe!*

After the solar eclipse received major coverage in *Science Probe!*, Dr. Walter R. Hearn of the American Scientific Affiliation published a review of *Science Probe!* in the February/March 1993 issue of the *Newsletter of the American Scientific Affiliation & Canadian Scientific & Christian Affiliation*: "To read about the ozone layer (with good references to the scientific literature) or learn how to build your own TOPS instrument to measure it, check out two major articles by Forrest in the Nov 1992 issue of *Science Probe!*. That issue could become a collector's item . . . . In *Nature* (1 Oct 1992) James Lovelock of Gaia fame wrote a very favorable review of *Science Probe!*; he wrote that one article (by Mark Hartwig) 'makes basic statistics so lucid that some professional scientists who read it may be able to distinguish precision from accuracy.'"[4]

The Lovelock review in *Nature* was a major development, but it arrived too late. The cover story of the next issue was to have been an escorted tour of the lunar surface by Harrison Schmitt, who had expressed interest in *Science Probe!* when I met with him during the eclipse cruise. When I asked if he would be willing to take our readers on a tour of the moon, he quickly agreed, and a few weeks later sent an outstanding manuscript that nicely fulfilled my goal. His article was slated for the cover story, but it was never published. Because the magazine was receiving insufficient support from advertisers, it was closed just days before Schmitt's tour of the moon was to have been published.

The Institute for Creation Research

After the ship returned to port, Minnie and I drove to San Diego to meet Dr. Henry M. Morris, cofounder of the Institute for Creation Research. Morris had multiple degrees, including a doctorate in hydraulic engineering, and he had been a professor at several universities.

The ICR and other organizations that advocate a literal interpretation of the Bible's book of Genesis are often ridiculed by the biological-evolution segment of the mainstream scientific community. The ICR's response was to organize public debates with leading evolutionists. Morris, who wrote many articles and books advocating biblical creation, participated in more than a hundred debates.

Dr. Duane Gish, codirector of the ICR, held a doctorate in biochemistry and became well known for his debates across the US and in many countries. While

---

4. Walter R. Hearn, "Mauna Loa Fallout," *Newsletter of the American Scientific Affiliation & Canadian Scientific & Christian Affiliation* 35, no. 1 (February/March 1993), https://www.asa3.org/ASA/topics/NewsLetter90s/FEBMAR93.html.

Morris and Gish based their debates on carefully presented scientific facts, their opponents often resorted to ridicule and ad hominem attacks. This tactic backfired when the skeptics refused to answer questions that even Darwin had asked about his theory, and the evolution community eventually urged its advocates to avoid debating Gish and Morris.

While some ICR staff scientists have performed high-quality science, their work is consistently condemned by the scientific community if it questions evolution or the age of the earth. ICR scientists are known as young-earth creationists, for they believe that the creation days mentioned in Genesis are literal twenty-four-hour days. Other advocates of creation cite Bible passages that suggest an alternative interpretation of the days in Genesis. For example, the apostle Peter wrote, "But, beloved, be not ignorant of this one thing, that one day is with the Lord as a thousand years, and a thousand years as one day."[5]

Phillip Johnson founded the modern intelligent design movement with his book *Darwin on Trial*, which I discussed in chapter 13. When he gave a public lecture in nearby New Braunfels in 1999, Minnie and I attended with our daughter Sarah. Since I knew Phil, we were seated with him and his wife, Kathie, at a table in the auditorium where lunch was served. People seated at tables around us might not have realized that the speaker they came to hear was so near, but that was Phil's low-key style.

Phil began his talk by answering the question he most often hears: "What do you believe about the age of earth?" He then said he has good friends who are on both sides of this issue. Some are young-earthers, who believe that the earth is no more than a few tens of thousands of years old. Others are old-earthers, who believe the earth is billions of years old. He then said he did not want to offend his friends, so he does not take a public position on this question. I am not a geologist or a theologian, and I have borrowed Phil's response many times since his talk.

The Historic Mount Pinatubo Eruption

On June 15, 1991, a month before the solar eclipse, Mount Pinatubo in the Philippines erupted with a massive plume of volcanic debris and gases. Some 20 million tons of sulfur dioxide were injected into the stratosphere, where it formed a veil around the planet and caused a slight temperature reduction. It also caused spectacular, hourlong twilight glows. The first Pinatubo twilight Minnie and I observed occurred the evening following the July 11 solar eclipse, while we were on the eclipse cruise. The Pinatubo cloud of aerosols (fine particles) had just reached

5. The Bible (King James Version), 2 Peter 3:8.

Baja, and the sky became brilliantly red after sunset.

My LED sun photometer measured sharp reductions in direct sunlight caused by the Pinatubo aerosols. The power generated by a bare solar cell was reduced by around 15 percent. However, the magnitude of the sun's ultraviolet rays increased sharply, a byproduct of the significant reduction in stratospheric ozone measured by TOPS-1 and TOPS-2. After several months of measurements, Eric assisted me with a paper about these findings that we titled "Atmospheric Turbidity and the Ozone Layer After the Eruption of Mt. Pinatubo." The paper began:

> *Debris injected into the stratosphere by the paroxysmal eruption of Mt. Pinatubo in the Philippines last June 15 has caused spectacularly colorful, extended twilights and greatly increased stratospheric turbidity in both hemispheres. In Texas during October 1991 we measured a mean atmospheric optical thickness (AOT) at 1003 nanometers (nm) similar to that observed at Mt. Wilson, California, in August 1912 after the historic eruption of Katmai (0.082 and 0.088, respectively). There is great interest in the effect of volcanic cloud-induced heterogeneous chemistry on stratospheric ozone during winter (3-5), especially since most of the Pinatubo debris lies within the ozone layer. The mean amount of ozone over our location was 4.2 percent lower in December 1991 than in December 1990, a decrease which exceeds the 3.4 percent dip of mean global ozone at 20-40 degrees N from December 1981 to December 1982 following the eruption of El Chichón in March 1982.*[6]

We included three carefully prepared charts showing our optical depth and ozone measurements and acknowledged twelve well-known ozone experts and a statistics expert for helpful discussions. We also cited papers that disclosed ozone reduction after the 1912 eruption of Katmai in Alaska and after the 1982 eruption of El Chichón in Mexico.

On January 14, 1992, we sent the paper to *Nature*. Based on the quick acceptance of my LED sun photometer paper and our eclipse paper by other leading journals, I was optimistic that our paper might survive peer review at the prestigious publication. However, *Nature* sent the paper to two reviewers, and the first reviewer's most critical remarks were these: "The paper raises important questions in relation to the eruption of Mount Pinatubo. . . . The paper is very superficial . . . . The study of Mims and Mims is not conclusive. More detailed analysis based on

6. Forrest M. Mims and Eric Mims, "Atmospheric Turbidity and the Ozone Layer After the Eruption of Mt. Pinatubo" (unpublished paper).

data gathered over a large fraction of the northern hemisphere will hopefully lead to more solid conclusions concerning the potential destruction of stratospheric ozone following Mt. Pinatubo's eruption. In conclusion, I do not recommend that the paper be published."[7]

The second review included these comments: "Comparisons are made between 1990 and 1991 with a non-standard instrument for which the calibration is essentially unknown. Although an ozone reduction may have occurred, it has not been established by this paper. . . . This subject will be the focus of many analyses in the future and should not be cheapened by this trivial effort. I cannot recommend publishing in *Nature* under any circumstances."[8]

Neither reviewer mentioned our optical-depth data, which clearly showed the significant blockage of sunlight caused by the Pinatubo aerosol cloud. That was puzzling, for our paper was among the first to mention colorful twilight glows caused by the Pinatubo aerosols. We even included four photos showing brilliant Pinatubo twilight glows. (One of my Pinatubo twilight photos, the brightest Pinatubo twilight photo in a Google image search, was eventually used in other scientific publications.) But instead of recommending we drop the ozone section of the paper, which the reviewers had criticized, *Nature* assistant editor Philip Ball politely rejected the entire paper we had worked so hard to write.

We learned much from this experience. Both reviewers noted that our measurements were made from a single site, but this is not uncommon. The 1912 ozone decline after Katmai was measured at the Smithsonian Astrophysical Observatory site at Mount Wilson, and many papers in today's satellite era have been based solely on data from the Mauna Loa Observatory and other fixed sites.

Despite our paper's rejection, the effects we measured were correct. The Pinatubo aerosol cloud was significantly increasing the haziness of the sky while also reducing the concentration of ozone in the stratosphere. The lesson from this is what I taught our three children when they did their science fair projects: If you are confident of your data but scientists are not, believe your data and not the scientists. What happened next proved Eric and I were right all along.

Months after *Nature* rejected our paper, other papers describing the decrease in stratospheric ozone appeared in print. The first was "Observations of Reduced Ozone Concentrations in the Tropical Stratosphere After the Eruption of Mt. Pinatubo," by NASA's Dr. William Grant and ten coauthors. This paper, sent to *Geophysical Research Letters* on March 31, 1992, and published on June 2, 1992,

7. Anonymous review of "Atmospheric Turbidity and the Ozone Layer After the Eruption of Mt. Pinatubo."
8. Anonymous review of "Atmospheric Turbidity and the Ozone Layer After the Eruption of Mt. Pinatubo."

proposed what Eric and I had proposed to *Nature* on January 14, seventy-eight days before Grant's paper was submitted. It concluded, "Two independent data sets, one of ozone from ozonesonde measurements, and one of aerosols from an airborne lidar [light detection and ranging] system, suggest that significant ozone decreases may have occurred as a result of the injection of debris by the Mt. Pinatubo volcano in June 1991."[9]

The next major paper was "Record Low Global Ozone in 1992," by NASA's Dr. James F. Gleason and thirteen coauthors. This paper was submitted to *Science*, *Nature*'s main competitor, on March 11, 1993, and published on April 23, 1993, fifteen months after I had sent our paper to *Nature*. This paper's conclusion also proposed what Eric and I had proposed in our rejected paper: "The cause of the 1992 low ozone values is uncertain. Although the mechanism for ozone decrease is unknown, the understandable first guess would be that the decrease is related to the continuing presence of aerosol from the Mount Pinatubo eruption."[10]

Meanwhile, *Nature* finally recognized what the *Science* authors and many other atmospheric scientists had discovered when it published on March 25, 1993, a week after the *Science* paper, "Role of Sulphur Photochemistry in Tropical Ozone Changes after the Eruption of Mount Pinatubo." The first sentence of the abstract of this paper by Dr. Slimane Bekki, Dr. Ralf Toumi, and Dr. John A. Pyle of the University of Cambridge echoed what Eric and I had sent to *Nature* fourteen months earlier: "RECENT observations suggest that the eruption of Mount Pinatubo in June 1991 has had a considerable effect on ozone concentrations in the tropical stratosphere. . . ."[11]

These three publications about ozone decline after Pinatubo and those that followed have received many hundreds of citations in the scholarly literature. The *Science* paper alone has received 438 citations, and it continues to be cited. I began to think that if TOPS ozone measurements and sun photometer readings were to be taken seriously, a global network would be required. But I had no way of funding such a network.

NASA's Satellite Ozone Measurements Disagree with TOPS

Recall that TOPS was described by one of *Nature*'s reviewers of the Pinatubo paper as a "nonstandard instrument for which the calibration is essentially unknown."

---

9. William Grant et al., "Observations of Reduced Ozone Concentrations in the Tropical Stratosphere After the Eruption of Mt. Pinatubo," *Geophysical Research Letters*, June 2, 1992, 1109.
10. J. F. Gleason et al., "Record Low Global Ozone in 1992," *Science*, 260, no. 5107 (April 23, 1993), 523.
11. S. Bekki et al., "Role of Sulphur Photochemistry in Tropical Ozone Changes after the Eruption of Mount Pinatubo," *Nature*, 362 (March 25, 1993), 331.

This criticism was very much on my mind when the ozone measured by TOPS-1 and TOPS-2 eventually began to differ from that measured by TOMS, NASA's famous ozone instrument aboard the *Nimbus-7* satellite. Were the nonstandard TOPS data wrong? Or was the satellite data drifting away from that measured by my homemade instruments? In 1914, Alexander Graham Bell wrote in *National Geographic*, "Don't keep forever on the public road, going only where others have gone and following one after the other like a flock of sheep. Leave the beaten track occasionally and dive into the woods. Every time you do so you will be certain to find something that you have never seen before."[12]

I decided to dive into the woods, for, as Bell also wrote in his 1914 article, "A thermometer is an instrument for measuring heat, and whenever you can measure a phenomenon you have a basis upon which may be built a science; in fact, all science is dependent upon measurement."[13]

TOPS was like Bell's thermometer. It was homemade, but it used high-quality ultraviolet filters provided by Barr Associates that were leftovers from a Smithsonian project. Was TOPS providing more accurate ozone readings than TOMS?

During the spring of 1992, ozone data measured by both TOPS-1 and TOPS-2 agreed well with one another but continued to differ by several percentage points from the ozone readings measured by TOMS. On June 12, 1992, I loaded my Ford pickup truck with instruments and a computer, a monitor, and a printer and drove to Charlottesville, Virginia, where the Quadrennial Ozone Symposium was being held at the University of Virginia. The plan was to present a poster paper on the waves in the ozone layer, which Eric and I had detected during the total solar eclipse of July 11, 1991. I also had another objective.

Surely, I thought, the ozone scientists at this meeting would be intrigued by my findings about the satellite. Therefore, I showed many scientists a chart I made of the growing difference between ozone measured by TOPS and NASA's TOMS aboard the *Nimbus-7*. Most were courteous, a few were not, and only one agreed I was correct. He was a French researcher who specialized in measuring ozone from high-altitude balloons. "Of course, you are correct!" he happily exclaimed on seeing my chart. "NASA is wrong!"

The scientists who disagreed with my claim thought they had a valid reason, for the stratosphere was still loaded with aerosols from the volcanic eruption of Mount Pinatubo. I assumed they did not know that TOMS was more subject to aerosol errors than TOPS, for the two wavelengths measured by TOPS (300 nm and

---

12. Alexander Graham Bell, "Discovery and Invention," *National Geographic* 25, no. 1 (January 1914), 650.
13. Bell, 652.

305 nm) were much more closely spaced than those used by TOMS. Then there was my underlying concern based on the reaction of the reviewer of the paper rejected by *Nature* that TOPS was a "nonstandard instrument for which the calibration is essentially unknown."

The wavelengths I used were suggested in a paper by ozone expert Dr. John J. DeLuisi, who later became a friend. He was at the symposium and agreed that my findings merited further study.

Key evidence in favor of the stability of TOPS was its calibration. Instead of using the complex ozone algorithm, I devised a much simpler method that compared five months of TOPS data with the ozone amount over my site measured by TOMS. The resulting TOPS ozone equation provided ozone data within 1 percent or so of the TOMS ozone. Thus, TOPS ozone data were initially in lockstep with TOMS data.

A subsequent difference in the data could be caused by one or both instruments, but there was only one TOMS, and I had built TOPS-2. After the difference in ozone amounts measured by TOPS-1 and TOMS appeared, TOPS-1 and TOPS-2 still agreed. This strongly suggested that the TOMS ozone data were in error.

When I called one of the ozone experts at the Goddard Space Flight Center to report a possible TOMS error, I was reminded that TOMS was part of a major, very costly program, while my instrument was merely a homemade project. I then mentioned that I had two TOPS that showed the same result, while they had only one ozone satellite, but that failed to persuade the TOMS expert.

During the closing days of the conference, I met NASA scientist William Grant, lead author of the post-Pinatubo ozone paper in *Geophysical Research Letters*. Grant was planning to become an independent scientist, and he showed me boxes of some of his books in the trunk of his car.

Some Russian scientists were staying in a dormitory room near mine. Their room was furnished with a new color television set they had purchased, and several bottles of vodka. My interest was in their M-124 ozonometer, which was used in the Russian ozone network. When I asked how much one cost, they said they would sell me one for $750 ($1,607 in 2023). Based on this instrument's specifications, I knew it could not be as accurate as TOPS, but I asked to purchase two of them. They immediately agreed.

Several weeks after my return to Texas, a UPS truck delivered a large, crumpled cardboard box that contained the two M-124s. I hoped that they would confirm that TOPS-1 and TOPS-2 were working properly and that NASA's TOMS satellite ozone instrument was providing erroneous data.

1. Col. Forrest M. Mims Jr. introduced his son to basic science and electronics.
US Air Force

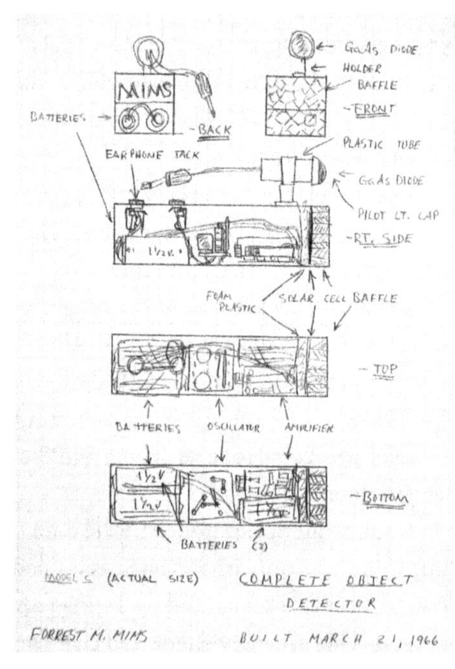

2. Mims described his travel aid for the blind in his first notebook on March 21, 1966.

3. Mims built this language-translating computer while a high school senior in 1961.

4. Mims' travel aid for the blind and the handle of his great-grandfather's cane.

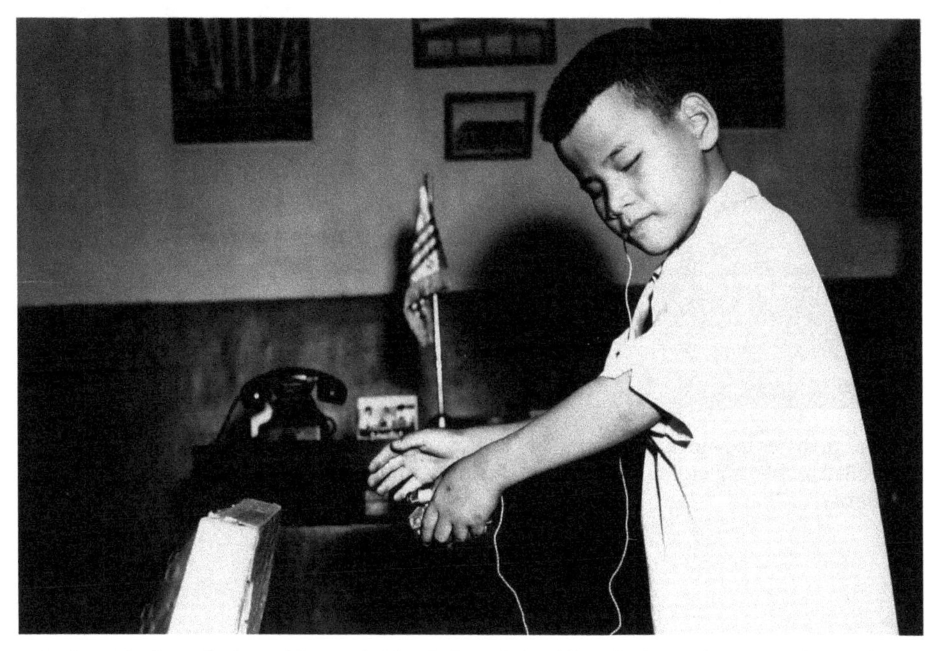

5. Le Que Manh exploring with travel aid at Saigon School for Blind Boys in 1967. *Robert Ellison*

6. Mims' eyeglass-mounted travel aid for the blind received a 1972 Industrial Research 100 Award.

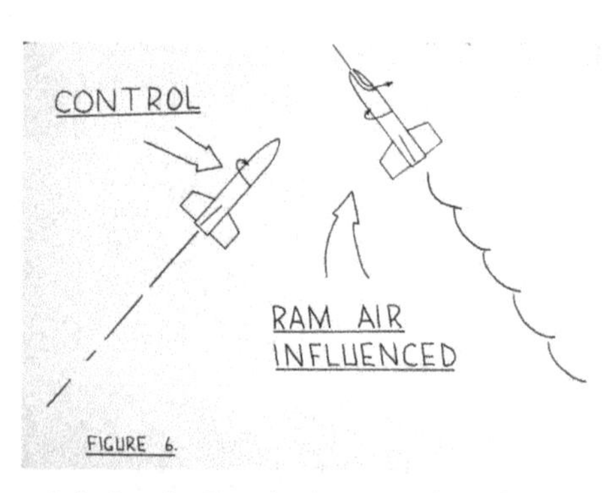

7. Spiral path of a spinning ram-air rocket with a nose port.

8. Mims prepares to launch a ram air–controlled rocket in Vietnam in May 1967. *Air Force*

9. Ram-air guided test rocket launch from race track near Saigon in spring 1967.

10. Most of Mims' Saigon recovery teams were very young; this gentleman was an exception.

11. Strap-on wind tunnel attached to Mims' 1966 Chevy Malibu in January 1968.

12. Minnie Chavez Mims during her Colorado honeymoon with Forrest in June 1968.

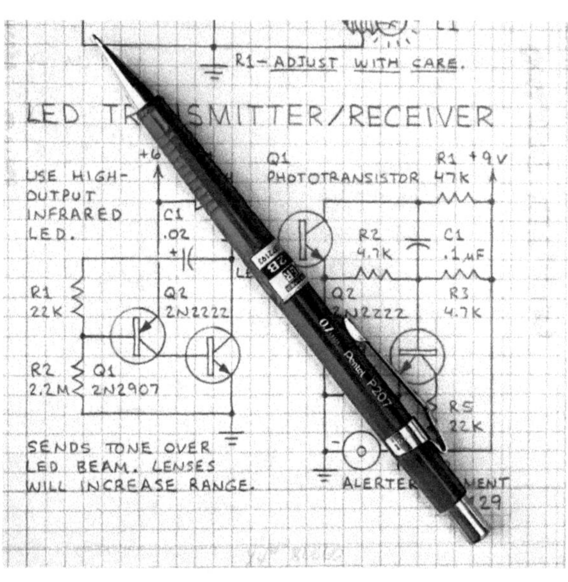

13. Mechanical pencil used by Mims to print and illustrate seventeen Radio Shack books that have sold 7.5 million copies.

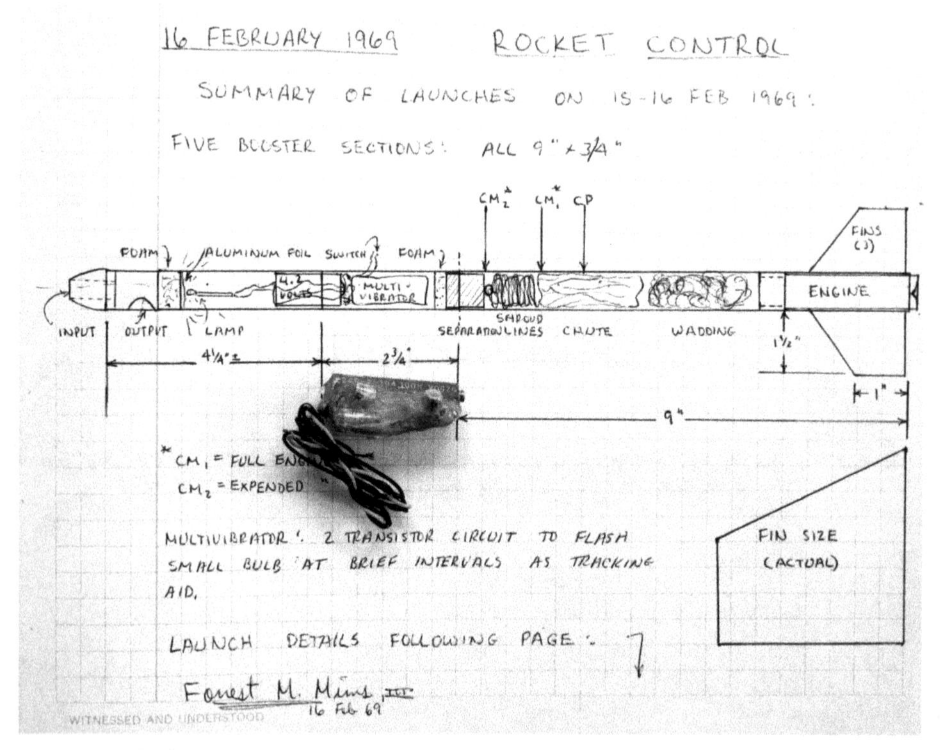

14. Rocket light flasher that began Mims' writing career and launched MITS in September 1969.

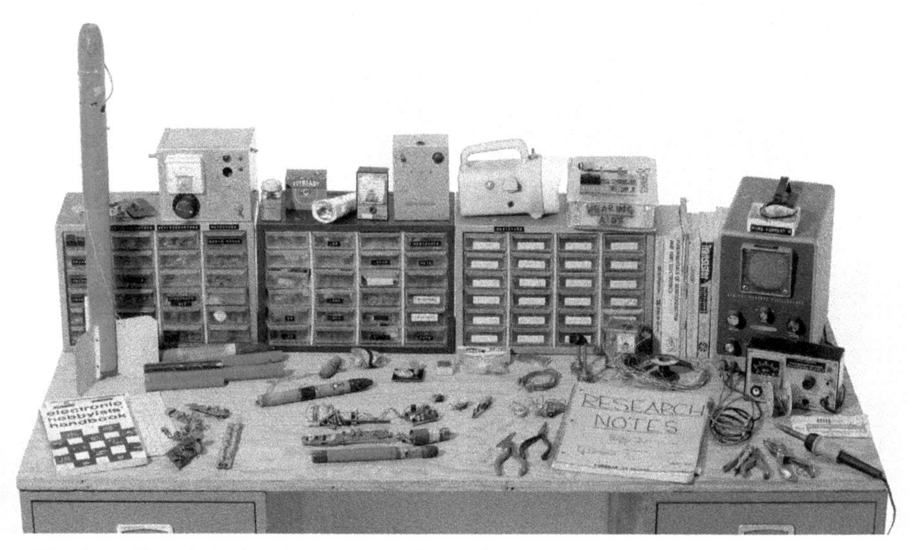

15. Mims' workbench during the 1970s. *Mark Langford*

16. Mims promoted the Altair and *Siliconnections* during a 1985 book tour.

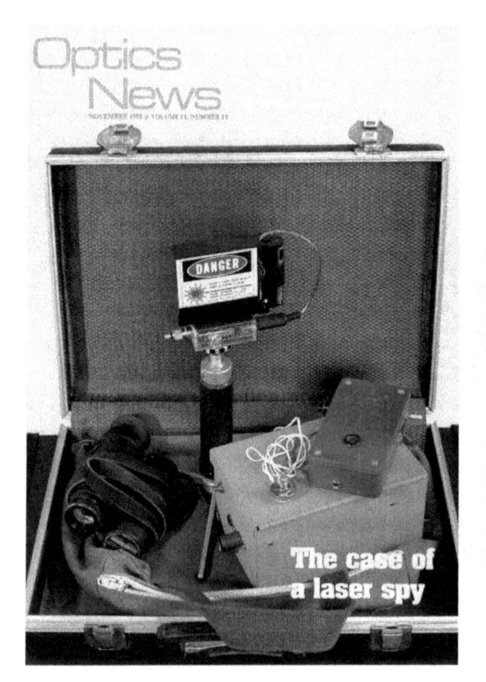

17. *Optics News* (November 1985) featured a cover story by Mims on "Surreptitious interception of conversations with lasers" (pp. 6-11).

18. Mims used a Microtops II to measure air quality for NASA at Alta Floresta, Brazil, in August 1997. *Bradley White*

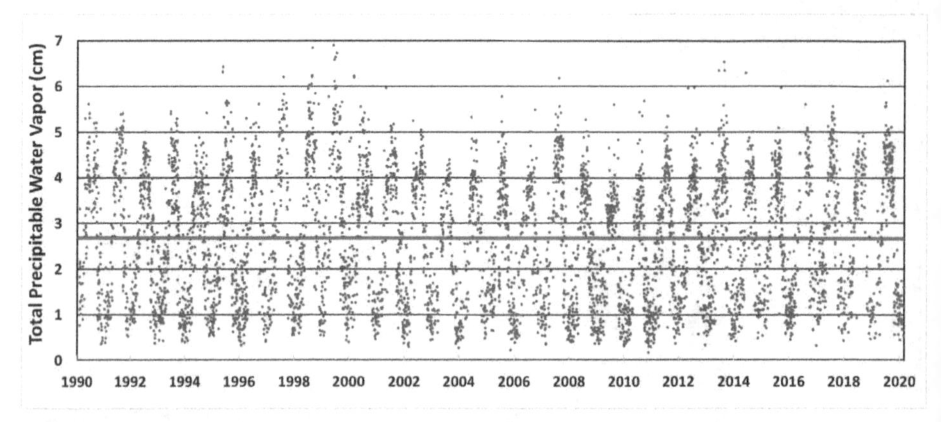

19. Thirty years (1990–2020) of total water vapor published by Mims in *Bulletin of the American Meteorological Society.*

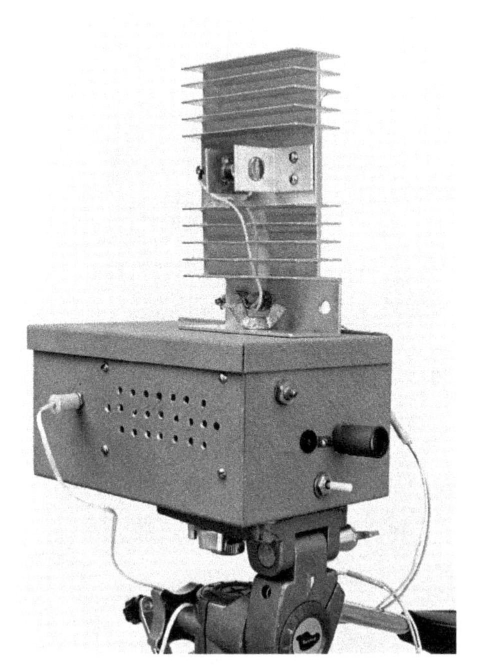

20. Laser-eavesdropping system built for *National Enquirer* in 1976.

21. Bell Labs attorney retrieving box of book manuscripts from Mims' attic during 1980 search of Mims' office.

22. Bell Labs patent attorney reviewing Mims' laboratory notebooks in 1980.

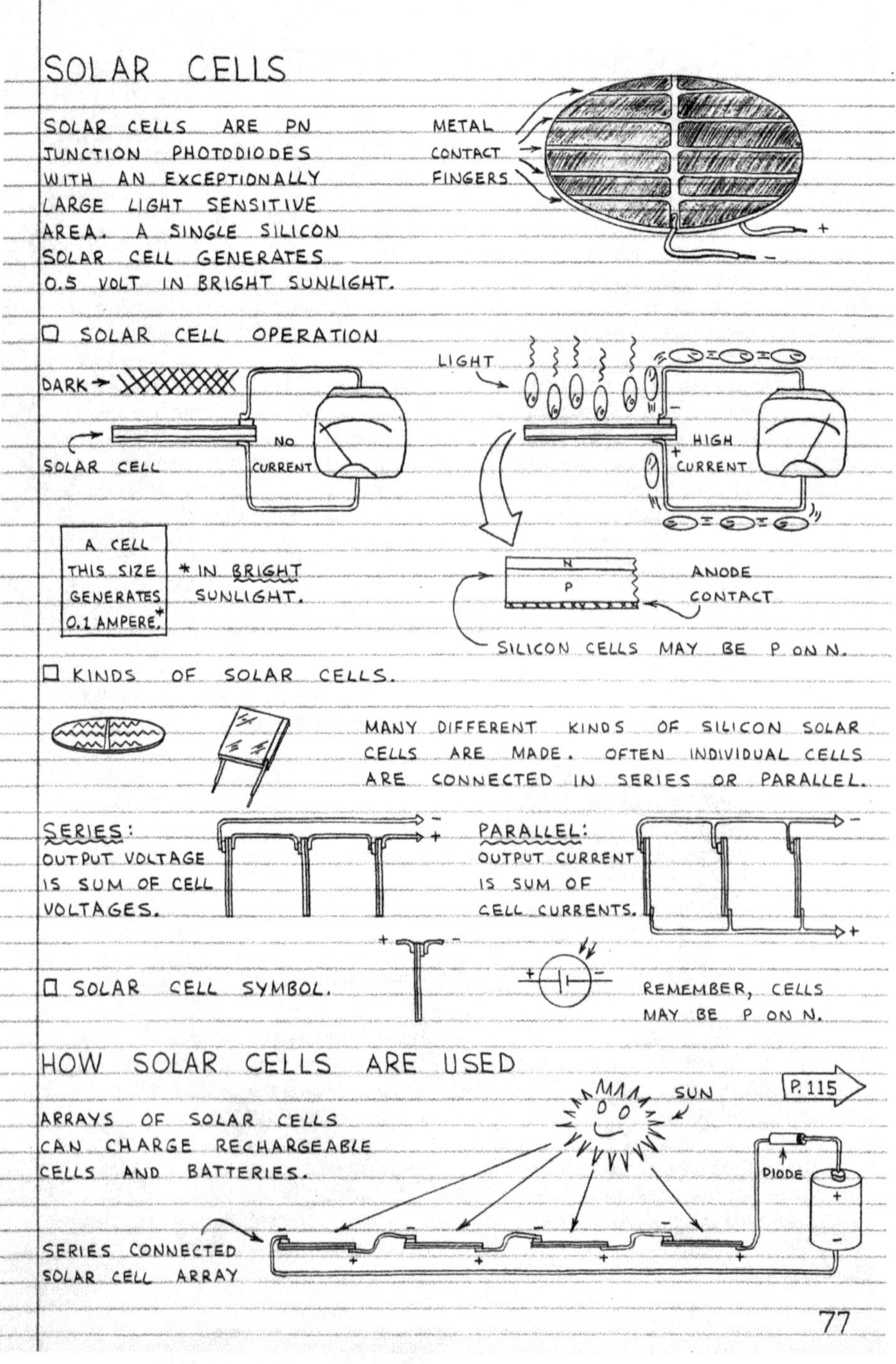

23. Page from Mims' best-selling book *Getting Started in Electronics.*

24. Mims (left), Bob Zaller, Paul Allen, and Ed Roberts (seated) at StartUp on November 18, 2006. *Minnie Mims*

25. Mims transported his aerial-photography blimp via bicycle in 1990. *Eric Mims*

26. Mims' homemade TOPS ozone monitors found errors in ozone retrievals by NASA's ozone satellite in 1992.

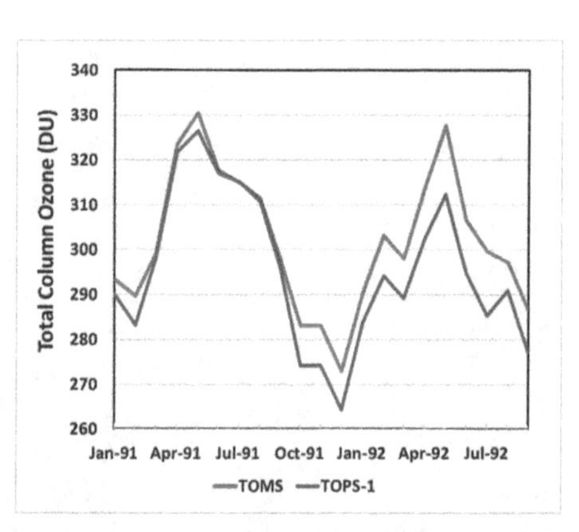

27. This graph from Mims' first paper in *Nature* (1993) shows the error (red) in ozone measured by NASA's Nimbus-7 satellite.

28. Mims displayed the first issue of *Scientific American* in a talk about the *Scientific American* affair at the National Press Club on March 28, 1991.

29. Mims points at a layer of pulp in orange juice to explain the ozone layer to Sir Edmund Hillary during the 1993 Rolex Awards Ceremony.

30. Forrest and Minnie at the National Geographic Photophone Centennial on February 19, 1980.
Bruce Dale

31. Mims used a water-proof sensor to measure solar UV at Hapuna Bay, Hawaii, in 1994. *Eric Mims*

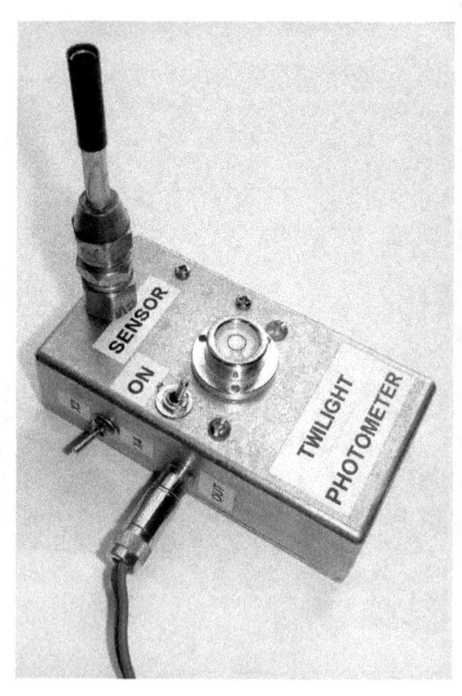

32. LED twilight photometer built by Mims for his 2013 column in *Make:* magazine.

33. Russian M-124 and TOPS-1 and -2 atop the world-standard ozone instrument when satellite error was verified in 1992. *Grae Roth*

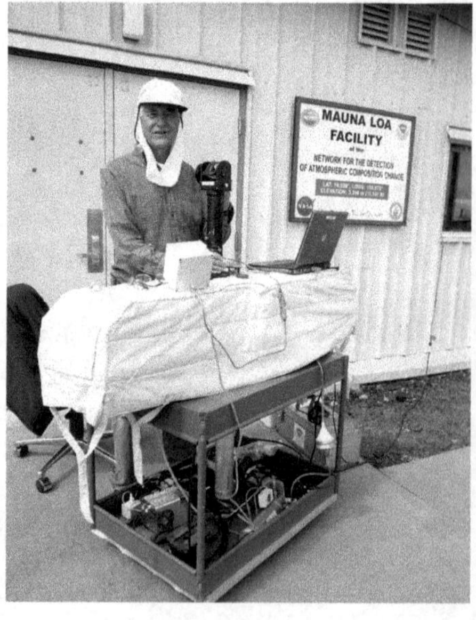

34. Mims calibrated the world-standard ozone instrument for NOAA during a 64-day stay at Mauna Loa Observatory in 2016.

35. Mims and a Microtops II at Smithsonian Institution in Washington, DC, where sun photometers were first used 125 years ago. *Minnie Mims*

36. Mims and his instruments on the 25th anniversary of his atmospheric measurements. *Minnie Mims*

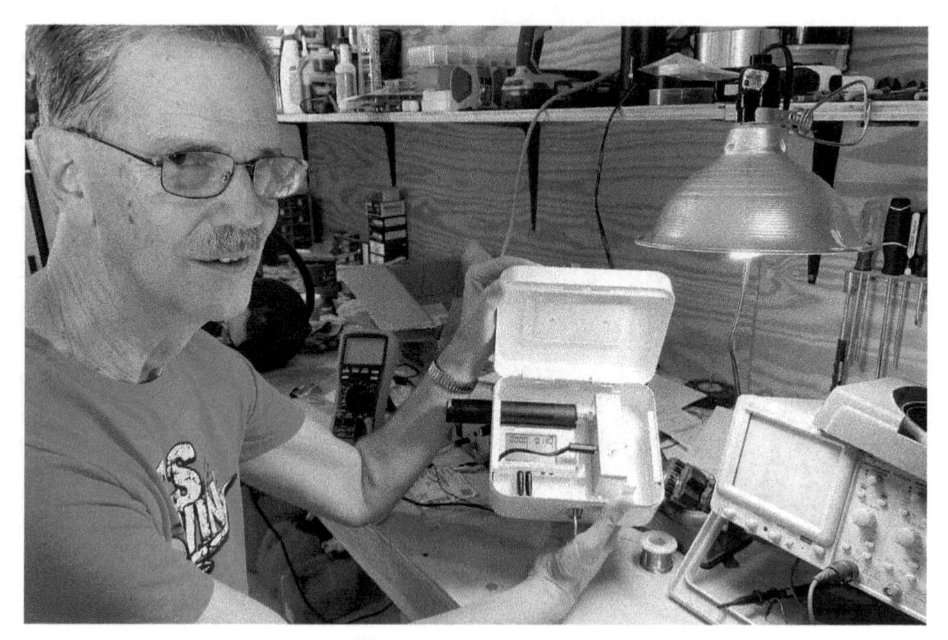

37. Scott Hagerup and one of five LED twilight photometers he built for Mims under a 2022 NASA assignment.

38. Mims created Sunny Sam in 2018 to measure solar UV on a person's face during a Rolex-sponsored project in Hawaii.

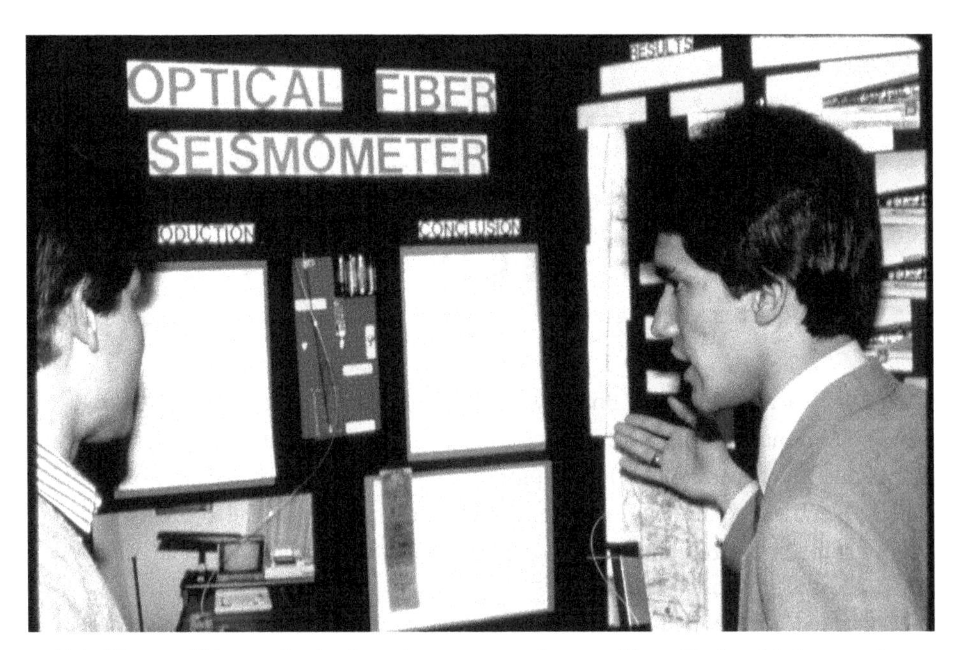

39. Eric Mims and his award-winning seismometer at the 1995 Alamo Regional Science and Engineering Fair.

40. Vicki Mims with one of her award-winning science fair projects.

41. Sarah Mims and her mother Minnie at the 2004 Siemens competition at the University of Texas.

15
THE ROLEX AWARD

In 1986, I learned about the Rolex Awards for Enterprise, a program sponsored by the famous Swiss watch company to recognize people with unique projects in applied sciences and invention, exploration and discovery, or the environment. Their objective is "empowering exceptional individuals," and the Rolex Awards "support inspiring individuals who carry out innovative projects that advance human knowledge or well-being."

The travel aid for the blind was on my mind, and although I lacked the funds to develop a better prototype than my 1972 eyeglass version, I sent for the awards application, which required a detailed explanation of the project and how it would be more fully implemented should it win an award. Several months later, Rolex sent notice that the project was among the finalists and would be more fully evaluated during a visit by a Rolex representative.

The Rolex person was very thorough during her interview and in her evaluation of the project. She was already well informed about my technical background, and she concentrated on what I proposed should I win an award. Months later, Rolex sent notice that my project had received a runner-up award: a gold-and-steel Rolex watch.

The 1993 Rolex Award

By 1992, it was clear that the TOPS ozone instrument could measure the amount of ozone in the ozone layer to within a few percentage points of that measured by NASA's ozone satellite, so I decided to again apply for a Rolex Award. My proposal was for a network of ground stations that would each employ an improved version of TOPS. The application process required just as much detail as the travel aid application. The main difference was that I was able to provide two years of ozone data that clearly showed the comparison with satellite data and the annual cycle of the magnitude of ozone overhead.

As before, Rolex sent a representative to interview me in detail about the proposed project. He took a serious interest in the project, but so had his predecessor back in 1986, so I was not particularly optimistic that the project would win. I was especially concerned that the widespread publicity about the *Scientific American* affair would ruin my chance.

Several months later, the kitchen phone rang. Western Union was calling, and a male voice read a congratulatory telegram from Rolex that my project was one of the five winners of the 1993 Rolex Awards. Minnie was also in the kitchen, and after hanging up the phone, I told her, "I am going to be back in *Scientific American*! The TOPS project has won a Rolex Award!"

I then explained that Rolex always announces the Rolex Award winners in a full-page ad in *Scientific American*, and maybe TOPS would be in the Rolex announcement photograph. I then told Minnie that Rolex would fly us to Geneva for the awards ceremony and give me a gold Rolex and the US equivalent of $35,000 ($72,958 in 2023) to implement the ozone network. The awards ceremony was scheduled for April 30, 1993. Meanwhile, Rolex arranged for a London film crew to make a documentary about my ozone project.

Looking back, it seems strange that I was still preoccupied with *Scientific American*. But there was good reason, for how to build TOPS was to have been the subject of one of the projects in The Amateur Scientist before Jonathan Piel told me I would never again appear in the pages of his magazine. Ironically, had Piel allowed me to continue the column, I would have been much too busy to apply for a Rolex Award.

The Ozone Satellite Error Is Confirmed

As related in chapter 14, during the Quadrennial Ozone Symposium at the University of Virginia in June 1992, I showed many scientists charts that displayed the increasing difference between ozone measurements by the TOMS instrument aboard the *Nimbus-7* satellite and by both TOPS instruments. While most were unimpressed, I became confident TOPS had found an error in the satellite data when TOPS data agreed with Dobson 83, the world-standard ozone instrument, at the Mauna Loa Observatory while I was twice there in August 1992.

I kept the TOMS team at the Goddard Space Flight Center fully informed about my findings, and on December 2, 1992, NASA ozone scientist Dr. Richard McPeters informed me he would soon send a fax about its evaluation of my claim that the TOMS satellite data were several percentage points too high. By a surprising coincidence, the Rolex-commissioned documentary crew arrived that evening. While driving producer Jeff Harvey and his documentary crew to my office so they could see whether there would be space to set up their equipment, I told them about the possible satellite error and that NASA might soon reply to my findings with a fax.

Before picking up the film crew the next morning, I sent Dr. Walter Komhyr, an ozone expert with the National Oceanic and Atmospheric Administration, a graph comparing TOMS and TOPS-1 ozone data and wrote, "The mean for TOPS-1 for all of 1991 is .02% higher than the mean for TOMS. The mean for TOPS-1 for all of 1992 is 3.4% lower than the mean for TOMS."[1]

I then picked up the film crew at their hotel. When we returned to my office, a curled-up fax was lying on the desk. When I reached for it, Jeff Harvey shouted, "Don't touch that fax!" I asked why, and he said that he had made many documentaries about Nobel Laureates and Rolex Award winners but had never been present for an actual discovery. If I had found a satellite ozone error with a homemade instrument, he wanted to record the confirmation as it happened.

The five long minutes required to set up their camera and lights and do a soundcheck seemed like an hour. While they adjusted their equipment, I stared at the rolled-up fax and wondered whether it was from NASA. Even if it were, it could easily have denied my premise, much as it had been previously denied by NASA and many scientists at the Quadrennial Ozone Symposium six months before. I knew TOPS data were much less affected by volcanic aerosols than TOMS, which had to mean the satellite data were in error. But how would I explain that to the documentary crew if NASA denied my claim?

Finally, everything was ready, and Harvey asked me to leave the office and then walk back inside. I did so while the cameraman began recording. When I was back inside, Harvey looked at me with a smile, motioned with his hand toward my desk, and said, "You may read your fax."

The fax was from NASA's Rich McPeters, and I was elated to read his confirmation that TOPS had indeed found an error in the satellite ozone readings:

To: Forrest Mims
From: Rich McPeters
Date: December 3, 1992

Message: The I83 [World Standard ozone instrument] comparisons show a strong scan angle dependence for the TOMS data—indicative of an error caused by Pinatubo aerosols. The actual calibration error appears to be 0–2%. We have to do some aerosol simulations to be sure which scan positions have the least error.[2]

The producer was just as excited as I was, and we were especially surprised

1. Forrest M. Mims, letter to Walter Komhyr, December 3, 1992.
2. Richard McPeters, fax to Forrest M. Mims, December 3, 1992.

moments later when McPeters called while the camera was still running. When Harvey asked if he could speak to McPeters about the satellite data, McPeters agreed, and I gave the phone to Harvey. McPeters agreed to be interviewed, and some of their conversation was included in the Rolex documentary.

The documentary was shown during the Rolex Awards ceremony and is still online. Just as Harvey had hoped, the scene showing me opening and reading the fax was unrehearsed and shows that unexpected agreement with my discovery as it happened.

I notified Joe Abernathy of the *Houston Chronicle* about what the TOPS data had uncovered. Abernathy's 1990 story in the *Chronicle* had been the first to report how *Scientific American* had terminated my assignment to write The Amateur Scientist, so he took my ozone report seriously, sent a photographer to my office for a photo of me holding TOPS-1, and began researching the story. On December 7, 1992, in a lead article headlined "Texan Vindicated—Amateur Scientist Proves NASA Wrong on Size of the Ozone Hole Over Antarctica," Abernathy wrote:

> *NASA scientists have confirmed findings by an amateur scientist in Texas, showing that the space agency's measurements of ozone depletion are in error.*
>
> *This error means that the ozone hole over Antarctica this year was bigger than announced, and also calls into question the ongoing validity of NASA's data, said Dr. Richard McPeters, project scientist for Nimbus 7, the satellite that the agency uses to measure ozone.*
>
> *The National Aeronautics and Space Administration will announce the problems with the satellite data today at the American Geophysical Union meeting in San Francisco.*
>
> *The error was found by Forrest M. Mims III, the Seguin-based editor of Science Probe!, an amateur scientists journal. His tests were conducted with an ozone-measuring instrument he designed himself.*
>
> *"We claim we're still doing a good job, but we're worried about how accurate we are," said McPeters. "There was some worry that we might be off 4 or 5 percent. These ozone trends might only be 4 or 5 percent over 10 years, so it looks big."*
>
> *"Some of the implications of that include how deep the ozone hole was this year, and whether or not the ozone depletion in the Northern Hemisphere is continuing," McPeters noted.*[3]

Earlier, I had prepared a short paper for *Nature* about the ozone error, and on November 27, 1992, I sent the draft to both McPeters and Walter Komhyr and asked

3. Joe Abernathy, "Texan Vindicated—Amateur Scientist Proves NASA Wrong on Size of the Ozone Hole Over Antarctica," *Houston Chronicle*, December 7, 1992.

them to join as coauthors. I was not optimistic Komhyr would participate, for when I had visited him at Boulder in June 1991 to show him TOPS, he had not been impressed: When I had handed him the homemade instrument, he had frowned and slammed its base down on his desk.

Komhyr, the world's foremost ozone-monitoring expert, had been responsible for establishing Dobson 83 as the world-standard ozone instrument. My little TOPS looked more like a toy than a legitimate scientific instrument. I had not said anything at the time, but I had been worried he might have damaged the optical alignment of the instrument that had found an error in data from the world's only ozone satellite. I was surprised, therefore, when Komhyr took the draft paper seriously enough to send a lengthy list of suggestions for improving it.

After the documentary crew left on December 3, I corresponded with both McPeters and Komhyr about the paper. Ultimately, they declined to join me as coauthors, and on December 4, I sent it to *Nature*. Some people thought I selected the publication because of its prestige. That played a role, for I wanted ozone operators around the world to know about the satellite error. But the main reason was that *Nature* does not charge page fees to publish accepted papers. When Maxine Clarke, *Nature*'s executive editor, notified me that they would consider publishing my piece only as a news story, I responded on December 11, 1992:

> *I discovered the TOMS error at considerable personal expense and with no outside funding whatsoever. . . . Moreover, I have suspected the error since the spring of last year when I noticed a clear departure between simultaneous measurements by my two instruments and the Nimbus 7/TOMS. But these data were politely discounted by all of the many ozone scientists to whom I showed data at the 1992 Quadrennial Ozone Symposium last June. Finally, to my knowledge, my ozone station, which is the only one in the entire southern United States, correlates its observations with the satellite considerably faster than government stations. . . .*
>
> *My letter to 'Nature' is intended to announce and explain the error. Certainly a news article in Nature is highly appropriate. But it is highly unlikely that such an article will be cited as a reference in the many papers that will now require revision. Had I known that my letter would be viewed as a news tip rather than a Scientific Correspondence, I would have sent it to 'Science' or published my findings in 'Science Probe!'. Therefore, I respectfully request that you reconsider any plans to publish this new development solely as a news item.*[4]

4. Forrest M. Mims, letter to Maxine Clarke, December 11, 1992.

Nature replied that NASA should publish the error, for *Nimbus-7* was their satellite. I responded by reminding the editors that NASA did not always announce problems with its satellites. Meanwhile, two teams of ozone scientists stationed in Antarctica had noticed that TOMS ozone readings over their sites were unexpectedly higher than their ground readings. Unfortunately, unwelcome news arrived from prominent ozone scientist Dr. Rumen D. Bojkov, of the World Meteorological Organization, after he received a note I sent him about the satellite error: "I have grave doubts about the simplistic analogue which you are promulgating as the reason for the differences between Dobson 83 and TOMS is close to reality [sic]."[5]

Bojkov had not been impressed by my poster at the Quadrennial Ozone Symposium on waves in the ozone layer during the 1991 solar eclipse. Nor had he been impressed by TOPS-1. But his "grave doubts" were premature, and on December 17, 1992, I replied to him with new information from McPeters, who had written, "There is an error resulting from the presence of Pinatubo aerosols . . . . In mid-1992, because of the lack of a solar calibration update [caused by satellite drift], the TOMS values drifted higher by 1–2% (this is a true calibration error and will be corrected before final release of the data for this year)."[6]

A month later, good news arrived from *Nature*. After John Maddox, *Nature*'s editor, intervened, my paper was sent to a peer reviewer. The reviewer quickly approved the paper, and on January 15, *Nature* informed me it would be published as a Scientific Correspondence. On February 11, 1993, *Nature* published "Satellite Monitoring Error."[7] Some of NASA's ozone scientists had second thoughts, and on July 15, 1993, *Nature* published "Errors and Ozone Measurement," a response to my paper by Richard McPeters and James Gleason that itself began with an error: "Mimms [sic] recently compared ozone measured using his hand-held TOPS instrument with ozone data from the Total Ozone Mapping Spectrometer (TOMS) on *Nimbus 7* and noted possible errors in TOMS data. The TOMS ozone data used in Mimms's [sic] analysis were preliminary and should not have been used for quantitative analysis."[8]

They then explained how orbital drift had prevented calibration of TOMS from February 12 to September 30, 1992, and that the data between these dates was preliminary and "should be used with caution and that only 'final' data should be

5. Rumen D. Bojkov, letter to Forrest M. Mims, December 17, 1992.
6. Richard McPeters, letter to Forrest M. Mims, December 14, 1992.
7. F. M. Mims III, "Satellite Monitoring Error," *Nature* 361 (1993), 505.
8. Richard D. McPeters and James F. Gleason, "Errors and Ozone Measurement," *Nature* 364 (1993), 198.

used for quantitative ozone analysis." However, they also acknowledged the TOMS error: "Mimms's [sic] comparison of his TOPS measurement with TOMS provided a valuable early indication that the extrapolated TOMS calibration was in error. The satellite ozone measurement programme will always need ground measurements to monitor the calibration of the satellite instruments. But we must emphasize that the 'real-time' data we produce and make generally available should be used with caution and that only 'final' data should be used for quantitative ozone analysis."[9]

While the ozone data generously provided by NASA was never designated as preliminary, this response specifically acknowledged that my finding was a "valuable early indication"[10] of the TOMS error, an error that was also obvious in comparisons of TOMS data with official ground-based instruments.

The problem with the official ground-based instruments is that their data often went unreported for days or even weeks and the ozone layer disappeared every weekend, when many stations did not measure it. My data were available immediately.

"Earth Science on a Shoestring Budget"

A week after the Rolex documentary film crew departed, a letter arrived from NASA's Goddard Space Flight Center, inviting me to give a presentation titled "Earth Science on a Shoestring Budget" at its March 29, 1993, Goddard Engineering Colloquium. This would be a major opportunity to meet Arlin Krueger and other GSFC ozone scientists who had provided important advice about my plans for TOPS back in 1989.

Before the colloquium, half a dozen ozone scientists took me to lunch. I looked forward to asking them technical questions about my ozone-measurements program. But they wanted to hear all about the *Scientific American* affair. None of them challenged my skepticism about Darwinism or my answers to their questions about evolution, which is a typical reaction by nonbiology scientists who ask about my views. They even laughed at some of the anecdotes I related about *Scientific American* during the American Scientific Affiliation meeting at the University of the Nations. Unfortunately, there was not time to discuss ozone.

During the talk, I said nothing about finding the TOMS ozone error with my "shoestring" instruments or the paper in *Nature*, even though that seemed to be the reason they had invited me to speak. The Rolex Award, which was decided before confirmation of the satellite error, had not been officially announced, so I was unable to mention it. After describing some of my science, however, I encouraged the GSFC to greatly expand its connections with students and serious

9. McPeters and Gleason.
10. McPeters and Gleason.

amateur scientists and asked, "Who knows what new knowledge might be gained if more data in your tape archives and on your CD-ROMs can be examined by hundreds or even thousands of enthusiastic students and serious amateurs?"

I also listed several other suggestions for using volunteers to analyze satellite data. Following the colloquium, GSFC scientists Dr. Brent Holben and Tom Eck asked me to meet with them to discuss the AERONET (Aerosol Robotic Network) program, their global network of Cimel robotic sun photometers. They wanted to know whether Cimels could be modified to measure the ozone layer, and I explained why that would be impractical. That brief meeting led to major, unexpected field studies for NASA related later and installation of an AERONET Cimel sun photometer in my field.

In Switzerland for the Rolex Award

Minnie and I left for Zurich on April 23, 1993, flying first class for the first time. While the American Airlines segment of the flight provided more space and service than coach class, the Swissair segment across the Atlantic was far more impressive.

In Zurich, we boarded a train for Arosa, near the eastern border of Switzerland. My goal was to visit the Light Climatic Observatory Arosa, where the world's longest series of ozone measurements have been made since 1926. But we arrived on a snowy Sunday when the observatory was closed. The next morning, we took a train to Chur, where we boarded the Glacier Express to Brig. This was by far the most scenic segment of our brief tour of Switzerland, for we passed through snowy mountain passes and valleys coated with spring flowers.

From Chur, we went to Interlaken and saw from a distance the famous Sphinx Observatory, between the Mönch and Jungfrau peaks. The observatory includes astronomical telescopes and various instruments for measuring the sun's ultraviolet and many other atmospheric parameters from an elevation of 11,716 feet (3,571 meters), a couple thousand feet lower than Hawai'i's Mauna Kea's astronomical observatories at 13,796 feet (4,205 meters) and around 500 feet higher than Hawai'i's Mauna Loa Observatory at 11,141 feet (3,397 meters). Unfortunately, there was no time to visit the observatory.

The next morning, we took the train to the Swiss capital of Bern, where we were impressed by the buildings and a chess board with life-size pieces. While walking along a street, we noticed a large tower with an elaborate clock face and moving figures. I told Minnie that Einstein's theory of relativity had been influenced by a clock tower when he lived in Bern.

A few steps later, we were surprised to see a small sign by a door at Kramgasse No. 49 that read, EINSTEIN HAUS. This is where Einstein had lived and had developed his famous theory from 1903 to 1905 while working as an examiner at the Swiss patent office. We visited the tiny apartment and bought some postcards printed with quotes attributed to him.

Meeting Sir Edmund Hillary

The next day, we traveled by train to Geneva and checked in at the five-star Le Richemond Hotel, overlooking Lake Geneva. The following morning, I put on my brand-new suit—still my only suit—and Minnie and I followed Rolex's printed instructions to meet our Rolex Award escort in the hotel's lobby.

That is when I was amazed to meet Sir Edmund Hillary, who had become my childhood hero after I read in the July 1954 issue of *National Geographic* his account of being, with Tenzing Norgay, first to climb Mount Everest. I had not known that Sir Edmund was a juror on the Rolex judging panel and would be our escort to the awards ceremony and a private dinner thereafter.

Sir Edmund, who had once worked with his father and brother as a beekeeper in New Zealand, was the first man to reach both poles and the summit of earth's highest mountain. He had received awards and recognition from around the world. Yet Minnie and I found him to be both humble and friendly. I wanted to ask questions about Mount Everest, but that was ancient history for him; he wanted to ask me about the ozone layer.

During the Rolex Award ceremony, each of the five winners received a gold Oyster Perpetual Chronometer, a work of precision craftsmanship down to its ten-piece waterproof winding and setting stem. We also received a beautifully inscribed parchment scroll and a check for 50,000 Swiss francs ($33,444.81; $69,716 in 2023) for our projects.

During our time in Geneva, we spent quality time with some of the Rolex staff. We also received a tour of the Rolex manufacturing building, where its famous watches are assembled by skilled technicians wearing white lab coats in state-of-the-art facilities complete with sophisticated equipment that subjected the Rolex Oysters we received to the equivalent of being immersed to a depth of 361 feet (110 meters).

We also visited the Cathédrale Saint-Pierre, where Protestant reformer John Calvin had preached his famous hourlong sermons from 1536 to 1564. Minnie loves bells, and she was rewarded with a series of loud chimes as we climbed the 157 steps of the cathedral's tower.

We closed our visit by visiting a store that sold a wide variety of cuckoo clocks, one of which we purchased. I have heard people, none of whom could make one of these remarkable timekeeping machines, mock cuckoo clocks. That clock still hangs on our kitchen wall, where it reminds me of the gold Rolex locked away and worn only on special occasions.

Doing Science on Swissair

Following our time in Geneva, we took a train to the airport, where we were seated in the spacious first-class section of a Swissair McDonnell Douglas MD-11. Rolex judge Dr. Charles F. Brush III, past president of the Explorers Club, was seated a few rows behind us. When the plane reached 31,000 feet (9.4 km), I did what I often do while flying and pulled three instruments from their bag and pointed them at the sun, one after another, to measure the total water vapor and oxygen between our altitude and the top of the atmosphere.

After the plane climbed to 35,000 feet (10.7 km) east of Iceland, Minnie became concerned that the flight attendants might wonder what I was doing, but she agreed to record the next set of 24 measurements while I quietly called out the numbers. Later, I passed the time by making several sets of additional measurements. This time, Minnie noticed that the flight attendants were talking about us, so she declined to record the data.

Halfway across the Atlantic, my notebook held 284 measurements, and I became less concerned about the furtive glances of the flight attendants. But during a measurement session that began at 4:51 p.m., a flight attendant walked over and asked what I was doing. When I explained that the instruments were measuring water vapor outside the plane, she frowned and headed for the cockpit. Minnie looked straight ahead and whispered, "I told you so!"

A minute later, the flight attendant returned, held out her hands, and firmly stated that the captain had ordered me to "please present your instruments" so he could inspect them. I laid them on her open palms, along with a Rolex Award brochure about my ozone research, and she hurriedly walked to the cockpit. While Brush was watching these developments from behind us, Minnie again whispered, this time a little louder, "I told you so!"

Several minutes later, the flight attendant arrived with a friendly smile and returned the instruments. The Rolex brochure had worked, for she said the captain had invited me to visit him in the cockpit. I could hardly believe that my dream of someday seeing what Daddy had seen during his countless jet flights was about to come true.

The flight attendant escorted me to the cockpit and introduced me to the captain, who motioned me to sit in the observer's seat behind and between him and the first officer. After I fastened the seat belt and gawked at the view through the front windshield and the big side windows, the captain asked about my ozone research that had led to the Rolex Award. After I replied, I asked him about the plane and its controls. The captain pointed to the radar display and explained we were flying much farther south than usual due to a 140-mile-per-hour jet stream.

Both men told me about the times they saw spectacular meteors, northern lights, and huge thunderstorms. All this time, we were flying alongside snow-white contrails left behind by aircraft ahead of us. A close-up view of those artificially produced cirrus clouds was a once-in-a-lifetime experience for me; seeing contrails was routine for those pilots.

Halfway through the cockpit visit, the first officer exclaimed, "*Le Concorde! Le Concorde!*" and pointed with his finger. Suddenly, a supersonic Concorde flashed by above and to our right on its way to Europe. The pilot radioed several words of greeting to the speedy Concorde, and its captain responded to the lowly MD-11 with an aloof, single syllable that I did not understand.

Our discussion next turned to water vapor and pioneering Swiss research involving ozone and solar ultraviolet. Both the captain and the first officer said they were impressed by Americans who asked about their background but annoyed by those who confused Switzerland with Sweden. Finally, they discussed the deposits still being left on their windshield by material from the volcanic eruption of Mount Pinatubo in the Philippines two years after the June 15, 1991, eruption—sulfuric acid from the sulfate in the Pinatubo aerosol plume.

The flight attendants were all smiles when I returned to my seat after a thirty-three-minute stay in the cockpit. Minnie, glad I was not going to be arrested, said that the Rolex judges had asked what had happened, so I told them. Meanwhile, I continued making measurements until we landed in New York eight hours and seventeen minutes after we left Geneva.

Our Rolex adventure ended in New York with a visit to the Explorers Club arranged by Brush, who had been president of the club from 1978 to 1981. Brush was an explorer and archaeologist who had led a series of high-altitude expeditions in the Andes. In 1983, he and Dr. Johan Reinhard had explored macrofaunae while scuba diving in one of the world's highest ponds, the crater lake of Chile's Licancabur volcano, at an elevation of 19,450 feet (5,928 m). At the time, I knew none of this, and during our time together, he never mentioned any of his many exploits.

I told Minnie that visiting the Explorers Club was significant, for the club was famous for not admitting women. I did not know then that when Brush was president, he had led the way for the acceptance of women during a secret vote of the members in 1981.

Brush had been sometimes described as eccentric, but that is not the man we saw. He gave us and other guests a tour of the club's famous rooms and enthusiastically pointed out their historic artifacts. He was intrigued by the ozone measurements I made from the club's patio and, again, never mentioned his impressive exploration achievements. In this regard, he reminded me of Sir Edmund Hillary, who was much more interested in my ozone research than in recounting his conquest of Everest.

Some people enjoy relating their adventures in stories and memoirs like the one you are reading. But Sir Edmund and Brush reminded me of astronauts and other famous people I have met whose fame has led to humility. Instead of recounting their achievements and discoveries, they prefer to learn something new from the next generation.

The Impact of the Rolex Award

The Rolex Award played a huge role in my budding science career. It transformed me from mainly writing electronics projects for magazine articles and books into doing nearly full-time atmospheric science. The Rolex Award also provided the money to hire my friend Scott Hagerup to develop a microprocessor-controlled, five-channel version of TOPS. Scott is an exceptionally talented electrical engineer who had become my friend at the church we and our families attend.

During a planning meeting with Scott, I described the microprocessor version of TOPS I envisioned: The instrument should be handheld and battery powered. It should detect five wavelengths of light—three in the ultraviolet for detecting ozone (300 nm, 305 nm, and 312 nm), one for water vapor (940 nm), and one for haze (1000 nm). Each of the five detectors should look at the sky though narrow metal tubes. Data should be saved in the instrument's memory.

While Scott went to work designing the electronics, I made many sunlight measurements to determine the amplification required for each of the five channels. I also found a source for the highly specialized UV detectors the instrument would require.

Within a few months, Scott designed, built, and programmed the prototype microprocessor-controlled instrument we called Microtops. It worked so well that Ace Radio Control agreed to manufacture twenty Microtops units I could send around the world.

My son, Eric, and I took all twenty Microtops to the Mauna Loa Observatory to calibrate them. While there were many volunteers for the ozone network I planned to establish, however, it was very difficult to find ways to send the instruments to foreign countries. I finally gave a few to my international students at the University of the Nations and a local woman who was working in South America. Unfortunately, the UV filters in the instruments soon began to deteriorate. I soon learned that this was a common problem for such filters and that much more costly filters would be required. By then, however, nearly all the funds I had received with the Rolex Award had been spent.

The filter-degradation problem and international shipping challenges had defeated my Rolex goal of establishing a global ozone network. Meanwhile, my many measurements with several Microtops attracted the interest of the Environmental Protection Agency, which I had criticized for failing to measure the very high levels of solar UV that resulted from the reduction in the ozone layer following the volcanic eruption of Mount Pinatubo.

The EPA knew that my homemade TOPS instruments had found an error in NASA's ozone satellite, and in 1994, the University of Georgia, its ozone contractor, asked to place a $120,000 ($243,089 in 2023) Brewer automated ozone instrument at my site for sixty days. The plan was to compare ozone measurements by the Brewer with those made by SuperTOPS, a Microtops with double filters to provide better UV resolution. This comparison yielded excellent results, with SuperTOPS agreeing with the Brewer to within 1 percent. When clouds interfered with a Brewer measurement, SuperTOPS always yielded much better data. Why? I waited for the clouds to move; the automated Brewer did not.

The EPA comparison attracted the attention of Saul Berger of the Solar Light Company, which secured the rights to manufacture an advanced version of Microtops in return for a royalty for Scott and me. The much more sophisticated Solar Light version of the instrument, designed by Marian Morys, is called Microtops II. It includes the ability to calculate total ozone, total water vapor, UV-B, and the optical depth of the sky (haze) and store up to 800 of these measurements in its memory.

The design and details of Microtops II were eventually published in a peer-reviewed scientific journal.[11] This paper, which I cowrote with Morys and other authors, has received 471 citations in research publications by scientists around the world, the most of any of my scientific papers.

Microtops II, which can be carried in a large coat pocket, is operated by opening a forward-facing hatch and pointing the instrument toward the sun. The instrument is carefully moved until the sun forms a bright spot centered on a crosshair behind a small, circular window. A button is then pressed, and the instrument makes a rapid sequence of scans of all five channels and stores the data in its memory. The instrument then automatically converts the data into total ozone, total water vapor, and aerosol optical depth (haze).

11. Marian Morys et al., "Design, Calibration, and Performance of Microtops II Handheld Ozone Monitor and Sun Photometer," *Journal of Geophysical Research* 106 (2001), 14,573–82.

Solar Light has sold a few thousand Microtops IIs. One version is designed to measure the ozone and water-vapor layers and haze. A second version replaces the three UV channels that measure the ozone layer with haze channels to better measure natural and anthropogenic air pollution.

While the global-network goal in my Rolex proposal failed, since 1995, Microtops IIs have been used around the world at fixed sites, aboard ships, and on scientific expeditions to make the same kind of measurements I began in the field outside my office on February 4, 1990. The expensive, high-quality UV filters in Microtops II are much longer lasting than the inexpensive filters in the original Microtops.

My original Microtops II has been used almost daily at my Texas site, during my second Brazil campaign for NASA in 1997, and during annual calibrations at the Mauna Loa Observatory from 1997 to 2018. Its calibration has never been updated, for I want to know how long the instrument can reliably work at a location where calibrations are impossible. As explained in the next chapter, its ozone measurements still closely agree with those by the world-standard ozone instrument, Dobson 83.

As of this writing, an online search on "Microtops" yields 511,000 results. These results are far more significant than my original Rolex plan, which would have merely duplicated a small part of the existing global ozone network.

The International Symposium on the Impact of Increased UV-B

Nineteen ninety-three was a busy year for me. Because of unprecedented reductions in stratospheric ozone caused by the eruption of Mount Pinatubo in 1991, the intensity of solar ultraviolet during my daily measurements was significantly higher than usual. I notified the EPA and urged it to begin making UV measurements. The agency's response was to retain me to represent the EPA along with several of its scientists at the second Pan Pacific Cooperative Symposium on Impact of Increased UV-B Exposure on Human Health and Ecosystem.

The meeting was to be held at Kitakyushu, Japan, in October 1993. The EPA would pay my travel expenses and provide a consulting fee of $600 ($1,250.70 in 2023). I spent two weeks preparing a slide show for my talk and an accompanying paper for the conference proceedings titled "Exceptionally Low Ozone and High Solar UV-B Radiation at Texas."

Upon arrival at Kyushu, Japan, on October 12, I was met by several polite students from the University of Occupational and Environmental Health Japan. They provided conference details and drove me to a hotel.

The following morning, I met the EPA staff, who informed me the trip was in trouble. Their Washington, DC, office had sent notice that the EPA inspector general declared that their attendance at a foreign conference had not been properly approved, and that we would be responsible for our registration, hotel, and meal expenses. My $600 consulting fee would not be paid. While our plane tickets were prepaid by the symposium and were already in hand, this created a brief crisis that was resolved when conference officials and the EPA worked out a compromise. I was told that the university or the Japanese government had agreed to take care of our expenses.

The conference talks, given in a large auditorium at the university, were well attended. What I had heard about the challenge of attending university in Japan was affirmed during the presentations, for each time the lighting was turned down for slide presentations, virtually all the students and professors seated around me quickly began dozing.

Fortunately, I observed this before my presentation, which I adjusted to present my slides at several intervals during the talk. This strategy worked, for when the lights were on between clusters of slides, all those I could see in the audience were looking straight at me. But when the lights were turned down for my slides, many heads began to tilt.

The brief time in Japan provided an opportunity to experience some of the country's culture. During a free afternoon, I joined Dr. Richard McKenzie of New Zealand and Dr. Jan van der Leun from the Netherlands. I am six feet tall, and they are much taller, so I was surprised that none of the locals we encountered while walking along a busy sidewalk looked up at us. But when I turned around, all the people we had passed had stopped to stare at us from behind. This lesson in size was repeated at a temple we visited, for none of the slip-on slippers we were supposed to wear fit our big feet.

The conference provided buses for our trips between the university and the hotel, and we were told to be at the bus stops no later than the specified departure times. One morning, the uniformed driver closed the door with his gloved hand just as a man arrived and beat on the door. The driver left him behind. While boarding the plane to Tokyo after the conference ended, a flight attendant moved a Russian passenger and me—the largest men on the plane—to the emergency-exit seat rows. When the plane was towed away from the gate, the uniformed mechanics and baggage handlers formed a neat line and bowed.

The papers presented at the symposium were published in 1993. My paper, which described the stability of TOPS-1 and TOPS-2 and included five graphs of UV and ozone data, ended with this comment: "The coincidences of this unprecedented, ongoing decline in total ozone, the failure of the Nimbus-7/TOMS, and the operational problems with the Meteor-3/TOMS emphasize the importance of ground-based ozone and solar UV-B observations."[12]

This conclusion brings me back to this chapter's theme, for the Rolex Award led directly to Microtops II, which measures the ozone layer at least as accurately as satellite instruments. Sun-photometer versions of Microtops II provide highly accurate measurements of total water vapor and both natural haze and air pollution. While my science career was inspired by my blind great-grandfather and the *Scientific American* affair, it received a major boost from the 1993 Rolex Award. The hundreds of papers that cite Microtops II measurements around the world and many of my subsequent papers would have never appeared if my ozone project had not received that recognition.

12. "Exceptionally Low Ozone and High Solar Ultraviolet-B Radiation at Central Texas," The 13th UOEH International Symposium & The 2nd Pan Pacific Cooperative Symposium on Impact of Increased UV-B Exposure on Human Health and Ecosystem, October 13–15, 1993, 126–135.

16

HAWAI'I AND THE MAUNA LOA OBSERVATORY

While working on the ultraviolet project for *Scientific American*, I learned about the importance of Hawai'i's Mauna Loa Observatory and became convinced it would be the best place on the planet to calibrate my homemade instruments. I was envious of Russell Ruthen, who edited my columns for *Scientific American*, when he told me Hawai'i was the destination for his upcoming honeymoon trip. I had no idea that the *Scientific American* controversy would soon lead to teaching a science course in Hawai'i and a multidecade relationship with the Mauna Loa Observatory (MLO). It is a story worth telling.

The American Scientific Affiliation

The American Scientific Affiliation, mentioned previously, is an association of scientists who are Christians, and I had heard from some of them during the public-relations-nightmare phase of the *Scientific American* affair. The purpose of the ASA is "to investigate any area relating Christian faith and science" and "to make known the results of such investigations for comment and criticism by the Christian community and by the scientific community."[1]

While I believe that God played a direct role in the creation of life, many ASA members are outspoken advocates of theistic evolution, So, I was surprised to receive a call from Walter R. Hearn, the longtime news editor for the ASA, shortly after returning to Texas from the Quadrennial Ozone Symposium.

Hearn, who earned a doctorate in biochemistry, had a distinguished career doing research and university teaching. He reflected on his call to me in "Players," an essay for the ASA's Fall 2014 issue of *God and Nature*, which focused on the theme "History of Science & Christianity":

1. About the ASA, American Scientific Affiliation, https://www.asa3.org/ASA/aboutASA.html.

> *In 1990 I was in Houston for a week's visit at the time when the Houston Chronicle broke the story of Forrest Mims not being hired by* Scientific American *to write its "Amateur Scientist" column, evidently because of his creationist beliefs. I drove to his home in Seguin (near San Antonio) and explored his backyard laboratory filled with electronic gear. Although shy about his lack of formal training in science, Forrest had written many electronics "how-to" books for Radio Shack and seemed to me to have more empirical spirit than many academic scientists I knew. With my encouragement, ASA had him speak at our 1992 annual meeting in Hawaii.*[2]

The meeting with Hearn occurred in the Silicon Farmhouse, where twenty-four books and more than a thousand articles and papers have been written and where these words are being typed. While Hearn was very interested in the atmospheric science in which I had become deeply involved, his professorial side emerged when he began asking probing questions about my beliefs. He could see that I was an authentic amateur scientist, and he wanted to know if I was also a practicing Christian.

I must have passed the test, for he later asked if I would be willing to give a talk about the *Scientific American* affair at the 1992 ASA convention, to be held at the University of the Nations campus in Kailua-Kona, Hawai'i. I replied that I knew that ASA members were required to have a degree in a scientific field (just as Air Force Weapons Laboratory staff were) and agreed to speak only after he said that was not a requirement for guest speakers.

A few months later, I received an invitation to give a plenary talk at the 1992 ASA annual meeting, to be held July 31–August 3, followed by a field trip to the astronomical observatories atop Mauna Kea on August 4. Hearn, Phillip Johnson, and Mark Hartwig and his wife, Janelle, would be at the meeting, along with an array of prominent scholars.

Speaking at the ASA meeting was an important opportunity to meet some of the leading advocates of theistic evolution and the budding intelligent design community and listen as they engaged in friendly debate. I also hoped to visit the Mauna Loa Observatory to calibrate my growing collection of homemade atmospheric instruments.

I was surprised by the reception to my plenary talk, for the audience loudly laughed at quotes I read from *Scientific American*'s correspondence and cheered my responses. Walter Hearn described my plenary talk on August 2:

---

2. Walt Hearn, "Players," *God and Nature*, Fall 2014, http://godandnature.asa3.org/essay-players-by-walt-hearn.html.

> *On Sunday afternoon there was no doubt about the excitement engendered by guest speaker Forrest Mims. Telling the tale of his rejection by Scientific American, Mims sincerely thanked ASA members for writing letters and offering encouragement at a critical time when he was being discriminated against because of his beliefs.*
>
> *Mims, who now edits Science Probe!, is a science writer, an instrument designer, and an "amateur scientist" whose unbounded enthusiasm for scientific experimentation proved infectious.*[3]

While I was meeting with some of the conference attendees the next day, Minnie and our daughters, Vicki and Sarah, were watching the scenery from the second floor of the GO Building at UofN when Dr. Howard Malmstadt, the famous chemist who was UofN's provost, engaged Minnie in a friendly conversation about our stay on the Big Island and our church involvement back home. Soon after his conversation with Minnie, he asked if I was willing to become a visiting instructor at UofN. Later, Minnie and I speculated that Howard had interviewed her first to make sure our relationship was sound.

Howard and I soon became close friends. He had served in the US Navy during World War II and had been involved in the development of radar. He was famous for developing a line of electronic instructional kits sold by Heathkit to universities, and he wrote 10 textbooks and more than 150 scientific papers.

While Howard was chair of the chemistry department at the University of Illinois at Urbana-Champaign in 1978, he was asked to help found the Pacific and Asia Christian University, which was renamed University of the Nations in 1989. Howard left behind a prestigious academic career and moved to Kailua-Kona to begin developing the new university. He was followed by other prominent academics, including Dr. Derek Chignell, who had chaired the chemistry department at Wheaton College for twenty years, and South African veterinarian Dr. John Kuhne.

These highly experienced academics devised a novel format of four semesters per year in which students attended a single course each semester. The courses were led by on-site faculty and visiting lecturers who taught classes for a single week. The lecturers included physicians, scientists, authors, magazine editors, missionaries, and others. I had experienced nothing like this during my college days, and I had not worked alongside academics as sharp as Howard and Derek since my time at the Laser Division of the Weapons Lab under Roger Mark.

---

3. Walter Hearn, "Thinking Ahead," *American Scientific Affiliation Newsletter* 34, no. 5 (October–November 1992), https://www.asa3.org/ASA/topics/NewsLetter90s/OCTNOV92.html.

Teaching at University of the Nations

The novel format of UofN made accreditation impossible, but no one seemed to mind. I soon learned that no one needed to mind, for UofN was in a league of its own. I was assigned to the School of Humanities and Sciences, where I worked closely with Derek while teaching basic electronics and experimental science for a week each summer from 1993 to 2010 at the UofN campus at Kailua-Kona and five sessions at their campus in Lausanne, Switzerland. The Hawai'i version of the course included an overnight field trip to visit the Mauna Loa Observatory and the astronomical observatories atop Mauna Kea.

The Mauna Loa Observatory was dedicated in June 1956 with a traditional Hawai'ian prayer circle and rededicated similarly during various anniversaries thereafter. After my May 2007 science class toured MLO, the sixteen students formed a circle in front of the original 1956 building where previous anniversaries had been observed, and Derek led them in a prayer of dedication for the observatory and its scientists as they began MLO's second fifty years.

My students came from around the world, with one class having twenty-five students from sixteen countries. Most students were in their late teens or early twenties, and some had college experience. A few students were much older, including James Watt, who had served as secretary of the interior under President Ronald Reagan from 1981 to 1983. Although Watt was controversial, I was intrigued by his range of knowledge and impressed by his willingness to participate in a daily duty required of UofN students, all of whom were much younger.

Most serve in the kitchen or perform lawn maintenance. Watt's chore was the daily raising and lowering of each of the thirty-eight flags surrounding a circular pool at the center of the campus. This was a time-consuming task that Watt approached with a smile. After the final day of class, I drove Watt on a tour of the southern half of Hawai'i Island. That's when he became the teacher as he provided some of the best advice I have received about organizing my writing career while doing science.

Almost all my students were humanities majors, and many were terrified about being required to take a science short course, so I tempered their concern at the beginning of the first day of class by introducing myself to each student seated around me in an arc of desks. I then sat on the floor to examine the contents of my pockets: a compass to check the flight direction of the plane that carried me to Hawai'i (or Switzerland), a bubble level to observe the up-and-down motion of the plane, an orange filter to better see the landscape through a hazy sky, a polarizing filter to better view the sky and the ground outside the plane, a pocket notebook, and a small pencil. I then opened my carry-on bag and extracted the Geiger counter I used to measure the background radiation, which was 35–40 times higher at

35,000 feet (10.7 km) than on the ground. This allowed the elevation of the plane to be estimated.

After this show-and-tell exhibit, I told them my college major had been in government. Suddenly, the frowns and concerned looks became smiles and even laughter. From then on, most students paid very close attention to my lectures, for they were presented in common language and not the jargon of science.

Each of my students was required to do a basic science project over the next four days, present a talk about the project the morning of the fifth day, and display the results on a poster to the entire student body at supper the evening of the fifth day. It was always a treat to watch the humanities students enthusiastically present their science projects to their fellow students gathered near their posters.

My students asked many questions in and out of class. When several students saw me walking past the nearby Bubba Gump Shrimp Co. restaurant in Kona after class one evening, one shouted, "Run, Forrest! Run!" I assumed this was a line from the movie *Forrest Gump*, which I did not see until I was writing this chapter. That's when I learned that Forrest Gump and I had both been born in 1944, had both served in Vietnam in 1967, and had both met some famous people.

The UofN teaching assignment was a rigorous week of teaching, preparing for the next day, and helping students make graphs and presentations for their science projects. After a difficult first year, I made more than a hundred overhead transparencies and detailed lecture notes for the 1994 class, which went much more smoothly than that first year. A decade later, I transitioned from overheads to a laptop and a PowerPoint slide show that allowed slides to be easily added. My 5-pound box of overheads was replaced by a pocket flash drive the size of a pack of chewing gum.

The academic faculty at UofN were among the friendliest and brightest scientists and scholars I have ever known. They were also humble. Meals at UofN in Kona are served under outdoor shelters, and during an evening meal in 2002, when I was seated across from Howard, two female students from the nursing school asked if they could join our table. We introduced ourselves, but Howard did not bother to give his title. He simply identified himself by his first name.

When I told the women that Howard was the university's provost, they continued the conversation as before while still referring to him as Howard. They knew nothing about Howard's role in founding UofN, or that he was widely considered one of the world's most influential analytical chemists of the past fifty years. Instead of telling them about himself, he asked about their plans. All the faculty I met followed Howard's example.

When the president of another university appointed me as a Distinguished

Lecturer in Physical Science and invited me to give a series of lectures, I had to listen to two professors scream at one another about a personal dispute after they, several students, and I were leaving my opening lecture. I never went back. Nothing remotely like this occurred at UofN, where for seventeen years I taught my short course once a year at the Kona campus and for several years at the Lausanne, Switzerland, campus.

Hawai'i's Mauna Loa Observatory

A few months before leaving for Hawai'i to speak at the American Scientific Affiliation meeting, I telephoned the Mauna Loa Observatory and spoke with secretary Judy Pereira. She told me that University of Colorado student Gretchen "Grae" Roth had been hired to calibrate Dobson 83, the world-standard ozone instrument, for NOAA and NASA that summer. She was staying with a friend at Kailua-Kona and might be able to drive me to MLO for a few visits.

During the ASA conference, I telephoned Grae to ask whether she would be able to take me on a few of her early-morning trips to the observatory. She agreed to pick me up at the hotel at 4:30 a.m. on August 7. The night before, I carefully packed TOPS-1, TOPS-2, and the sun photometers along with a clipboard and data sheets I had printed back home.

Grae arrived on time, and, after a quick stop for breakfast and coffee to be consumed along the way, she drove us 31 miles (50 km) along nearly empty Highway 190 to Saddle Road, a narrow, curvy road with so much pavement damage that cars drove along the center of what remained of the asphalt. After 25 miles (40 km), we turned right onto the then unmarked MLO road. A squiggly white line marked the center of the single-lane road, at least where there was asphalt between gaps of raw lava. The white line, which resembled the trace of a seismometer indicating earth tremors, had been formed by punching a hole in a series of cans of white paint held at the back of the MLO pickup by one of the day crew.

The road to the observatory terminates at a series of small buildings on either side of the pavement 11,141 feet (3,397 m) above sea level. Back then, the main structure was the original concrete-block building from 1956, where for the first decade, teams of two or three staff had lived for nearly a week at a time before rotating off the mountain. It was a low-budget installation.

Fortunately, I experience only minimal effects while working at high altitudes, for as soon as Grae and I arrived at MLO, there was no time to become acclimated. We had to start our respective measurements as quickly as possible. The Dobson 83 calibration protocol required that Grae make a series of ozone measurements from when the sun was a few degrees over the horizon to nearly noon. I followed her protocol for both TOPS and the sun photometers, including the one I had showed Jonathan Piel in 1989.

As the sun rose higher in the sky, its angle shifted more slowly. This provided time between measurement cycles for us to chat about what we were each doing. I was especially interested in how Grae operated world-famous Dobson 83, having no idea that I would be calibrating the same instrument twenty-six years later while living at MLO for two months.

The observatory's day crew arrived at midmorning. After unloading their truck, with clipboards in hand, they began making their rounds, checking the status of the many instruments deployed around the site and in various buildings.

Grae drove me to MLO a second time, on August 9. During a break there, I told her that TOPS had begun measuring less ozone than the satellite instrument several months before and asked what she was measuring. Grae showed me her data for that day, which were then preliminary but which agreed closely with TOPS. This seemed to confirm that TOPS was working well and had indeed found an error in NASA's ozone satellite. But I would need the satellite data for the two days at the observatory to be sure.

The 1992 visits to MLO had fulfilled a major goal, and I did not then realize that my tenure there was just beginning. The UofN teaching assignment provided seventeen years of free passage to Hawai'i. From 1993 to 1997, after the course was over, I stayed at the UofN campus for three days and left for MLO every morning at 3 a.m. I calibrated many instruments from sunrise to noon and then returned to Kona. From 1998 to 2018, I spent 235 nights at the observatory to avoid the tiring, predawn drive up the mountain.

The Russian M-124 Ozone Instrument

I was hopeful that the two M-124 ozonometers purchased for $750 each from Russian scientists at the Quadrennial Ozone Symposium in June 1992 would support my claim that NASA's ozone satellite was returning erroneous data. Therefore, I took one of the M-124s to Hawai'i during my first visit in 1992 and each of the next two years. During the August 9, 1992, MLO visit, I photographed Grae operating Dobson 83, and she photographed me holding TOPS-2 while TOPS-1 and the M-124 were perched atop Dobson 83.

The M-124 is not handheld like TOPS. It is much larger and heavier and is designed to be placed on a level table during measurements. Preparing the M-124 I took to MLO was not easy, for it uses very different batteries than those available in the US. I used a string of 9-volt batteries to provide the proper voltage.

During nearly three years of ozone layer measurements from the Geronimo

Creek Observatory, which included the 1992–1994 MLO visits, I measured the ozone layer with the M-124 and both TOPS and compared their data with measurements from space by NASA's failing Total Ozone Mapping Spectrometer (TOMS) aboard *Nimbus-7* and its replacement, a TOMS instrument installed on Russia's *Meteor-3* satellite that provided questionable data due to its unusual orbit.

The results of this study are summarized in this previously unpublished table that clearly shows that the average ozone amount measured by the M-124 was not nearly as close to the ozone measured by the satellites and, especially, TOPS-1 and TOPS-2.

Device	Days	Nimbus-7	Days	Meteor-3
M-124	30	-11.7%	123	-10.4%
TOPS-1	126	-2.1%	552	-4.9%
TOPS-2	121	-1.1%	481	-3.2%

The mediocre performance of the M-124 was unsurprising, for its two UV filters are broadband and not the ultranarrow band filters used in TOPS. This alone meant the M-124 could not be as accurate as TOPS, especially during winter, when the sun is low in the sky.

Because of the poor performance of the M-124, I decided to abandon the paper I had planned. My comparison was based on only one M-124; perhaps it had not been properly calibrated before it was shipped. Also, while the *Nature* paper about the satellite error found with my homemade TOPS instruments had been well received by some ozone scientists, a few were resentful, and I did not want to provoke them. For this reason, I did not write a paper about the easily found errors in the *Meteor-3* data, about which the Goddard Space Flight Center was already aware.

Earlier, I had decided not to write a paper about a significant sensitivity error I found in NOAA's Advanced Very High Resolution Radiometer satellite instrument, which measures the presence of aerosols in the sky. After a conference talk in July 1993 by a NOAA scientist who claimed that the AVHRR was no longer detecting Pinatubo aerosols, another scientist and I disagreed with him during the comment period. My sun-photometer data clearly showed that Pinatubo aerosols were still present, and their pink glow in the sky could be seen by anyone during twilight. NOAA scientists arranged a comparison of their AVHRR data and my data, which proved that the AVHRR instrument had developed a sensitivity problem.

Years later, I wrote a paper about the excellent agreement in haze measured by NASA's *Terra* satellite and my LED sun photometer when *Terra* overpasses were very near the Geronimo Creek Observatory. This restored some of my lost confidence in multimillion-dollar satellite programs. But I still felt short-changed that I was required to pay taxes to support huge satellite programs run by well-paid government scientists when I could barely afford to support my personal research. Nor could I afford to pay the extravagant page fees most scientific journals charged, or to travel to scientific conferences.

When a major paper presenting global ozone results was published in a leading scientific journal, I was surprised that the paper included Soviet M-124 ozone data alongside much better data from Dobson spectrophotometers. I called the lead author, a NASA scientist I had met, and he agreed with me about the poor quality of the M-124 data and explained that it had been included "for political reasons." Yet the M-124 data were known to be questionable, as reported in Dr. Uwe Feister's fifteen-month study presented at the 1992 Quadrennial Ozone Symposium, which concluded, "Individual measurements with the M 124 instrument #200 cannot be considered as reliable and can, therefore, not be recommended for use in analyses of ozone data."[4]

Perhaps I should have written a paper on my M-124 findings. I am not a professionally trained scientist, but I have a degree in government. Including questionable data "for political reasons" in a major paper for a leading professional journal of science does not pass a basic integrity test.

The 1994 UV Survey

Because Eric was working and unable to accompany Minnie, Vicki, Sarah, and me during our first visit to Hawai'i in 1992, his college-graduation present was a trip to the Big Island after I completed teaching my course at the University of the Nations in June 1994. This was Minnie's gift, for she had planned to accompany me to Hawai'i that year. Instead, she insisted I take Eric.

He and I visited some popular tourist destinations, and I took advantage of his presence to assist me in calibrating twenty Microtops during his first trip to the Mauna Loa Observatory. Eric also assisted with what I called the Hawai'i UV Survey. His help was especially valuable when I was measuring the passage of solar ultraviolet through seawater at Hapuna Bay with a homemade sensor connected to

4. Uwe Feister, "Comparison Between Brewer Spectroradiometer, M 124 Filter Ozonometer, and Dobson Spectrophotometer" Quadrennial Ozone Symposium, University of Virginia, Charlottesville, Virginia, June, 1992.

a data logger and installed in a waterproof camera case. We had trouble keeping the device steady until we placed a lava rock over the camera case's strap and allowed the case with its UV instrument to float a few feet over the bottom.

The most important measurements during the UV survey were made with sixteen miniature UV monitors I placed in isolated locations around the island with a full view of the sky. Only one was missing when I recovered them several days later. While the purpose of the study was to investigate the difference in UV between sea level and MLO, where UV is typically 15 to 20 percent higher, a surprising finding was that scattering of sunlight from cumulus clouds in the pristine sky over MLO caused increases of up to 15 percent in solar UV-B.

A literature search did not disclose any similar studies, so I contacted UV expert Dr. John E. Frederick at the University of Chicago to ask whether he would join me on a paper about this finding. Frederick agreed and contributed a paragraph about the theory behind the finding. I then submitted the paper to *Nature*, which sent it to two referees. Surprisingly, both referees soon recommended that the paper be published. One wrote that while the phenomenon was known, it had not been previously published. "Cumulus Clouds and UV-B" appeared in the September 1994 issue of *Nature*.[5] As of this writing, it has 103 citations in the scholarly literature.

The 1994 UV survey led to many more UV measurements in Hawai'i. Clouds often arrive at MLO before solar noon. When this occurred on shower and grocery days, I drove down to Spencer Beach Park on the upper west coast and made sunlight measurements to compare with those made with the same instruments at the observatory. During my second visit to MLO, on August 9, 1992, simultaneous measurements made by Minnie at Kona (TOPS-2) and by me at the observatory (TOPS-1), and many such measurements since, showed that the total ozone column is around 9 Dobson units higher at sea level than at MLO. Years after I began comparing surface and MLO ozone measurements, NOAA did likewise with a pair of their big Dobson ozone instruments. Their results were remarkably similar to mine.

A Prominent Scientist Objects to My MLO Visits

Dr. Stan Anderson of Westmont College developed the ozone algorithm used in Microtops II, the sophisticated version of Microtops designed and developed by Marian Morys of Solar Light described earlier. Stan needed to know the ozone measured by world-standard ozone instrument, Dobson 83, at the Mauna Loa Observatory in 1994 when I was there making measurements.

My routine request for three days of data did not receive a routine response, so I

5. Forrest M. Mims III and John E. Frederick, "Cumulus Clouds and UV-B," *Nature* 371 (September 1994), 291.

wrote Dr. David J. Hofmann, a widely known atmospheric scientist who led NOAA's Climate Monitoring and Diagnostics Laboratory, in Boulder, Colorado. Hofmann was aware of my visits to MLO and my paper in *Nature* about the satellite ozone error discovered by my homemade TOPS instruments. On November 22, 1994, he wrote me that "it is Laboratory policy that observatory programs involving guest investigators or instruments be reviewed for scientific integrity as well as appropriateness to CMDL's mission. This is to avoid impacting an already crowded observatory and supporting measurements which are not considered appropriate by our scientists."

The letter continued with more about Hofmann's concerns about my routine data request. But on November 29, Hofmann relented and sent the requested Dobson 83 data by fax. I thanked him and immediately forwarded the data to Stan.

Despite Hofmann's objections, I continued my annual calibration visits to MLO, where none of the staff ever objected. I did not realize that Hofmann was kept fully informed about my visits in the observatory's monthly activity reports. Nor did I know that Hofmann would eventually reverse his attitude toward me and my research in an unexpected and very positive manner during my many years performing annual calibrations there. I will have more to say about this later.

Overnight Stays at MLO

In 1997, the Network for the Detection of Atmospheric Composition Change (NDACC) building was completed at the Mauna Loa Observatory. This structure and its elevated solar deck provides significantly more space for instruments that measure atmospheric gases, solar ultraviolet, visible and infrared sunlight, and the ozone layer. Both NOAA and NASA installed lidar systems in the NDACC building to measure the profiles of aerosols, water vapor, and ozone high over MLO. The building also includes a workshop, two tiny bedrooms, a kitchen, and a restroom.

When Dr. John E. Barnes became the observatory's station chief in 1998, he asked if I would like to stay overnight in the NDACC building to eliminate my long predawn drives from Kona. I was then calibrating two LED sun photometers for the GLOBE program at MLO each summer, which would provide official justification for spending nights at the observatory. The GLOBE connection was important, for it was an international program for students to make environmental measurements.

I gladly accepted John's offer and began spending nights at MLO after the 1999 UofN course. I have since spent more than 235 nights in the NDACC building. Below, I describe my daily routine at the observatory.

Morning at MLO

On a typical day, I arise at 5 a.m., quickly dress, and go outside to retrieve the twilight photometers, which I had placed behind the NDACC building the previous evening. Usually, the sky is clear, but occasionally, overnight rain leaves a thin layer of slippery ice. Back inside, I download the twilight data into a laptop computer while having oatmeal with raisins and milk for breakfast.

My homemade twilight photometers use LEDs to detect slight changes in the twilight glow straight overhead for an hour after sunset as the sun sinks below the horizon and an hour before sunrise. The signal from the twilight glow is saved in a device called a data logger. After the data is transferred into the laptop, a program produces a graph that shows the brightness of the twilight from the elevation of MLO to a point about 60 miles (100 km) or more in the sky overhead. Bulges in the graph indicate the presence of aerosols.

After checking the twilight data, it is time to warm the rental car, from where I will make the morning calibration measurements with my original LED sun photometer, two GLOBE sun photometers, five Microtops IIs, and a Radio Shack Sun & Sky Station. (The warm car is necessary to accommodate the temperature sensitivity of the instruments.) By 6 a.m., I am in the car and pointing instruments at the rising sun in quick succession while writing down the results on a form clamped to a clipboard.

All these instruments indicate the brightness of the sun on a digital readout. The goal is to calibrate the instruments by determining their extraterrestrial constant (ET)—what they would read if pointed at the sun from above the atmosphere. This is done by pointing the instruments at the sun as it rises through a sequence of angles that progressively reduce the thickness of the atmosphere (the air mass, or m) between the instrument and the sun. When the sun is 19 degrees above the horizon, $m = 3$. When the sun rises to 30 degrees above the horizon, $m = 2$, and so forth. If the sky is very clear, the points will form a straight line known as a Langley plot, named after the former secretary of the Smithsonian Institution who perfected the method. Extending the line to where $m = 0$ yields the ET.

Several times during the calibration, I photograph the solar aureole, the bright glow around the sun caused by aerosols in the sky. This is done with a camera fitted with a small black ball 6 inches from the lens, which I described how to assemble in my first article for *Make* magazine. The photo is taken after the camera position is adjusted so that the black ball forms a shadow over the camera's lens. I take solar aureole photos during all my Texas sky measurements, and they always show a glow around the black disk blocking the sun. The sky over MLO is often so clear that there is no solar aureole.

Clouds do not often interfere with morning calibrations at the observatory, but they do back home in Texas. Spring and fall cloud interruptions provide an opportunity to gaze at the many wildflowers sprinkled around the Geronimo Creek Observatory. I also note birds, butterflies, dragonflies, and wasps, and spiders drifting by under silken strands. These observations are summarized in an annual wildlife survey.

Noon at MLO

By midmorning, cumulus clouds often rise to the altitude where MLO is situated. When they arrive overhead, they dance around the sky in mesmerizing patterns. When clouds don't arrive at noon, I carry a case of instruments to the MLO solar deck and use a homemade radiometer (light meter) I made in 1995 to measure the intensity of sunlight from the full sky and the intensity of light from the full sky when the sun is blocked.

These measurements are made using UV-B sensors and LEDs sensitive to visible and near-infrared installed in aluminum tubes with Teflon caps. The sensors are plugged into two sockets on top of the radiometer. All these components were assembled or purchased from a $1,000 fund set aside from my 1993 Rolex Award.

During these measurements, the radiometer is placed on a flat, level surface on the solar deck, from where it first measures the full-sky irradiance of the sun and the sky. I then carefully hold a 1-inch-diameter black disk mounted on the end of a piano wire so that its shadow completely covers the Teflon diffuser of each sensor probe. The digital readout then indicates the diffuse sunlight scattered by the sky and any clouds present. The diffuse UV-B is typically half the full-sky UV-B, but this ratio decreases sharply with wavelength so that near-infrared diffuse sunlight is typically less than 10 percent of the full-sky irradiance.

The UV-B at MLO is typically 15 percent higher than that on the beaches far below. UV-B sterilizes the outdoor environment and is essential for producing vitamin-D in vertebrates, including you and me. The downside is that excessive UV-B causes erythema (sunburn), which can lead to skin cancer.

Afternoon at MLO

On nongrocery days, I stay at MLO to process data. Sometimes, I help the staff by leading planned tours and checking on instruments. The day crew usually leaves around 2:30 p.m., and I then work on the data. Sometimes the day crew leaves early, like the afternoon a powerful thunderstorm arrived. They knew that lightning had struck MLO, so they did not bother to say goodbye when they hopped in the observatory's truck and raced away while I photographed their taillights shortly before heavy rain and hail arrived.

Every two or three days, after the noon measurements, or earlier if clouds arrive, I load some key instruments into the car and head down to Spencer Beach Park, on the west side of Hawai'i Island. On the way, I usually stop at McDonald's in Waimea for a quick Big Mac. Spencer Beach Park is ideal for sea-level measurements, a fifteen-minute swim in the Pacific, and a solar-heated shower. I then head back to Waimea to buy groceries, top off my car's gas tank, and buy a Subway sandwich to eat on the drive back to MLO. The plan is to arrive back at the observatory in time to set up the twilight gear and photograph the sunset.

Night at MLO

Shortly before sunset, I program the data loggers for my twilight photometers and install them on the concrete pad behind the NDACC building to shield them from radio-frequency interference. I then walk to the west end of the site to photograph the sunset and the twilight glow.

After sunset, MLO becomes Hawai'i's loneliest outpost. Lights from several communities along the shoreline far below are visible, and the headlights of a dozen Mauna Kea tour vehicles can be seen slowly driving down from the summit and the visitor's center 21 miles (34 km) across the saddle from the observatory. Thereafter, the sky becomes so dark that the Milky Way resembles a cirrus cloud stretched across the coal-black sky.

When the sky is clear, NASA's lidar operator arrives after dark to fire up a high-powered ultraviolet laser that measures the profile of the ozone layer. This lidar and its support gear and computers are installed in the most elaborate work area at the observatory.

Once a week, the MLO lidar operator arrives to activate John Barnes's lidar, which is installed in a much more compact space than the NASA lidar. Its stunningly brilliant green beam provides profiles of dust particles and water vapor up to an altitude of about 25 miles (40 km).

Thanks to John's training, I have operated his lidar a dozen or so times. After my first session alone on June 27, 2014, John wrote, "Glad it worked Forrest. I now proclaim you a 'Lidar Guy.' Congratulations."[6] Skeptics who mock my lack of a science degree and believe I cannot do real science until I accept Darwinian evolution will be displeased to read page 1 of the June 2015 MLO Activities Report, which states, "5/27–6/8 Forrest Mims continued his solar calibrations, twilight instrument tests, ran the NOAA lidar, and helped with tours."

---

6. John Barnes, email to Forrest M. Mims, June 27, 2014.

By 11 p.m., the lidar operators are gone, and it is usually time to sleep. My first nights sleeping at MLO in 1999 were difficult, for the ultradry air makes breathing a challenge. Minnie suggested spraying saline solution into each nostril, and that worked.

After becoming accustomed to the sounds of pumps feeding air from atop the MLO tower to an array of instruments that measure carbon dioxide and other gases and periodic bangs of a ventilation door slamming shut, I was able to get a good night's sleep. After the first night, I usually sleep well. Only twice have I been awakened by earth tremors during the night.

Occasionally, something unusual occurs at night. Shortly before going to bed late one night, I went outside to test a new ultrabright LED flashlight by pointing it up the mountain. A bright light flashed back! That was more than puzzling, for there were no cars in the small parking area for hikers just below the observatory. I flashed several more times, and each time, my flash was returned.

Half an hour later, I tried again, but there was no return flash. Then I looked down the slope and saw three lights floating over the lava as they slowly drifted east. This was even stranger than the light flashes seen earlier. I had no idea what was happening until a voice from below shouted, "Can we come up?" Suddenly, I realized that the floating lights were headlamps worn by hikers.

Several minutes later, a dozen heavily laden soldiers arrived. Those soldiers were practicing high-altitude climbing for a forthcoming trip to the Himalayas to look for the remains of US fliers downed during World War II. They had planned to spend the night at the summit cabin, but one of their group had developed altitude sickness, and they had had to get her down to a lower elevation as soon as possible. They had no water, so I shared mine and allowed them to spend the night in the NDACC building's hallway. This raised a few concerns among MLO staff when I informed them the next day, but it was the right thing to do.

Another nocturnal incident deserves mention. I was packing my rental car at 3:30 a.m. during thick fog when I noticed movement about 50 feet (15 m) away. Presumably, I was the only person on the mountain, but when I nervously asked, "Can I help you?" a voice from an invisible source replied, "Yes."

The hair on the back of my head stood erect as I stared into the fog and saw a dark figure slowly moving my way. Seconds later, a man arrived who claimed he was a Mauna Kea tour guide who had always wanted to visit MLO, which made no sense considering the late hour and the dense fog. He said he could not find his way back to his vehicle, so I drove him down to the visitor's parking area. He thanked me and seemed believable, but he was driving a car, not a tour van.

Then there was the woman dressed in a white gown standing near the road when I was driving to the observatory one night. She did not motion for me to stop, so I assumed she was with nearby campers and was not Madame Pele, the Hawai'ian volcano goddess. MLO staff love stories like this, for very few of them have stayed overnight at MLO, and none of them have seen Madame Pele.

The MLO History Book

During my 2004 stay at the Mauna Loa Observatory, John Barnes told me that a fiftieth anniversary of the observatory in 2006 was being planned. Would I be willing to write a book to celebrate the occasion? I quickly agreed and began work on a proposal to send Russell Schnell, a former MLO director and outstanding atmospheric scientist who then directed all of NOAA's observatories. The plan was for NOAA to hire me to write the book and pay travel expenses; I would forgo a royalty.

After the proposal was accepted, none of us realized the magnitude of the task before us. I made two trips to the observatory and a trip to NOAA in Boulder to review hundreds of historical photographs and documents and to do video interviews with forty former staff and scientists.

The visitor's log at the observatory was especially important, for it included signatures of early MLO staff, astronauts, elected officials, reporters, and hundreds of scientists. The log provided specific dates for important events, including the dedication of the observatory in 1956, the first arrival of staff members, and the installation of the historic Scripps carbon dioxide analyzer in November 1958. The Boulder visit was also highly productive, for Dr. Russell D. Schnell allowed me to spend several hours in a document-storage room that included decades of MLO status reports.

Hawai'i's Mauna Loa Observatory: Fifty Years of Monitoring the Atmosphere was published by the University of Hawai'i Press in November 2011. The 462-page book includes many black-and-white photos and a hundred color plates. Atmospheric Science Librarians International (ASLI) gave the book its 2012 ASLI Choice Award in the historical category.

The book, commended for its "engaging perspective on the scientists, discoveries, and ground-breaking atmospheric measurements done at Mauna Loa Observatory,"[7] details the difficulties of working at MLO during the early years. It includes stories about the staff, funding shortages, a Mauna Loa eruption that cut power to the observatory for more than a month, snowstorms and hailstorms,

---

7. 2012 ASLI Choice Awards Winners, Atmospheric Science Librarians International, http://www.aslionline.org/wp/2012-asli-choice-awards-winners.

and prominent visitors. The famous Keeling Curve, the record of the atmosphere's increasing carbon dioxide measured by Dr. Charles David Keeling from 1958 onward, is covered in detail. While Keeling's original instrument has been replaced by more modern technology, visitors to the observatory can see it on display in the NDACC building's hallway.

Calibrating the World Standard Ozone Instrument

During my 2015 stay at the Mauna Loa Observatory, John Barnes, the station chief, asked if I would be available to calibrate Dobson 83, the world-standard ozone instrument, during the summer of 2016. Of course, I replied yes. Dobson 83 was awarded that status in 1962, the year I graduated high school, and it has been used to calibrate some 90 Dobson instruments around the world (plus TOPS and Microtops II).

Dobson 83 was the instrument Grae Roth was assigned to calibrate when she drove me to my first two visits to MLO in August 1992. As explained in chapter 16, that experience confirmed that my TOPS instruments were measuring the ozone layer more accurately than NASA's ozone satellite. Calibrating Dobson 83 would provide an ideal opportunity to compare ozone measurements by my Microtops II with those by the world standard.

On June 3, 2016, Matthew "Marty" Martinsen helped me prepare Dobson 83. I was grateful for his help, for the ancient instrument had given me an electrical shock the first time I plugged its power cord into an outlet. Marty was new to the observatory, but he had considerable experience operating a Dobson when he was station chief at NOAA's Barrow Atmospheric Baseline Observatory, near Barrow (now Utqiagvik), Alaska.

The sixty-four days with Dobson 83 did not go smoothly. Unusually frequent cloud cover often blocked measurements, and several days with extraordinarily strong wind were a major problem. Gusts rocked the heavy Dobson cart, and it was necessary to use duct tape to keep the laptop computer from blowing off the cart. Blown fuses, defective calibration lamps, and a worn-out adjustment screw caused delays.

Especially troubling were the many times the meter needle atop the Dobson began moving after I set it to zero. A decade earlier, NOAA employee Mark Clark had experienced the same problem while I was watching him calibrate Dobson 83. This problem was likely caused by radio signals, for MLO has become a relay station for military and other communications.

Then there was RIMPAC 2016, a major naval, air, and surface military operation by twenty-six nations that began on June 1, 2016, the day Dobson 83 arrived at MLO, and ended on August 4, two days after I completed the sixty-four-day calibration assignment. Because of the many military actions in the saddle between Mauna Kea and Mauna Loa and the fact that electronic warfare is a major aspect of RIMPAC, I included references to the operation in a detailed report to NOAA about the Dobson 83 problems.

I also included details about how the US Naval Research Laboratory's microwave instrument for measuring high-altitude water vapor had been seriously disrupted by unknown radio signals. Mike Gomez, who manages that expensive instrument, arrived on my final day at the observatory, and he showed me how interference was ruining his data.

While I obtained only a few interference-free calibration runs with Dobson 83 during that MLO assignment, 194 ozone readings with my original 1997 Microtops II, which was calibrated against Dobson 83 in 1997, were within 1.9 percent of those recorded by that venerable instrument. That agreement was within the standard used to validate other ozone instruments, and that alone made those sixty-four days at the observatory worthwhile.

During those sixty-four days, Kilauea was erupting. Aside from driving down to the beach for a shower and groceries every three days, my only other break was to visit the Kilauea eruption and see molten lava up close as it slowly flowed to the ocean.

17

TO BRAZIL FOR NASA

Following the 1993 colloquium "Earth Science on a Shoestring Budget" at the Goddard Space Flight Center, I stayed in touch with Brent Holben, director of AERONET. In 1995, Brent asked if I was available to participate in the Smoke, Clouds, and Radiation–Brazil (SCAR-B) field campaign. He explained that SCAR-B was a major collaboration between US and Brazilian scientists during August and September 1995 to study the impact of serious smoke pollution caused by the massive burning of the landscape across Brazil.

The TOMS ozone instrument aboard NASA's *Nimbus-7* satellite had finally failed, and NASA had no portable ozone-monitoring instruments. Brent asked if I would be willing to use a Microtops to measure the ozone layer from Brazil for three weeks during SCAR-B, and I enthusiastically agreed. Although my assignment was to simply measure how severe air pollution affects ozone, I had other plans: How much does smoke reduce sunlight? Does it also reduce the sun's ultraviolet rays?

Shortly after Brent asked me to participate in SCAR-B, I gave a talk at Colorado Christian University in Lakewood. One of the professors informed me that a student, Damian Kilday, wanted to accompany me to Brazil. He was willing to help carry my equipment and make measurements, and I gladly accepted his offer. NASA had provided a contract for $5,000 ($9,870 in 2023) for the trip, and that would be enough, so I thought, to cover airfare, meals, and housing for both of us.

A few weeks before departure, Brent informed me that the Brazilian military had withdrawn permission for SCAR-B. A well-traveled NASA scientist I knew said I should go on my own by posing as a tourist and making measurements as inconspicuously as possible. I was aware of Brazil's strict requirements for doing science in their country and wondered how my scientist friend would get me out of a Brazilian jail. Only days before the canceled departure date, however, Brazil's military relented. The expedition was on again, and Damian and I were off to Brazil.

Flying to Brazil

Damian and I left Dallas–Fort Worth International Airport at dusk on August 24, 1995, and crossed the Equator around 3 a.m. As we flew further south, we saw flickering strands of orange, red, and yellow that formed eerie outlines below. They were the fires whose smoke we would soon be measuring—and breathing.

When we entered São Paulo's airport terminal, Damian and I were apprehensive about making it through customs with all my equipment. Before the trip, I had emailed the scientist friend who had earlier advised me to conduct research in Brazil as a tourist and asked if the list of equipment I sent the Brazilian consulate would get us through customs. He replied that it probably would not, and the best thing I could do is dress like an American tourist and put my equipment in regular luggage and hope for the green light!

That is what we did. After a friendly greeting at passport control, we picked up our checked bags and walked through the door under the Nothing to Declare sign. Suddenly, two men in suits and carrying briefcases raced ahead of us. The customs inspectors took notice and stopped them for inspection with the red light and waved us through with the green light.

We were warned that the airport had a reputation for pickpockets and baggage theft, but we soon learned our paranoia was misplaced when we visited an airport bank to exchange my money. When I placed $540 in $20 bills in the slot below a thick glass window, the woman on the other side slipped five of the bills under the counter while trying to distract me by saying the first $20 bill was fake. When I firmly said that I gave her $540, the missing $20 bills somehow emerged from under the counter.

Smoky Cuiabá

Cuiabá, the capital of the state of Mato Grosso, is impressive in size, and its people were as friendly as Texans. They were also young. At fifty-one, I was the oldest person around when we walked the streets. We spent nights in a tiny room at the Hotel Mato Grosso. Each morning, we took a taxi to the National Institute for Space Research (INPE), near the city, where we met our NASA contacts and set up our instruments on the roof.

We spent six days on the INPE roof making hundreds of measurements of the ozone layer and ultraviolet and visible sunlight through the smoky sky. On some days, the smoke was so thick, we could barely see the skyline of Cuiabá. Occasionally, we took breaks to watch monkeys in the trees and find quartz crystals lying on the ground.

Every day at noon, the NASA scientists joined their Brazilian counterparts for a fish dinner at a nearby restaurant. Damian and I had to use most of our meager NASA budget for the expensive taxi rides between our hotel and the observatory, and this left us with no money to join the scientists at the restaurant. During their noon absences, I altered the measurement schedule of INPE's $120,000 Brewer spectrophotometer like the one I had learned to operate when the EPA placed one at my Texas site for two months. This allowed me to compare my solar-noon ozone and ultraviolet measurements with the Brewer data while we ate peanuts, crackers, and an orange for lunch. Minutes before the scientists returned from lunch, I reset the program to the INPE settings and returned to the roof. While my Microtops ozone measurements were within a few percentage points of the Brewer's, my UV-B measurements were around 20 percent higher.

Since Damian and I needed to save funds for a trip to Manaus, we were unable to accept the offer of an enthusiastic travel guide who wanted to take us on a tour of the Pantanal, the world's largest wetland. He proudly showed us a thick album filled with photos of terrified tourists trudging through piranha-infested water up to their chests.

We were surprised to see that biomass burning in Brazil is not confined to the countryside, for landscapers working around tall buildings in downtown Cuiabá were burning leaves they had raked into piles.

Also, one evening, we saw a crowd in a small square where a man had set up a steel hoop a yard in diameter in which a dozen or more knives were inserted. The blades of the knives were all pointed inward, which is why there was a hush when a young boy around twelve years old ran toward the hoop and dove through the ring of blades without receiving a scratch.

I brought a global positioning satellite (GPS) receiver to provide our exact location for Microtops measurements. Late one evening, Damian and I took the GPS to the monument that marks the geographic center of South America. Surveying methods in 1909, when Marshal Cândido Rondon pinpointed the spot, did not match the precision of the GPS receiver, and the coordinates engraved on the Centro Geodésico da América do Sul monument did not match those displayed by the GPS. Instead, the GPS took us to an abandoned bar atop a hill a couple thousand feet (600 m) away, where I used a rock to pound a rusty nail into the ground to mark the geographical center of South America.

When we told the NASA scientists about our discovery, Brent suggested it would not be prudent to make a public announcement until we were home.

Pantanal Adventure

On September 1, the scientists held a meeting to discuss how to coordinate an illegal burn planned by a farmer with an overflight of an instrument-equipped plane to measure the properties of fresh smoke. While a maid was busily popping bubbles in the plastic protecting one of Brent's instruments, one of the Brazilian scientists insisted they must not do anything to cause an illegal fire to be set and threatened to send for the police to arrest all the scientists. He calmed down when they explained that the farmer would burn his trees no matter what the SCAR-B team did.

Afterward, Damian and I accompanied Brent Holben, Tom Eck, and a driver on a trip into the Pantanal with their instruments and ours. This was the dry season, and the scary waters of the Panantal we had seen in the travel guide's photos were confined to ponds sprinkled across the landscape. Our objective was to make sky measurements when a NASA satellite and two instrument-equipped aircraft, a Convair C-131A and a Lockheed ER-2, flew overhead. After all the instruments were set up, we watched tropical birds and hundreds of alligator-like caimans lounging around a large pool while we waited for the planes to arrive.

When a fish jumped from a pond, half a dozen caimans plunged after it. While we were packing the instruments after the overflights, the group was entertained by the spectacle of a colleague fleeing a huge sow whose five piglets he had stumbled on while removing an instrument from a post. That is the first time I have had to drop a delicate instrument and run for my life.

The Rio Negro

After a week at Cuiabá, we and three sleepy NASA scientists boarded a midnight flight to the remote city of Porto Velho. Before landing, the plane circled low over the runway to make sure no cattle were present. After we stopped at the gate, the NASA guys were so exhausted, I could not wake them. A persistent flight attendant finally shook them awake and made them leave the plane so we could take off for Manaus.

After they stumbled down the stairs, we watched through a terminal window as they collapsed on the concrete floor to continue their sleep. Damian and I continued to Manaus, where we planned several days of measurements from a tree house hotel on the Rio Negro. We arrived in Manaus around 4 a.m., waited for sunrise to make sun measurements from the roof, and found a hotel, where we slept until late that afternoon.

The most significant scientific finding back in Cuiabá was that the thick smoke blocked most of the sun's ultraviolet radiation. Over a supper of funny-tasting pizza, I wondered what the instruments would reveal during the boat trip to our jungle hotel just below the Equator. Farther south, smoke from thousands of acres of burning forest shrouded the land of parrots and quartz. Would the sky over the mighty Rio Negro also be smoky?

Robber Monkeys and Black River Piranhas

The Rio Negro is so huge, it resembles an endless lake more than a river. We were cruising up the black river, just above its confluence with the Solimões River, where it becomes the mighty Amazon. The water resembles very dark tea, a byproduct of tannin from decayed vegetation.

We saw no smoke during the voyage, and the wide river provided an open sky perfect for making ultraviolet measurements. Surprisingly, the ultraviolet so near the Equator was much less intense than in Hawai'i and similar to a summer day in Texas. This was because the high humidity formed haze that blocked up to 20 percent of the sun's UV.

After a few hours, we turned south into the Rio Ariau, a tributary of the Rio Negro. Soon, we spotted the tall observation tower of the Ariau Towers hotel emerging from the bright green forest. The hotel (later much expanded and renamed the Ariau Amazon Towers Hotel, but now closed) was constructed almost entirely among the trees growing along the river.

Clusters of tiny rooms were connected by catwalks to a multistoried complex containing a registration area, a restaurant, and a conference room. Window screens and double doors kept the monkeys out. Our room included a prominent sign that read, WHEN LEAVING THE ROOM PLEASE LOCK THE DOOR, BECAUSE THE MONKEYS CAN CAUSE A CONSIDERABLE AMOUNT OF DAMAGE WHICH IS ALMOST THE SAME AS AN ELEPHANT.

My original plan was to mount ultraviolet instruments high atop the tall tower, which Damian and I promptly climbed. But it did not take long for monkeys to convince us to change plans and study the ultraviolet down in the rain forest. Experiencing the Amazon rain forest up close was an adventure of its own. The humidity was so high that our shirts and my notebook were quickly soaked.

The next day, we boarded the hotel's riverboat for a short cruise down the Ariau and across the Rio Negro. On the way, we watched smiling women washing the hotel's snow-white sheets and towels in the black water. As we approached the north bank of the Rio Negro, we saw a motorized canoe with six European tourists and a pilot. It was the first boat we had seen since leaving Manaus. The passengers looked terrified as they furiously bailed water from their tiny vessel. They had good reason to be working so hard, for the overloaded boat was within a few inches of sinking. Their pilot was providing them an Amazon experience they would never forget.

Our guide took us to a family that employed ancient slash-and-burn agriculture to raise crops. Corn was growing among the trunks of large trees they had cut to make room and allow sunlight for their crops.

Back at the treehouse hotel, three terrified Spanish women would not enter their room, because a long green-and-yellow snake was emerging from under their door. Sensing a photo opportunity, I told Damian the fun would begin when the hotel staff arrived to remove the trespasser. We were soon treated to pandemonium and ear-piercing screams as a porter tried to catch the snake while the women attempted to climb the walls.

That evening, a guide took us piranha fishing in a small boat. He passed around chunks of chopped chicken, which we speared on our hooks and dropped over the side. After a few minutes with no nibbles, there was a loud *whack-whack-whack*, and we looked back to see our guide slapping the water with his fishing pole. Within seconds, he had hooked a foot-long piranha. He set the hook by slinging the fish out of the water and between us in the boat. Its protruding, triangular teeth killed any desire to touch the fish. Instead, we began whacking the water with our poles.

Back at the hotel, the cook prepared our piranhas for supper while we watched the monkeys and macaws. One of the monkeys took a liking to Damian and jumped into his arms. I quickly left, for we had been warned that the monkeys enjoyed emptying visitors' pockets and tossing their loot into the river below.

Damian was more fortunate. Instead of stealing from him, his new friend gave him gifts, which he discovered as we prepared to eat supper when some of those gifts began jumping from his shirt onto his plate. Suddenly, Damian shouted, "Fleas! Fleas!" And flee we did, as those of us at the table raced to another one on the opposite side of the dining area.

After three days at the jungle hotel, we left the sweltering Amazon and Manaus and returned to smoky Cuiabá for a few more days of observations. Soon, we were flying home with thousands of measurements stored in three miniature computers. It was night, and a look through the plane's window revealed familiar orange-and-yellow patterns scintillating far below. South of the Amazon, Brazil was still burning.

The 1995 Brazil Smoke Report

Back home, six weeks were required to condense all the data into a thick technical report for NASA under purchase order No. S-59036-Z. (The number is provided because a Wikipedia editor once deleted another editor's reports about my Brazil trips because there was no proof NASA had sent me to Brazil.) The report should have been called "Smoke vs. Sunlight." Instead, I called it "Aerosol Optical Depth, Ultraviolet-B, and Total Sky Irradiance during SCAR-B." The report included dozens of graphs and tables of data, and acknowledgments of the many scientists who had made suggestions about my measurement protocols.

The trip would have been impossible without the assistance of Damian Kilday, and I acknowledged how he "performed virtually all LED sun photometer and Microtops-10 observations. He also performed various other observations and assisted in logistical arrangements and data entry and analysis in the field."[1]

Among the principal findings was that at noon on a smoky day in Cuiabá with the same amount of ozone overhead as during a clear day at Hapuna Bay in Hawai'i, the solar UV-B at Cuiabá was only 16.6 percent of that at Hapuna Bay. This seemed significant enough to warrant some speculations about the impact of smoke-reduced solar UV on the environment. In a paragraph titled "Human Health," I wrote:

> *Various newspaper and anecdotal reports indicate a significant increase in respiratory ailments in Rondonia during the burning season of 1995 (see, for example, "Fires Raging in Amazon Region," New York Times News Service, 13 October 1995.). Can the intensity of respiratory difficulties be correlated with the aerosol optical depth of the atmosphere? Does the significant reduction in UV-B permit the survival of airborne pathogenic microorganisms? What is the population of microorganisms on clear and on smoky days? . . . it is important that any future version of SCAR-B include an investigation of airborne, surface and soil bacteria and other microorganisms under varying conditions of AOT, hence solar irradiance . . . .*[2]

I closed the report by offering to organize and carry out this suggestion. Brent Holben received the report on November 11, 1995, and wrote, "Excellent job. Your report is very much appreciated and the biological emphasis raises questions that few people from NASA have attempted to study . . . ."[3]

I hoped that the report would receive wide circulation within NASA, but that did not occur. While the INPE team also measured UV-B with its Brewer instrument at Cuiabá, its measurements and mine and their possible connection with infectious respiratory disease were not a listed goal for SCAR-B. Yet looking back, SCAR-B provided the experience that would be helpful for a future Brazil campaign.

---

1. F. M. Mims III, *Aerosol Optical Depth, Ultraviolet-B, and Total Sky Irradiance During SCAR-B*, final report for NASA purchase order No. S-59036-Z, 1995.
2. Mims, *Aerosol Optical Depth.*
3. Brent Holben, email to Forrest M. Mims, November 11, 1995.

The International Radiation Symposium

In August 1996, Saul Berger of Solar Light paid travel expenses for me to attend the International Radiation Symposium (IRS) at the University of Alaska Fairbanks to give a talk on my 1995 ultraviolet findings in Brazil. This was an important opportunity to promote my hypothesis about smoke, pathogenic bacteria, and human health. The abstract read:

> *Smoke from biomass burning caused very significant AOT [haze] and up to a 74% and 81% reduction, respectively, in UV-A and UV-B . . . [at] Cuiabá, hundreds of kilometers from the most widespread burning. An increased incidence of respiratory, cardiopulmonary, and other disease is associated with severe air pollution, but the responsible biological mechanisms are unknown. The bactericidal and viricidal effects of solar UV-B are well known, and significantly reduced UV-B resulting from severe air pollution in cities and regions where UV-B levels are ordinarily high might enhance the survivability of pathogenic organisms in air and water and on surfaces exposed to sunlight.*[4]

The conference also provided an opportunity to again meet some of the world's leading UV experts. They included the prominent scientist who had invited my paper and who had made fun of an idea I had recently suggested during a meeting in Boulder with NOAA scientists. My idea was for students to measure UV using simple instruments based on the design I presented in *Scientific American*. His response was, "But won't they get peanut butter on the instruments?"

The group laughed at his response, but they did not laugh at mine. Shortly before the meeting, NOAA ozone expert Robert Evans had taken me to the roof of NOAA's building to see some instruments. I was surprised that the wind was blowing a fine mist from an air-conditioning system over a row of their expensive UV sensors, so I responded to the peanut butter comment by saying that students would never make UV measurements downwind from a misty air conditioner like the one on their roof. The group's laughter was replaced by total silence.

4. F. Mims, "Biological Effects of Diminished UV and Visible Sunlight Caused by Severe Air Pollution" (International Radiation Symposium, Fairbanks, Alaska, August 19–24, 1996).

18

SCIENCE ADVENTURES AND A MISADVENTURE

My science projects and field trips kept life interesting. Preparing papers for scientific journals was far more difficult than writing magazine articles and Radio Shack books. But those papers were key to establishing the goal set after the *Scientific American* affair: A self-taught individual who rejects Darwinian evolution and opposes abortion can invent scientific instruments and use them to make discoveries.

I was impressed that few of the many professional scientists I met were concerned about my beliefs. Some even privately agreed with them. Their main interest was my science, and that was demonstrated by many collaborations and assignments, including several years as chair of the Environmental Science Section of the Texas Academy of Science.

As you may recall from my discovery of an error in data from NASA's ozone satellite (discussed in chapter 14), professional scientists occasionally disagreed with my findings. While NASA's ozone team members were gentlemen, I've had some negative experiences with some environmental scientists employed as government regulators. A classic example follows.

The Defective Ozone Monitor

As related earlier, after nearly failing first-year algebra at Texas A&M in 1962, I changed my major from physics to government. I told my father that governments need people who understand science, and I learned why during fifteen years of voluntary service on a regional government air quality committee.

In 2002, one of San Antonio's three EPA ozone monitors (CAMS 23) measured record-high ozone. While the regulators took the unprecedented readings seriously, I wondered how San Antonio could have higher ozone levels than Houston. I also wondered why only one of the three monitors indicated record-high ozone.

After examining the data, on September 25, 2002, I wrote the chief ozone scientist at the Texas Natural Resource Conservation Commission (TNRCC) in Austin, "Ozone measured by CAMS 23 during June–September 2002 could have a mean positive bias of at least 11.4 ppb when compared to the same months of 1997–2001."[1]

The scientist said my analysis did not follow an approved method. Nevertheless, on October 1, 2002, the TNRCC sent two technicians to CAMS 23 to audit the station. They found that CAMS 23 had a positive bias of 13.6 percent, 2.2 percent higher than what I had discovered. They also found that CAMS 23 indicated +7 percent ozone concentration when exposed to air with no ozone.

I erroneously assumed that the defective data would be abandoned, but on October 4, 2002, TNRCC chairman Robert J. Huston notified the Alamo Area Council of Governments (AACOG), "... we have an internal goal of achieving +/− 15 percent accuracy in measurements compared with an EPA requirement of +/− 20 percent. All of the CAMS 23 data for 2002 meet the [Texas Commission on Environmental Quality]'s target confidence interval . . . .Thus, we have concluded that there is no basis to invalidate any of the 2002 data."[2]

This decision raised a major question in my mind about the quality of ozone data being used to regulate businesses and the public. After Seguin's city council voted unanimously to place me on AACOG's Air Improvement Resources (AIR) Technical Committee, I learned that the TCEQ and the EPA required AACOG's modeler to include the erroneous CAMS 23 data in his modeling, since it was within the EPA's +/−20 percent tolerance. This decision was incredibly wrong, for it seriously impacted air quality planning.

On September 25, 2003, I wrote to the EPA, ". . . The ozone analyzers used by the EPA and State agencies have a typical accuracy of +/−1 ppb. . . . The EPA allows a calibration tolerance in ozone measurements of +/− 20 percent. When I disclose this to elected officials, the media and various scientists, the response is always laughter, especially when I then state that third grade students can provide higher quality ozone measurements (+/− 10 percent) than the EPA using improved paper test strips developed with funding from NSF (National Science Foundation) and NASA."[3]

The EPA did not appreciate my letter and continued to require that AACOG models use all data within its +/− 20 percent tolerance. I continued to protest, and the EPA eventually reduced the allowable error range to +/− 7 percent. Others must have also protested, for the EPA now requires that the tolerance of ozone measurements must be within +/− 3 percent.

1. Forrest M. Mims, letter to the Texas Natural Resource Conservation Commission, September 25, 2002.
2. TNRCC chairman Robert J. Huston, letter to the Alamo Area Council of Governments, October 4, 2002.
3. Forrest M. Mims, letter to the Environmental Protection Agency, September 25, 2003.

After serving a year as Seguin's representative on AACOG's AIR Technical Committee, I began representing Guadalupe County. This became a voluntary fifteen-year adventure in government science that corroborated what I had told my father in 1962: Governments need people who understand science.

For several years, I was met with a hostile reception at AIR Tech meetings, even though I was the only member who measured air quality. I was also the only nongovernment employee on the committee, whose leadership consisted of employees from San Antonio and Bexar County, some of whom laughed and joked about cities outside San Antonio as "one-stoplight towns." The citizens and taxpayers who paid their salaries were laughingly referred to as Joe Six-Pack and John Q. Citizen.

AACOG's small professional air-quality staff had better manners, especially Peter Bella, their supervisor, and modeler Steven Smeltzer. While they were good at doing emission surveys and modeling, they were poorly informed about the transport of polluted air into South Texas from coal-burning power plants in the Ohio and Tennessee Valleys. So, I introduced them to the US Naval Research Laboratory's website, which forecasts the daily presence of pollutants, dust, and smoke across the country.

During one meeting, I was ridiculed by a member for informing the committee that biomass smoke from Mexico that arrives in Texas during spring can cause ozone. He was unaware of scientific studies and papers that affirm this. Eventually, however, the committee members realized that my concerns were valid, for every ozone violation in San Antonio occurred during days when polluted air arrived from sources hundreds of miles away.

Solar Eclipses

Finding waves in the ozone layer during the 1991 total solar eclipse (an event discussed in chapter 14) greatly enhanced my interest in these rare celestial events, and I made plans to observe and measure the May 10, 1994, annular eclipse. Significant cloud cover was forecast along the eclipse track across New Mexico and Texas, and considerable clouds were evident when I drove into Lubbock, Texas, the day before the eclipse.

I found the office of the National Weather Service in Lubbock and explained the measurements I planned. A friendly meteorologist took me to the computer-filled work area and showed me satellite imagery of the cloud conditions. After the staff discussed the situation, they said my best chance would be Las Cruces, New Mexico. I thanked them and headed south.

The Lubbock forecasters were right. The sky was nearly free of clouds over Las Cruces the next day, and I collected excellent data during the entire eclipse.

The track of the May 20, 2012, annular eclipse was directly over Albuquerque, so Minnie and I headed there. We reserved a room at a hotel overlooking the west side of town, where I was able to get nice photos of the various stages of the eclipse.

The track of the partial solar eclipse of October 23, 2014, was over Cloudcroft, New Mexico, as well as White Sands. Minnie and I photographed the eclipse from White Sands, and the following day, I showed a few of my photos during a talk I gave in Cloudcroft during the 2014 Solar Eclipse Conference.

The tracks of the next two solar eclipses across the US will be very near my place in Central Texas. The first is the annular eclipse of October 14, 2023; the second is the total eclipse of April 8, 2024. As of this writing, Minnie and I plan to be there.

Mysterious Twilight Glows

During September 1995, I began noticing bright, long-lasting twilight glows nearly as spectacular as the brilliant twilights that occurred in 1991–92 following the volcanic eruption of Mount Pinatubo in the Philippines. I notified some of my contacts at NASA's Goddard Space Flight Center, but they were not seeing such twilights and suggested they were caused by plumes from coal-burning power plants along the border of Texas and Mexico.

They were not impressed when I told them that the long duration of the glows suggested an aerosol layer in the stratosphere many times higher than a power plant's plume. The absence of any major volcanic eruptions supported their skepticism, but they worked in windowless offices and drove home from work under polluted skies, so they did not see what I was seeing.

The internet was still in its infancy in 1995, but I managed to find an online bulletin board called Usenet, where I posted a notice about the twilight glows. Soon, I heard from Robert Roosen and Carolyn Meinel, both of whom were also observing long-lasting twilight glows.

Roosen's messages were a major surprise, for two of his papers about the old Smithsonian Astrophysical Observatory sun measurements played important roles in my atmospheric-monitoring program. Carolyn Meinel was the daughter of astronomers Aden and Marjorie Meinel, authors of *Sunsets, Twilights, and Evening Skies*, one of my favorite books. By December 28, 1995, I completed "Stratospheric Aerosol Cloud of Unknown Origin," a paper that listed Meinel, Roosen, and several other coauthors.

I sent the paper to *Nature*, but its anonymous reviewer was unimpressed. He criticized the twilight lengths we had all measured as anecdotal and noted the absence of satellite data confirming what we were seeing. He was unaware of satellite errors like the many I had identified, and he recommended that the paper be rejected.

Meanwhile, Meinel, Roosen, and I continued contacting satellite and lidar scientists to determine whether any of them had found evidence of new stratospheric aerosols. We finally succeeded when Cuban atmospheric scientists Dr. René Estevan and Dr. Juan Carlos Antuña-Marrero reported that the old Russian lidar they used to probe the sky had detected an aerosol layer ranging in altitude from 9 to 11 miles (14 to 17 km). I revised our paper with this new information, but *Nature* was not interested, so we chose the Smithsonian Institution's *Bulletin of the Global Volcanism Network*, which published our findings and two of my charts supporting what we found in its Atmospheric Effects section.[4]

The lesson in this story is the same I taught each of our three children during their science fair years and have mentioned previously: If you are confident of your data but scientists are not, believe your data and not the scientists. This lesson was further affirmed while writing this memoir when I reestablished contact with Carolyn Meinel. The folder with Carolyn's information includes a March 25, 1996, letter to Dr. Larry Thomason, a NASA satellite scientist, in which I wrote, "There are no known volcanic events that can explain the enormous increase in optical depth observed from Cuba [with its lidar]. I have proposed as a source for these aerosols the emission of gaseous aerosol precursors from increased (perhaps record) biomass burning in the tropics."[5]

None of the scientists I approached agreed with this hypothesis, including one who was visiting the Mauna Loa Observatory. I told him that that the smoke over Brazil during my 1995 and 1997 campaigns was so thick that some of it must have reached the stratosphere. He said that was impossible, which convinced me it was probable. This was on Carolyn's mind, and on September 26, 2019, she wrote to me, "It turned out the twilight glows [in 1995] were caused by extreme forest fires in the Amazon injecting particulates into the stratosphere. Recently a paper was published on the mechanism whereby carbon particulates from such fires are lofted into the stratosphere. See attached. I thought you might like to read about what you and I were among the first to observe."[6]

More of Brazil's rain forest was burned in 1995 than in any year before or since, and I was there for three weeks in August during the peak of the burning season. I immediately downloaded the paper, which began, "In 2017, western Canadian

---

4. Forrest M. Mims et al., "Stratospheric Aerosol Cloud of Unknown Origin," *Bulletin of the Global Volcanism Network* 21, no. 2 (February 2, 1996).
5. Forrest M. Mims, letter to Larry Thomason, March 25, 1996.
6. Carolyn Meinel, letter to Forrest M. Mims, September 26, 2019.

wildfires injected smoke into the stratosphere that was detectable by satellites for more than 8 months. The smoke plume rose from 12 to 23 kilometers within 2 months owing to solar heating of black carbon, extending the lifetime and latitudinal spread."[7] This paper by Pengfei Yu and thirteen coauthors, published in 2019, likely solved the mystery of those brilliant 1995–96 twilight glows Carolyn Meinel, Robert Roosen, and I had observed twenty-four years earlier.

The World Trade Center

During a visit to New York City on January 28, 1996, I visited the World Trade Center on a chilly, windy afternoon to make measurements from the base of the towers and the observation deck atop Tower 2. I took the long elevator trip to the 107th floor, a glass-enclosed observation floor near the top of the huge building, which was made from 100,000 tons of steel and 212,000 tons of concrete. An escalator led from the 107th floor to an outdoor observation deck 1,377 feet (420 m) above an unforgettable view of New York City.

The observation deck was closed due to wind, but the sky was clear and deep blue, and I was determined to measure its clarity. Yvonne Correa, the young woman in charge of access to the deck that day, was intrigued by my research, and agreed to escort me up the closed escalator, unlock the heavy door, and watch from inside as I walked onto the windy deck.

I completed the measurements in less than ten minutes, took a quick look at the city spread out below, and hurried back to the door, where the smiling supervisor was waiting. I showed her the entries in my notebook and entered her name. I then rushed back to the elevator to return to the plaza below to repeat the measurements at street level exactly twenty minutes later. The measurements showed that there was very little ozone in the air at street level that day. There was also relatively little haze, with the sun at street level being only around 3 percent dimmer than at the observation deck.

After the 9/11 attack on the World Trade Center, I wondered whether Yvonne Correa had been one of the victims. Fortunately, her name was not on the list of casualties.

From Hawai'i to Alaska

During May 1996, I taught the experimental science course at the University of the Nations and stayed over to calibrate my instruments at the Mauna Loa Observatory.

7. Pengfei Yu et al., "Black Carbon Lofts Wildfire Smoke High into the Stratosphere to Form a Persistent Plume," *Science* 365, no. 6453 (August 9, 2019), 587.

In June, Bill Aldridge, executive director of the National Science Teachers Association, and I met with Radio Shack staff in Fort Worth to discuss educational opportunities. My hope was that Radio Shack would offer a wide range of books, kits, supplies, and display boards for the one million or so students who do science fair projects each year. Radio Shack's book and parts buyer liked the idea, but the higher-ups showed no interest then or at any of our follow-up meetings. Perhaps Radio Shack might have survived had the company expanded the best-selling books I wrote for them with a wide range of educational products.

The Compass in the Sky

If you wear polarized sunglasses and look up at the sky after sunset, you will see a compass in the sky in the form of a broad, dark band crossing the entire sky from approximately north to south. I spent considerable time learning to measure the polarization of the sky, which is a strong indicator of air pollution and which helps ants, bees, wasps, and birds navigate from one place to another.

After sunset one evening during my visit to Fairbanks, Alaska, I hiked to a nearby lake and watched geese taking off for their fall migration. I put on a pair of polarized sunglasses and saw that the geese were flying perfectly parallel with the compass in the sky. The sun had set at 237 degrees southwest, which meant the sky compass was pointed at the western US.

That is also where my research compass was about to be pointed, for two weeks later, I would be measuring forest fire smoke in several western states. During the IRS meeting at the University of Alaska Fairbanks, NASA's Dr. Pawan K. Bhartia took serious interest in my ultraviolet–disease hypothesis and mosquito project. Bhartia had approved the 1995 trip to Brazil, and he offered a $5,000 assignment for me to measure smoke over Brazil when NASA's new ozone satellite was overhead. Christina Hsu of Hughes STX, a NASA science contractor, had developed a clever way to detect smoke from the new ozone satellite, and my role would be to provide measurements from the ground while the satellite was passing overhead.

Unfortunately, there was little time to organize a major trip to Brazil and arrange for the required invitation from a scientist or institution, so I proposed an alternate plan while watching thick layers of smoke from major forest fires in Montana while flying back to Texas. I wrote AERONET director Brent Holben, "There are HUGE forest fires in the west. While flying from Seattle to Memphis yesterday, I photographed smoke from 37,000 feet that completely obscured the terrain. . . ."[8]

8. Forrest M. Mims, letter to Brent Holben, August 25, 1996.

Instead of flying to South America with little preparation, I wrote, why not study smoke from domestic forest fires? The NASA team approved on August 30, and on September 3, I flew to Reno, Nevada, with my instruments, rented a car, and spent the night at nearby Sparks, an appropriately named place for a fire chaser. The next day, I headed east to get in position under the massive smoke plume from the Ackerson Complex Fire, which eventually consumed 59,606 acres of forest, most of which was in Yosemite National Park. The smoke covered the entire sky, but it was too thin to provide useful data. To get closer to the fire, I headed for Bridgeport, California. The smoke was thicker, but nothing like what I had seen in Brazil.

Ordinarily, I plan trips well in advance, but forest fires do not allow planning, so I left my bags by the office door, ready to leave for the airport as soon as notice was received about the next big fire. Two days later, NASA called. I ran to the house to hug Minnie, grabbed the bags, jumped into my pickup, and drove to San Antonio International Airport. When I explained the purpose of the trip, the ticket agents were immensely helpful with last-minute reservations. I visited fires in Utah, Wyoming, and Montana, but their smoke was too dispersed for useful data.

NASA's purchase order No. S-78417-Z was finally approved on September 10, the day before I returned home from the second smoke trip. The work statement included a list of specific consulting services, tasks, measurements, and studies I had already been performing during the two smoke trips.

Forrest M. Mims Jr. (1923–1996)

Daddy and I enjoyed trading stories about our time in Alaska and Vietnam, and he provided sound advice about how best to respond to *Scientific American* during the busy months following that affair. He advised me to turn down suggestions that I pose as a tourist while doing smoke research in Brazil. He also accompanied me on a field trip to make measurements in New Mexico.

But Daddy's time had come, and late on September 27, 1996, some of us were standing by the hospital bed where he had laid comatose for several days when he unexpectedly opened his eyes and turned his head from side to side to look at us. He then raised his head from the pillow and, with his mouth and eyes wide open, as if in awe, looked up at the ceiling and slowly fell back on the pillow. My nephew Daniel Jimenez closed his eyes.

Daddy's death was devastating, and I was overwhelmed with grief. He was much more than a father; he was my closest friend and adviser who had introduced me to science and mechanics, taught discipline and loyalty, fully practiced his gifts in architecture and artistry, and lived his Christian faith.

The grief that followed the loss of my mother two years before was lessened by Daddy's presence. He was always there when advice was needed. He continued the family Thanksgiving, Christmas, and New Year dinners. And he was present when Sarah was baptized on Christmas Eve of 1995. Now he was gone.

As his oldest son, I had prepared a eulogy for Daddy I was expected to read during the funeral service. I asked Eric to come to the podium and take over if I could not continue, and halfway through, I could speak no more. But when I looked toward Eric, I felt the pastor's hand on my shoulder and heard him say, "Go ahead, son." The pastor sounded just like Daddy before his throat surgery, and I was able to continue. After the service, I asked Minnie, Eric, and Daniel whether they had noticed when the pastor placed his hand on my shoulder. They all said he had been seated the entire time I was at the podium.

The Out-of-Control Controlled Fire

Before Daddy's death, NASA had been sending notice about major fires. But by the time I arrived, there was not enough smoke, so I revised our protocol and established direct contact with the National Interagency Fire Center to find out about the next big fire. On October 8, a call arrived, and within several hours, I was in the air, on the way to the Fayette Fire, near Pinedale, Wyoming.

A controlled burn by the Wyoming Game & Fish Department on a windy day had gotten out of control and was burning a fire truck and half the timber on Half Moon Mountain. This fire was different from all the others, in that its smoke formed a thick blanket directly over a long dirt road. I was in position, collecting good data, two days after the fire began.

After packing up the instruments and driving several miles (5 km), I pulled off the road to take some final photos of the smoke plume. Daddy had provided good advice before the first two forest fire trips, and I could hardly wait to tell him about the success of this trip. But then I remembered he was gone.

The 1996 Smoke Report

The cover of my 1996 smoke report for NASA featured a dramatic image of the smoke plume from the Fayette Fire. The title read, *Aerosol Optical Thickness, Total Ozone, UV-B, Diffuse/Total Solar Irradiance, and Sky Polarization Through Forest Fire Smoke and Stratospheric Aerosols During TOMS Overpasses.*

Though I was unaccompanied while visiting seven forest fires for this study, the entire project was very much a team effort. While compiling all the data for the report, which included additional studies not covered here, I was surprised by the number of people who deserved inclusion in the acknowledgments. I also added a dedication: "This report is dedicated to the memory of my father, Forrest M. Mims, Jr. (1923–96), who took great interest in this study and provided helpful suggestions about its implementation."[9]

"The Creationist Left Last Night!"

Shortly after the third smoke trip of 1996, I was asked to give a talk at the National Conference on Student and Scientist Partnerships (NCSSP), funded by the National Science Foundation (NSF) under the leadership of Dr. M. Patricia Morse. The meeting, held in Washington, DC, October 23–25, was sponsored by Dr. Susan Doubler of the Technical Education Research Center and Robert Tinker of the Concord Consortium.

I had experienced several science adventures in Washington, DC, and I looked forward to the trip. The best Washington adventure was the 1980 Photophone Centennial. Three others were when I measured the clarity of the sky from where the first such measurements were conducted by the Smithsonian Institution's Samuel Langley a century before.

Still another adventure occurred when I walked by the White House and noticed circular holes in the masonry structures that separated sections of the steel fence. Could those holes conceal a laser intrusion alarm? To find out, that night, I returned to the White House with my infrared scope and began looking at one of the suspicious holes. Before you could ask, "What are you doing?" a security guard approached from behind and asked, "What are you doing?" I told him, and he politely, but firmly, said I should not be doing what I was trying to do.

On the second morning of the NCSSP meeting, Morse gave opening remarks, and I followed with a keynote talk on measuring haze with sun photometers. Having just returned from the Wyoming fire, I was able to provide a firsthand description of how NASA validates satellite measurements of smoke from the ground. This generated considerable interest, and, following my talk, the conference leaders flashed on the screen an overhead transparency they had just made. It read, PhD by the Authority invested in We by the SSP.

The fake doctorate was signed by Robert Tinker, Sue Doubler, and M. Patricia Morse, all of whom held real doctorates. They then called me forward and gave me a paper version of my new honorary diploma, which all three had also signed.

Before this conference, none of the many professional scientists I had personally encountered after the *Scientific American* affair had been critical of my position on

---

9. F. M. Mims III, *Aerosol Optical Thickness, Total Ozone, UV-B, Diffuse/Total Solar Irradiance, and Sky Polarization Through Forest Fire Smoke and Stratospheric Aerosols During TOMS Overpasses*, final report for NASA purchase order No. S-78417-Z, 1996.

Darwinian evolution, and some privately stated or wrote that they also had serious questions about evolution. But something new and unexpected occurred during this conference, and I sent Mark Hartwig the details:

The day the conference began, Dr. [M. Patricia] Morse had very enthusiastically greeted me and stated one of her staffers had "all your books." She was very complimentary about my science education work.

During my talk, I politely disagreed with a point Morse had made about not liking the competition of science fairs. I simply pointed out that scientists compete for prizes—like the Nobel Prize. Teachers also compete for recognition. Students should do likewise, because it's the real world and because they deserve recognition. Neal Lane, then NSF director, was present. I met him later, and he was quite friendly. He seemed to know about the SCIENTIFIC AMERICAN matter. At least that is what I sensed.

In a planning meeting for the final session to be held the last morning at the very elite NAS [National Academy of Science] auditorium, it was decided I would speak first and show my mosquito demonstration (color filters and larvae on an overhead projector, pretty neat to watch the larvae swim for the red and blue—which they cannot see and perceive as black). There was then a discussion about using retired engineers and scientists to teach science around the country.

Morse stated her strong opposition to this idea, since they might end up hiring some creationists. "You just don't know," she said. A woman to my right then stated that some Native Americans have creationist traditions. Morse said that made no difference to her. She was quite firm about this. Then Barbara Tinker. . . stated that there are even some well-known science educators who are creationists. She said we might even know some of them without realizing it. She and [her husband] Bob knew all about SCIENTIFIC AMERICAN (we had discussed it). Clearly she was covering for me, but I do not hide behind a lady's skirts, even Barbara Tinker's. So, I looked across the table at Morse and simply stated: "I lost my position at SCIENTIFIC AMERICAN because I reject Darwinian evolution."

Morse became white as a sheet, and her expression froze on her face. She began to shake and said she would have to leave. First, she said that the arrangement of speakers for the next morning was inappropriate, but no one replied. She specifically did not want me to speak. Everyone was silent. After all, she was the NSF rep for the entire conference. She put up the tax dollars to pay our expenses. Then I believe it was Susan Doubler who politely disagreed and pointed out the interest generated by my mosquito demonstration.

Morse then quickly left the room. I told Barbara I would be right back and followed her out the door. Morse was nearly in tears and shaking. I spoke to her about the discrimination she had experienced as a young female scientist (she had talked about this) and said, "I know exactly how you feel about discrimination." I then hugged her. . . . Mark, she did not merely look like she was shaking. She was shaking. She then seemed to recover somewhat, said she would be okay, and walked down the hallway. I returned to the room and the meeting continued—but with a somewhat more subdued tone. (I was the only male. Leave it to us to make a woman cry.)

The incident the next morning occurred as I was seated on the front row [of the NAS auditorium] reviewing overheads and making notes. The larvae demonstration was all set up and you could see them swimming on a huge screen. (Very fancy auditorium.) Morse walked by followed by [Rodger] Bybee [Executive Director, Center for Science, Mathematics and Engineering Education, NAS] and did not say a word. As they walked within feet of me, Bybee loudly stated, "I hear the creationist left last night!" (I wrote his exact words down on my notecard, which is around here somewhere. I do not know who told him, as the meeting the previous evening had been fairly late.) Morse spun around and pushed him away, mumbling something under her breath. They went outside. A few minutes later, they came back. Bybee stared right at me as he walked by and then went up on the stage. Morse went down a side aisle. Bybee kept looking my direction during the introductions. It was then that I decided to abandon my simple talk and go fairly technical about some of my instrument developments and findings in Brazil and so forth. I had just come from a NASA sponsored trip to measure some big forest fires out west, so I discussed that also. I emphasized the biological effects of smoke on mosquitoes and discussed my suppressed UV-bacteria-disease hypothesis. Fortunately, I had some very nice overheads to support this more technical talk.

After the talk I spoke to Morse, who seemed quite composed. When I asked about doing student work using Sun photometers, she said I should propose that to TERC for the GLOBE program. I tried to ask a follow-up question, and she pointed toward Dan Barstow [of] TERC, said I should speak to him and then walked away. The exchange was much more formal than before she found out about my tainted background.

The proceedings of the conference were published in "National Conference on Student & Scientist Partnerships," (National Science Foundation, 1997). My paper "Science that Happens through Partnership" is on pp. 58–61. They even included pictures of me and the mosquitoes (p. 59) and Microtops and me (p. 60). But they forgot to mention the incident late on the evening of 24 October 1996 in a little conference room at the Georgetown Holiday Inn.

. . . More than a year later Barbara Tinker told me that Morse was very remorseful about her reaction. Barbara said something about it being the biggest mistake of her career. Well, maybe this was said to Barbara. But Morse never told me. Someday I would like to visit her (and Jonathan Piel) to renew acquaintances and see what these folks think some years down the road. This approach certainly worked well with Martin Gardner, who agreed to correct some of his erroneous writings about me. Not long ago he even asked my advice about an article he was writing on the life of Hugo Gernsbeck [sic].[10]

I followed through on Morse's suggestion that I talk to TERC about using LED sun photometers in the GLOBE program for students around the world, and that became an active eight-year project with Dr. David R. Brooks. Thus, the Morse experience had an upside. It also reinforced lessons I had earlier learned from Jonathan Piel and later from various academics: Some true believers in scientific paradigms resemble devoted religious believers defending their faith; they are especially conflicted by someone who rejects a key scientific paradigm while still managing to be published in leading journals of science.

More important than those lessons was that Morse was the only science educator in the planning meeting who wanted to cancel my presentation at the NAS auditorium. During all the publicity about the *Scientific American* affair, I was concerned that Radio Shack, my book and magazine publishers, and the newly developed contacts at NASA and NOAA would want nothing more to do with me. But there was never a hint of this.

The readers of my books and articles and the editors, journalists, scientists, engineers, and fellow authors with whom I dealt never made an issue of the *Scientific American* affair. Nor did Rolex, the National Science Foundation, the National Science Teachers Association, and the various professional organizations I had joined. A few curious professionals asked questions about what happened, and I was roundly attacked by some skeptics. But only Jonathan Piel and M. Patricia Morse banned or attempted to ban me from a professional position or duty.

I closed out 1996 teaching my science short course at the University of the Nations campus near Lausanne, Switzerland, to an open-minded collection of university students from several countries. As with all my UofN courses, I ended with the *Scientific American* story. And, as always, the international students ranked it as the highlight of the course.

10. Forrest M. Mims, report to Mark D. Hartwig, May 23, 2000.

19
BACK TO BRAZIL

Following SCAR-B and the IRS paper, I became increasingly concerned by the apparent connection between ultraviolet reduced by smoke and human health. Brent Holben and some of his Goddard Space Flight Center colleagues were intrigued by my idea. While they had no authority to assign me to conduct a health study, they could send me back to Brazil to repeat the measurements I had made in 1995. A survey of the relationship of airborne bacteria to the solar UV that managed to leak through the smoky skies would be a bonus. Brent suggested that the proposed study be conducted at Alta Floresta, a remote town in the Amazon basin 400 miles (644 km) north of Cuiabá.

Unfortunately, NASA was not allowed to conduct smoke research in Brazil in 1997, so I again sought advice from my well-traveled scientist friend who had advised me in 1995 to visit Brazil as a tourist. That was a bad idea in 1995 and a worse one in 1997, for a prominent Brazilian scientist had warned me to avoid entering the country as a tourist when doing science was the real purpose.

Therefore, I wrote Dr. Paulo Artaxo, a prominent atmospheric scientist at the University of São Paulo, to ask permission to do smoke research in Brazil. Artaxo had participated in SCAR-B, and he agreed to send the invitation required by Brazilian law.

The 1995 trip would have been impossible without Damian Kilday's help, and it was essential to find a student who could take Damian's place. Fortunately, Bradley White volunteered for the assignment. Like Damian, Brad had just completed high school, and he was willing to delay his first semester at Texas Lutheran University a few weeks to make the trip.

While I was acquiring a visa and preparing the instruments, Brad was getting a passport and an armful of shots. Meanwhile, Brent arranged for a $5,000 ($9,385 in 2023) check under GSFC purchase order No. S-97728-Z. On August 19, Brad and I lifted off from San Antonio International Airport on the first leg of our journey to the Amazon basin. After arriving in Atlanta, we watched through a big window as the baggage was unloaded. Suddenly, I spotted my main bag full of fragile instruments being tossed 15 feet (5 m) onto the concrete. It did not fall accidentally; it was thrown.

When I told the gate attendants about the delicate gear in the flying baggage, they feigned shock and outrage. In all their years working by that huge window, they claimed, they had never seen baggage handlers toss baggage onto the concrete. They pretended not to notice as still more bags came sailing through the air and landed atop mine.

After a maintenance delay of several hours, we left for Brazil. When we passed over the Amazon, most of the passengers were asleep. But I could not sleep; as in 1995, my eyes were glued to the window, watching scintillating orange outlines far below.

As in 1995, getting through customs was on my mind when we arrived at São Paulo. Most of the instruments I build worry security and customs people, especially in foreign countries like Switzerland, Brazil, and New York City. I was especially concerned about this trip, for I had brought a plastic seedling nursery, potting soil, seeds, and more than fifty mold and bacteria growth plates, so I made sure that Artaxo's letter was within easy reach when we left the plane. Thanks to the delay in Atlanta, the customs people were not expecting us. We walked through without even seeing an inspector.

Since the late arrival caused us to miss our connecting flight, we spent most of the day at the airport in São Paulo. This gave us time to change our money, and, remembering what had happened in 1995, I chose a different bank. While Brad guarded the bags, I walked up to a lady behind a bulletproof window and placed $500 in $20 bills in the tray. She counted the money and announced, "$400." I paused and firmly said, "No, $500!"

The lady began counting again. When I pretended to look away, in a replay of what had happened in 1995, her right hand disappeared under the counter and miraculously reappeared with five $20 bills. With a straight face, she announced, "$500."

That evening, we arrived in Cuiabá too late for our flight to Alta Floresta, a town so remote it did not appear on many maps, so we took a midnight taxi ride to the Hotel Mato Grosso, where Damian and I had stayed in 1995. The next morning, we boarded a small turboprop for the flight to the smoke-covered wilderness of the southern Amazon basin.

As we flew north, the land below gradually turned to forest interrupted by enormous clearings that had been burned to form cattle pastures. Soon, the smoke beneath the plane became so thick that the ground was obscured. We were flying over a vast blanket of gray that extended to the horizon.

After half an hour, we descended through the gloomy gray and landed at Sinop, a small city halfway to Alta Floresta. Several acres of trees near the runway were in flames. After refueling, we took off for Alta Floresta. The smoke was so thick when we arrived that I thought the captain might be unable to land. But the landing was fine, and later, we learned that the smoke could be much, much worse. We were finally at our destination, and the GPS reported we were 3,890 miles (6,260 km) from my home in Texas.

Alta Floresta is in a gold mining area 9 degrees below the Equator, though most of the gold mines have played out and have been replaced by agriculture and logging. Tourism is important to the Floresta Amazônica Hotel, on the outskirts of town, which is where Brad and I set up shop. Visitors at the hotel can stroll along flower-lined walks while watching monkeys, toucans, and colorful macaws within a few yards of their rooms.

We arrived during the peak of the smoke season, so only a few rooms were occupied. After making some sun measurements, I poured potting soil into each of the seventy-two receptacles in a miniature Easy Grow Greenhouse—eighteen seeds each of wheat, corn, beans, and lettuce.

The Alta Floresta Routine

Early the next morning, we began what was to become a daily ritual: After a quick swim in the hotel pool, we set up instruments to measure and store sunlight and temperature. Then we pointed four sun photometers at the sun to measure how much sunlight was blocked by the smoke. Three of these were my homemade instruments, and one was a Microtops II, the sophisticated version of Microtops designed by the Solar Light Company. We also measured the ozone layer. This routine was repeated every fifteen minutes, so there was no time for breakfast. Nor could we take time for lunch, since midday is when the most important measurements were made.

The all-important study of bacteria floating in the air was sandwiched between the many sun and sky measurements. Every ninety minutes, we placed an open agar tray on a post for twenty minutes. We then sealed the tray and stashed it in our room so any bacteria and mold spores that fell on the agar would grow into colonies large enough to count.

A thousand or more new fires were started each morning, and the smoke was usually thickest by late afternoon. By then, the sun became a dim orange ball. While every day was smoky, some days were especially so. August 30 was the smokiest day at Cuiabá in 1995 and the smokiest day at Alta Floresta.

The smoke was so thick that it was necessary to employ a magnifying lens to find the sun in the viewfinder of the Microtops II ozone instrument. Around noon that day, less than a third of 1 percent of the green wavelengths of sunlight were penetrating through the smoke. The airport had to be closed, and the hotel's owner, Vitória da Riva Carvalho, and some of her workers were made ill by the smoke.

The most significant scientific finding on August 30 was that there was no measurable solar UV from sunrise to sunset. This reinforced my concern that infectious respiratory disease in Brazil during the burning season might be exacerbated by airborne pathogens protected from UV sunlight by dense smoke. This motivated Brad and me to wear dust masks when we were working. We did not wear them when people were around, however, because we did not have enough to share.

Late one afternoon, we went looking for trees to core to see if we could find any evidence of differences in the width of annual growth rings caused by sunlight suppressed by smoke. Recently cut trees were lying on the ground in an open area of the forest, and I went to work with my coring tool. Back home, it was easy to use the tool to drill into pine trees. But these were rain forest hardwoods. It did not take long to work up a considerable sweat while twisting the hollow bit into a downed trunk. Transporting biological specimens out of Brazil and back to the US requires special permission, so I measured the width of growth rings and photographed the cores instead.

Sunset was the highlight of each day. That is when Brad and I packed up the instruments and had our only meal of the day at the hotel's restaurant. Over steak that came from cattle raised on grass where rain forest had once stood, we relaxed while watching big geckos stalking insects under the lights on the wall. While later walking through the forest adjacent to the hotel, I was surprised that there was very little topsoil under the carpet of leaves on the ground. Instead, the ground was mainly laterite gravel. That explained abandoned cattle pastures in the area. A local explained that after a few years, the nutrients in the thin soil are exhausted and the pastures are abandoned.

Each night after supper, our flashlight beams formed yellow cones in the smoke, which blocked stars and transformed the moon into a faint orange disk. There were chores to do in our room. Tap water of uncertain origin had to be pumped by hand through a ceramic filter into our water bottles. Clothes had to be washed in the sink. Each night, I stayed up until midnight counting and photographing the bacteria and fungal colonies that had grown from bacteria and spores that had fallen on the trays two days before.

A Medical Emergency

Two days after we arrived at Alta Floresta, Brad badly cut his thumb. Fortunately, the hotel manager's son was there and offered to drive us to the hospital. It was nearing noon, the most important time of day for measurements, so I told Brad I would have to stay. He insisted I accompany them, so I reminded him that the principal trip rule was, "The data always comes first." That was a bad call on my part. Brad's thumb was badly cut, and I should have gone with him.

At the hospital, a physician sewed up and bandaged Brad's thumb. This occurred on a Sunday, and Brad said the doctor told him that if he had been in church, the accident would not have happened. The next day, I walked Brad back to the small hospital so his injury could be redressed. I was surprised by how few supplies the hospital room had in its single cabinet. The nurse was wearing a leg cast, for she had recently broken a leg while helping a disabled patient to the second floor on the steep ramp that substituted for stairs.

The Miniature Garden Project

We placed the miniature garden on a rooftop and watered it daily. The blue and red colors of sunlight responsible for plant growth are called photosynthetically active radiation. A PAR sensor was placed by the greenhouse and connected to a data logger that automatically stored the sunlight intensity once every twenty-four seconds. The height of the wheat and corn seedlings and the width of the bean-seedling leaves were measured twice each day for eight days after emergence.

Seedlings in shade grow more rapidly than those in full sunlight as they try to reach light. This is what occurred with the garden seedlings, which grew more on the smokiest days. While that seemed to make the smoke beneficial, reduced PAR during early growth might make plants less tolerant of full sunlight after the smoky burning season ends.

Vitória da Riva Carvalho's Dream

Vitória, the hotel's owner, took a serious interest in our findings. Deforestation was destroying the habitat for the birds many of her visitors came to see, few of whom arrived during the smoky burning season. As noted earlier, Vitória and others sometimes suffered ill effects from the smoke. Five days after we arrived, Vitória asked if she could speak with me about the situation.

"I have a dream," she began. "I am working to save the Cristalino area."

I have had plenty of dreams also, and I wanted to hear those of Vitória, who was frustrated that too few people were aware of what was happening to the rain forest. When I suggested that its destruction might be slowed if people were aware of the negative effects of the smoke on people, animals, and plants, she wanted people to see this for themselves. "We need to bring people here," she said. "The great problem here is, we don't have money."

Vitória also had another strategy: "My dream, apart from [your] research, is applied research about how to use the forest."

By this, she meant the potential of therapeutic plants. I suggested to Vitória that it might be good to hold a meeting with concerned local officials about the effects of the smoke. She asked for a list of possible effects, and I responded with a list of sixteen that included the well-known health issues, reduction of tourism, airport closures, and poor wildlife-photography conditions. To these, I added possible negative effects on crops and native vegetation and possible increases in pathogenic bacteria and viruses due to highly suppressed ultraviolet. Finally, I added that bird migration might be hampered by blocked views of stars and the absence of polarized skylight.

The Ad Hoc Smoke Meeting

Vitória went to work and invited thirteen professionals to a meeting at the hotel at 8:30 p.m. the next evening. The attendees at the ad hoc conference included a fisheries biologist, four physicians, a nurse, an attorney, a hotel consultant, and several others. As I wrote in my notebook, many questions were asked about the data I presented during the two-and-a-half-hour meeting, and there was considerable concern about the effects of smoke on children. Dr. Antonio Sinkos stated that before the burning season, he typically saw two or three cases of respiratory illness each week. During the burning season, he was seeing eight a day.

I suggested the need to conduct a health survey of the local population during the ongoing burning season. The attendees agreed that a local biology student could be hired for $350 ($657 in 2023) to conduct a study of hospital records for the past two burning seasons. I would have to pay for this on my own, because the NASA money was exhausted. Several days later, Vitória introduced me to Dr. Darriel Azevedo, a dentist who agreed to help find a student to compile medical records. We agreed to work out details after Brad and I returned from a trip to the Cristalino River.

After nearly two weeks of all-day measurements, garden tending, and bacteria counting, Brad and I could hardly wait for the field trip to the remote Cristalino Lodge. We were allowed to take only our backpacks, which we stuffed with instruments and water bottles.

During the bumpy, dusty ride through country that reminded our driver of scenes from western movies, I volunteered to open and close the farm gates through which we passed. After an hour or so, we arrived at the shady bank of a large river.

A few hours later, a boat arrived from the Cristalino Lodge with half a dozen bird-watchers. Their guide explained the ambivalent attitude Brazilians have about burning down their rain forest. He said those doing the burning view themselves as good citizens who are clearing a nonproductive wasteland to provide space for new pastures and crops. However, Brazil's tourist industry, he explained, despises the burning because the smoke drives away tourists and the burning is destroying what tourists came to see.

After Brad and I climbed into the boat along with three bird-watchers, we were transported through thick smoke to the tributary named the Cristalino River. Having seen only muddy brown and tea-black rivers in Brazil, I was looking forward to seeing a crystal-clear river. Finally, our boat turned into the tree-lined Cristalino. It was the clearest black river I had ever seen. You could see an inch (2.5 cm) into its tea-colored water.

The river provides the only access to the Cristalino Lodge, which consists of several one-room bungalows, a meeting room, a dining area, and a kitchen. As Brad and I walked to our bungalow in an isolated clearing, two foxes wandered into the dense forest. After supper, the lodge staff told us that electrical power would be available for the next two hours. Back in our room, I tried to recharge the battery in my laptop. There was no response, so I checked the electrical outlet with a voltmeter. There was power, but it was only 60 volts, not the expected 110 volts.

That evening, Brad and I watched bright orange and yellow-green flashes from fireflies along the riverbank. When I pointed my flashlight to our right, a pair of bright orange eyes popped out of the darkness a hundred feet (30 m) away. This was worthy of investigation, so we got into a canoe and paddled over for a closer view.

I had the flashlight, so I was in front. As we neared our quarry, leafy branches hanging over the water blocked our way. Oblivious to the spider webs and the wasps that probably lived there, we pressed on. Suddenly, the flashlight revealed what was behind those eyes: the biggest caiman I had ever seen.

The caiman's big head was lying at the edge of the water. The rest of the creature was sprawled out on the bank under the branches. Because Brad was still hung up in the limbs at the back of the canoe, he did not see the caiman and kept pushing forward.

"Brad," I whispered. "Stop!"

He did not hear me and kept pushing branches away while trying to paddle.

"Brad, we have to stop!" I quietly insisted. "Now!"

Just as I thought the caiman might lunge at the canoe, Brad spotted the giant. It was impossible to turn around, so I stayed motionless while Brad backed us out through the tangle of branches as quietly as possible.

Sun and Sky Measurements Above the Rain Forest

The next morning, Brad and I hopped in a canoe along with a backpack full of instruments and a camera, and a guide in a motorboat towed us a few miles upriver. Our goal was to climb a hill to measure sunlight above the forest canopy. There was much more life along the river than in the forest, and along the way, we saw many birds, clouds of colorful butterflies, and a large tapir. Eventually, our guide angled us toward the mud bank and pointed toward the hill. He then pointed at the sky and said, "*Muita fumaça*" ("Much smoke"). He then roared away.

Brad and I eventually found a trail to the exposed granite atop the hill. At the end of the trail, which was a bit steep in places, we were rewarded with a panoramic view of the rain forest below. We quickly ate bananas while waiting for the GPS to home in our location for the Microtops II measurements. I then made a full range of measurements between annoying attacks by squadrons of sweat bees and flies. Our protocol was to swish a butterfly net back and forth to capture the insects so that we were free to do a sequence of measurements with minimal distraction. A few minutes later, it was time to use the butterfly net again.

After fourteen minutes of measurement-net-measurement cycles atop that very hot hill, we packed the instruments, released the insects, and hurriedly left the summit. Of special interest was that there was measurable UV on the hill. Back at the lodge, I had measured zero UV at noon under the adjacent forest canopy. Trees, of course, block considerable sunlight, as I knew well from many hundreds of measurements in the woods behind our Texas home. But I had never measured a complete absence of UV at noon, and I wondered then, and still do, whether there are environmental consequences from exceptionally low levels of UV on a tropical forest floor at the peak of the burning season.

Back at the river, we placed the instruments in waterproof bags, pushed the canoe into the water, and began paddling back to the lodge. A small island we passed was inhabited by hundreds of loudly buzzing bees or wasps, so we stopped paddling and did not say a word until we drifted beyond them. We then shot through a noisy section of relatively tame whitewater in the black river and arrived back at the lodge in time for lunch with the birders.

The Bird-Watchers

One of the birders was a guide from Australia who I will call Jim. He was a compassionate man, for the fellow bird-watcher he was guiding was a paraplegic I will call Bob. It was impressive to watch Jim pushing Bob in his wheelchair, binoculars glued to his face, along trails through the forest. The third bird-watcher was an Englishman I will call Richard.

While the three birders, Brad, and I were waiting for supper the first evening, we learned about another side to Jim. We were all concerned about the massive burning of the forest and the resultant smoke, but Jim's solution was so surprising that Brad began taking notes. "If there's one thing we need," Jim proclaimed, "it's an environmental assassination squad."

He continued this theme for some time before moving to Chinese businessmen in Malaysia, whom he said were playing key roles in deforestation in the tropics. He was not shy about his feelings for the Chinese, whom he described as unprincipled, unethical, and generally worthy of ridicule.

The next day, at lunch, Jim continued talking about the massive destruction of Brazil's tropical forests. "It's the worst environmental disaster I've come across," Jim said. He then spoke with disgust about the dead bird they had seen that morning while Brad and I were making measurements from the granite hill. Jim said that anyone who kills a bird should himself be killed, to which Richard replied, "I shall be careful to avoid stepping on an ant." Jim then said helicopter gunships should be used to stop the burning.

Earlier, a rare bird had walked under Bob's wheelchair. When Bob had asked where it had gone, Jim had replied, "Put your binoculars down and look between your legs." Meanwhile, Richard was walking about while quietly telling his tape recorder about the birds he was viewing. When I told Brad how funny this was, he slyly replied that I did the same thing while doing sun and sky measurements. Brad was right, and I developed considerable respect for Richard and his tape recorder.

After lunch, I was rethinking my goal to swim in the Cristalino River now that it had been revealed to be another black river. But the afternoon was warm and the river was cool, so I put on swimming trunks, walked onto the floating boat dock, and dove in.

Somewhere between the dock and the water, I remembered the schools of piranhas, the vicious caimans, and the deadly parasites waiting below, so I quickly swam back to the dock and exited the water. Because nothing had bitten me, I twice repeated the dive-in-and-exit-quickly routine. Because of his bandaged hand, however, Brad could not enter the water. I would not have either, had I first read Theodore Roosevelt's 1914 description of the vicious piranha in *Through the Brazilian Wilderness*, which begins, "They are the most ferocious fish in the world. Even the most formidable fish, the sharks or the barracudas, usually attack things smaller than themselves. But the piranhas habitually attack things much larger than themselves."

Lessons from a Tree Hugger

The second morning, colorful butterflies decorated the tall, flowering trees by the river, where birds were singing, bees were buzzing, and an iridescent hummingbird was collecting silk for its nest from a tent caterpillar colony. Easily seen growth rings in a recently cut tree trunk revealed thin rings over the past decade, possibly a byproduct of reduced sunlight caused by smoke.

An American a few years older than Brad worked as a guide at the river lodge. He told us he was God, but his real name was Jen. One morning, Jen took us on a canoe trip to see some huge Brazil nut trees. As we were leaving the biggest of the giants, we looked back and saw Jen praying to the tree and asking it to forgive us for disturbing it. Back at the river, we waited for the birders to float by so we could join them. Meanwhile, Jen waded out into the water and began waving his arms for us to join him. It was hot, but I was too worried about piranhas and parasites to join him.

An ugly rash had developed around Jen's waist, perhaps caused by river organisms. This constantly worried him, so I suggested he use his divine power to heal himself. Jen's happy countenance faded, and thereafter he stopped claiming to be God. Brad and I said we would help him get some ointment when we returned to Alta Floresta.

When the birders arrived, we followed them as they floated very *s-l-o-w-l-y* while observing, recording, and photographing various specimens, including a pair of razor-billed curassows. Over lunch at the lodge, Jim and I debated various aspects of evolution. At supper, the conversation turned from Jim's admiration of Charles Darwin to a more disturbing topic: Perhaps Jim was noticing the Brazilians who prepared our meals, for he suddenly started talking about his contempt for mixed-race Aboriginal Australians. "They're garbage people," he blurted. He then laughed about a proposal he had heard about bulldozing them and their houses. I changed the subject to the *Scientific American* affair. Jim agreed they should have kept me on board but then said, "Maybe you shouldn't have applied for the job."

Jim's attitude toward Brad and me cooled when we asked him questions about the unexplained evolution of complex molecular motors in our cells, for which he had no answer. It was bridge-burning time, so I asked Jim how he liked his steak.

"Fine," he replied. "It's really quite good."

"It's rain forest steak," I said. "The cattle were raised on pastures that were once rain forest filled with birds."

Jim had no reply. He glared off into space as he wolfed down his steak, which was quite good.

Farewell, Cristalino River

Our adventure on the Cristalino ended much too soon. The third morning was spent packing our clothes and bacteria trays and conducting a final round of sunlight measurements. That afternoon, we boarded a motorboat that carried us back to where the hotel van was waiting to take us on the dusty trip back to Alta Floresta, where we washed our clothes and enjoyed a steak dinner.

The next two days were spent processing data and repeating the same cycle of all-day measurements we were doing before the Cristalino trip. We also met Dr. Jim Borrow, a physician from Seattle who expressed considerable interest in my smoke studies and the related health hypothesis. That night, hard rain fell, and I told Brad this would provide a clean sky by morning. I was wrong; the sky was as smoky as ever.

At 8 a.m. on departure morning, Brad and I met with Dr. Carlos Alberto Redondo and Gisele Cristina de Castro, a university student who had already begun the health study I had commissioned. She showed us some data, and we discussed extending the age range from twelve to fifty, as older people seemed to have many smoke-related health problems.

São Paulo or Bust

We departed Alta Floresta a day earlier than planned after Paulo Artaxo, who had provided the official invitation to do smoke research in Brazil, sent a fax asking me to give a seminar for his graduate students at the University of São Paulo. The first leg of the flight to São Paulo was rather scary. As I noted in the trip log, "Very <u>rough</u> ride from Alta Floresta to Cuiabá. Pilot warned of big jolt just as the plane slammed downward. Loud noise. Things flew up. My glasses landed in front of a passenger across the aisle. Camera bag struck my face."

The "loud noise" was more like a loud bang. The plane fell so rapidly that everything not tied down flew to the ceiling. Fortunately, everyone on board was wearing a seat belt. After the plane stabilized and while passengers gathered their things, I looked out the window to make sure the wing was still there.

After a one-night layover in Cuiabá, we continued to São Paulo. During a stop at Brasília International Airport, the captain and crew allowed me to leave the plane to make sun measurements from the ramp, an opportunity that would never happen back home.

After I returned to the plane, two injured field workers were brought on board. Both were wearing soiled, torn work clothes and had blood-stained bandages on their arms or legs. One man's head was bandaged, and both were obviously in pain. Despite their unpleasant appearance, the flight attendants were full of compassion for them. They treated them like royalty as they helped them into their seats for the flight to São Paulo and brought them snacks and drinks.

At the University of São Paulo, Artaxo and his students were interested in the health implications of the UV measurements, and one student was especially interested in the tree-ring studies. After the seminar, I asked Artaxo how many other scientists were in Brazil for the burning season.

"Just you two!" he replied.

In 1995, 101 scientists, academics, pilots, and officials had participated in SCAR-B. This time, NASA had not been allowed in Brazil, and the country's tiny number of atmospheric scientists were preoccupied with serious air pollution in São Paulo. Suddenly our hard-earned data seemed especially important. It was the only data.

At the airport, we remembered the woman who tried to cheat me when we changed money on arrival, so we went to a different bank to change our Brazilian money back to dollars. While an armed guard stood by watching, the male teller tried the same counting trick, which by now we were expecting. When Brad and I burst out laughing, the guard looked away while the teller frowned and recounted the money.

After a major airport-security problem involving my instruments (related in chapter 20), we boarded the plane home. During this trip, I had lost seven pounds, equal to my weight loss during the 1995 Brazil campaign. I also departed with an uncomfortable sensation in the right side of my chest that probably resulted from the smoke I had inhaled for the past three weeks.

The 1997 Brazil-Smoke Report

Each day in Alta Floresta, we made more than 18,000 measurements of temperature, water vapor, and various wavelengths of sunlight using automated radiometers and handheld sun photometers. The agar trays captured hundreds of bacteria and spores. The miniature garden provided good results. And we managed to core several rain forest trees and make hundreds of photographs.

The bacteria study provided significant results, for it supported my hypothesis that the increase in respiratory illnesses during the burning season might have been caused in part by the survival of pathogenic bacteria suppressed by normal levels of ultraviolet sunlight. Pathogenic bacteria usually lack the pigmentation that protects plant bacteria from natural levels of UV. The agar experiments clearly showed that the ratio of nonpigmented to pigmented bacteria was significantly higher on the smokiest day than the clearest day. Plotting the ratio for all seven days of the bacteria study against the solar UV-B index yielded a correlation coefficient (r^2) of 0.83, which supported the bacteria hypothesis.

The hospital records compiled by Gisele Cristina de Castro clearly showed a significant increase in influenza incidence and the death rate during the burning seasons of 1996 and 1997. A comparison of the total UV-B and influenza cases per day yielded a correlation coefficient (r^2) of 0.82, which also supported the bacteria hypothesis. While influenza is caused by a virus, it is vulnerable to UV.

The 1997 smoke report, *Scientific Studies During the 1997 Burning Season at Alta Floresta, Brazil*, was much more comprehensive than the 1995 smoke report. The report stated that it was prepared for Dr. Paulo Artaxo, and Vitória da Riva Carvalho of Floresta Amazônica Hotel e Turismo Ltda was acknowledged. But if I had known then what I know now, it would have been dedicated to her, for she successfully pursued her dream to preserve a substantial segment of rain forest along the Cristalino River.

I expanded on the possible connection between reduced UV-B caused by biomass smoke and infectious respiratory disease in several publications, including "Health Effects of Tropical Smoke." This letter closed by noting that satellite ozone instruments like TOMS can detect smoke and that "the close association of upper respiratory diseases with optical depth in Alta Floresta and the API [Air Pollution Index] in Sarawak suggest the possibility of using TOMS as an epidemiological tool to identify regions that might have smoke-related respiratory disease."[1]

In "Avian Influenza and UV-B Blocked by Biomass Smoke," a note published in *Environmental Health Perspectives*, I suggested, "Human cases of avian influenza in Thailand and Vietnam since December 2003 have peaked during both the rainy season and the burning season. Thus, periods of prolonged cloudiness and severe smoke pollution could play a role in initiating avian and other influenza outbreaks by attenuating the solar UV-B that might otherwise suppress influenza viruses in outdoor air exposed to sunlight."[2]

The Nature of Nature Conference

During the 1990s, I became acquainted with Dr. William Dembski through online conversations about intelligent design. Bill, who was familiar with what had happened to me at *Scientific American*, has an extraordinary academic background. His seven degrees culminated in a doctorate in mathematics (1988) from the University of Chicago and a doctorate in philosophy (1996) from the University

1. Forrest M. Mims III, "Health Effects of Tropical Smoke," *Nature* 390, no. 6657 (November 20, 1997), 222–3.
2. F. M. Mims, "Avian Influenza and UV-B Blocked by Biomass Smoke," *Environmental Health Perspectives* 113, no. 12 (December 1, 2005), A806–A807.

of Illinois at Chicago. He also has a master's degree in theology (1996) from the Princeton Theological Seminary.

Bill has written many books, including several that have influenced my thinking about the scientific aspects of the ID movement. While ID proponents do not necessarily claim a biblical explanation for the origin of life on earth, they do claim that neither chemical nor biological evolutionary theory explains life's origin. Bill makes this abundantly clear in *The Design Revolution: Answering the Toughest Questions About Intelligent Design.*[3]

Baylor University, founded in 1841, is the second-oldest institution of higher education in Texas. While Baylor was originally founded as a Baptist university, like many other Christian universities, it has drifted away from some of its theological roots. Therefore, being a Baptist, I was impressed when Baylor hired Bill in 1999 to lead the Michael Polanyi Center, ". . . at Baylor University researching the idea that life was created through 'intelligent design' instead of evolution."[4]

The first major initiative by the Polanyi Center was "The Nature of Nature: An Interdisciplinary Conference on the Role of Naturalism in Science," cosponsored by the John Templeton Foundation, *Touchstone* magazine, and Seattle's Discovery Institute. The conference, which took place April 12–15, 2000, was attended by several hundred scholars, academics, theologians, philosophers, authors, and students.

Dozens of talks were presented at the conference by a wide range of scholars, authors, and academics. My talk was titled "Solar Ultraviolet Radiation Is Finely Tuned to Enhance the Survival of Many Forms of Life." The summary in the conference program read, "While the current paradigm is that solar ultraviolet (UV) is harmful to plant and animal life, UV also plays a vital role in synthesizing vitamin D in mammals and suppressing parasitic and pathogenic microorganisms suspended in air and water."[5]

The theme of my talk was that the synthesis of vitamin D by UV-B is fine-tuned, and I showed a series of slides that made my point. For example, the vitamin D action spectrum ranges from 295 to 305 nm, and only about 0.4 percent of the sun's energy falls within this narrow band. About 70 percent of this energy is absorbed by the ozone layer. Thus, vitamin D synthesis appears to be fine-tuned. (That phrase is often used to suggest that a natural phenomenon has been designed.)

Because there were seven concurrent sessions the afternoon of my talk, I was surprised that the room was filled. When I finished, a scholarly-looking gentleman

---

3. William A. Dembski, *The Design Revolution: Answering the Toughest Questions About Intelligent Design* (Westmont, IL: Intervarsity Press, 2004).

4. "'Intelligent Design' Center at Baylor Gains Support from Review Committee," January 1, 1998. Reprinted by the Discovery Institute at https://www.discovery.org/a/487/.

5. The Nature of Nature Conference, Concurrent Session Speakers, April 12–15, 2000.

on the back row asked, "Does the fine-tuning argument rule out an evolutionary explanation?" After pausing a few seconds, I replied no. (I later wrote a short paper making a case for answering yes.)

This ended the session, and the back-row gentleman then rushed toward me at the podium. I hoped that he might have more questions; instead, he abruptly demanded, "Do you know where I can get a cup of hot tea?" I did not know, so he asked someone else. I did not realize then he was the famous Nobel Laureate Dr. Steven Weinberg, an atheist who has long had an interest in explaining fine-tuning arguments. That explains why he attended my talk.

During the conference, I learned that there was considerable objection to the meeting by Baylor faculty. On April 18, the Faculty Senate voted 27–2 to dissolve the Michael Polanyi Center. To this day, I remain amazed that those faculty members voted to disband the center that organized the historic conference in which Weinberg and Dr. Christian de Duve, another Nobel Prize winner, and scores of the country's most influential scholars, scientists, academics, and skeptics participated. Dr. Robert B. Sloan Jr., Baylor's president, responded to their vote with a lengthy statement that included the following analysis:

> *By dissolving the Center, as the Faculty Senate has proposed, we would in effect be imposing a form of censorship on the work of the Center. I believe there are matters of intellectual and academic integrity at stake here. Drs. Dembski and Gordon, both highly capable scholars with the credentials to support their qualifications to study the subjects that the Center was established to pursue, should be allowed to do their work. If their conclusions do not stand up to peer review, then so be it. But to quash their research and to mute their point of view because of political pressure and without sound intellectual cause is antithetical to everything for which a true university ought to stand. We should not be afraid to ask questions, even if they are politically incorrect. Indeed, I am proud of Baylor's willingness to ask questions which some are apparently afraid to entertain.*
>
> *. . . Whether or not there are patterns of design, information, and purpose in this universe that can be detected by scientific processes, I do not know. I do think, however, that it is an interesting question. Indeed many people regard it as an issue of significant intellectual import. Surely it is fair game in a place like Baylor to ask such questions. It is simply too easy to dismiss as "creation science" every attempt to relate belief in the Creator God to the human processes of understanding the created order.*[6]

6. Robert B. Sloan Jr., Michael Polanyi Center Statement, *Baylor Lariat*, 2000.

How could a Christian faculty disagree with this response? The attendees at the conference, including well-known skeptics like Weinberg, Dr. Michael Shermer, and others who disagree with the idea of design, did not, for they attended and participated. Baylor's Faculty Senate had become so eager to be accepted by the academic mainstream that it dismissed Bill Dembski under the umbrella of scientific materialism, exactly as the editor of *Scientific American* had reacted to me. Despite Sloan's support, however, the Faculty Senate's view prevailed, and the Michael Polanyi Center was closed.

During the faculty firestorm that followed the "Nature of Nature" conference, Bill called me several times to discuss strategies with a kindred spirit who had also been through a major controversy involving intelligent design. When I visited Bill at his home, I tried to lighten the conversation by advising him to keep an eye out for a crowd of torch-bearing Baylor professors marching on his home.

Bill remained at Baylor as an associate research professor in the university's Institute for Faith and Learning until 2005. He then moved onward and upward through a series of faculty appointments until becoming an independent entrepreneur and developer of educational software and websites. He also continued working as an independent writer, editor, and researcher.

During my career, Bill was tied in intellectual capacity only with Roger G. Mark, MD and PhD, during my days at the Air Force Weapons Lab's Laser Lab. Mark, a fellow believer, has made highly significant contributions to medicine and electronics, and Bill has done likewise in education, statistics, software design, and, of course, intelligent design. Bill's intellectual contributions probably exceed those of the entire Baylor Faculty Senate that voted to condemn the Michael Polanyi Center.

Bill's critics at Baylor will be forgotten, but the "Nature of Nature" conference will long be remembered for hosting the most impressive gathering of intellectuals in Baylor's long history. The conference record is preserved in *The Nature of Nature: Examining the Role of Naturalism in Science* (Intercollegiate Studies Institute, 2011). Included in this 964-page volume is a paper by Nobel Laureate Dr. Francis Crick, codiscoverer of the double-helix structure of DNA, and Dr. Christof Koch. Their paper closes with this challenge, "The explanation of consciousness is one of the major unsolved problems of modern science. After several thousand years of speculation, it would be very gratifying to find an answer to it."[7]

7. Francis Crick and Christof Koch, *The Nature of Nature: Examining the Role of Naturalism in Science* (Wilmington, DE: Intercollegiate Studies Institute, 2011), 762.

The Mosquito Project

The 1995 Brazil campaign enhanced my interest in studying mosquitoes. My first project was to devise an ultrasimple method to determine the spectral sensitivity of mosquito eyes. I placed a house mosquito (*Culex pipiens*) larva in a thin, water-filled tube and illuminated it with various colors of light using a spectrometer that divides visible light into a rainbow of its wavelengths. As I dialed the spectrometer's dial, light from blue to red swept across the larva, which reacted vigorously when illuminated by UV-A (344.6 nm) and green light (550.0 nm).

This had previously been discovered using a much more complex method, and I reported the results in "Biological Effects of Diminished UV and Visible Sunlight Caused by Severe Air Pollution."[8]

This paper also reported on my mosquito nursery project, in which I mounted from one to six layers of window screen over wood bases, each of which supported a plastic tray and a light-recording logger. The trays were placed in the woods below our house, where they collected rainwater and egg-laying mosquitoes. This study showed that female house mosquitoes deposited up to fifty times more eggs in the darkest nurseries. The obvious implication was that thick biomass smoke like that which I measured in Brazil might enhance the population of some mosquito species.

The Center for Tropical Diseases

My mosquito studies and the hypothesis about the possible connection of UV-B suppressed by thick smoke and infectious respiratory disease were of interest to Dr. David H. Walker, a noted physician with a wide range of interests who heads the University of Texas's World Health Organization Collaborating Center for Tropical Diseases and the Department of Pathology at the University of Texas Medical Branch (UTMB) in Galveston.

In 2000, Walker invited me to give a colloquium talk at the center titled "Diminished Solar Ultraviolet Caused by Severe Air Pollution in the Tropics May Enhance Populations of Certain Mosquitoes and Water and Airborne Microorganisms." The academics who attended my talk were interested in my ultraviolet hypothesis, the Brazil field trips, and my one-man mosquito studies. Two of the staff scientists and I later proposed a joint study of what my findings suggested, which we submitted to NOAA. Unfortunately, our proposal was not funded.

---

8. F. M. Mims, "Biological Effects of Diminished UV and Visible Sunlight Caused by Severe Air Pollution" (International Radiation Symposium, Fairbanks, Alaska, August 22, 1996).

Spaceship Earth

In 2002, Documentary Japan Inc. contacted me about doing a television documentary about my ozone research for *Spaceship Earth*, a Japanese science program on television. After I provided the information the company requested, it decided to film the first week in Hawai'i and the second week in Texas. The Mauna Loa Observatory was the emphasis of the Hawai'i segment, where the documentary crew filmed every aspect of my calibrations there. They also filmed my measurements elsewhere in Hawai'i.

The most memorable aspect of the Hawai'i filming was when I was driving down a steep section of the Chain of Craters Road. The producer, the sound man, and their translator were in the back seat, and the camera operator was seated next to me in the front seat. Suddenly, the camera operator opened his window and somehow managed to slip halfway out the window so he could film me through the windshield while the sound man held onto his belt.

After a week in Hawai'i, we flew to Texas, where the crew filmed me making my daily solar-noon measurements. When the producer saw a NASA decal on one of my instrument cases, he asked if I worked for NASA. When I said I had consulted for NASA on several occasions, he wanted proof, so he arranged for us to visit a studio in San Antonio to do a live video interview with two of the NASA scientists I knew at the Goddard Space Flight Center.

The crew also filmed me giving a talk at Seguin High School, addressing the Seguin City Council, testing children's sunglasses for ultraviolet blockage, and attending church. They assumed Minnie and I were rich because of the acreage around our house, and they especially enjoyed a supper Minnie prepared for them, which was preceded by our usual mealtime prayer.

When they were packing their gear on the final day at our place, the producer left his translator behind, walked over to me, and, in perfect English, thanked me for my participation. He then said that while he was an atheist, he was intrigued by my Christian faith and added, "I will try God." The documentary aired in Japan on March 4, 2003.

CNN's Twenty-Fifth Anniversary Video

A much briefer video documenting my science appeared in the Profiles Campaign, a series of thirty-second videos celebrating CNN's twenty-fifth anniversary. Producer Peter Mullett explained the series in an email that concluded, "In doing research, you came up as someone who we feel could be an interesting person to talk to, if you'd be willing."[9]

9. Peter Mullett, email to Forrest M. Mims, November 19, 2004.

A three-person CNN film crew and I met at Cypress Creek in Wimberley, Texas, on a pleasant January afternoon in 2005. Harry Helms had moved to Wimberley, and he showed up to watch. They had me sit on a big rock in the center of the fast-flowing stream and make several statements in support of CNN following their introduction.

After briefly introducing myself, I made a series of brief statements about my science. The one they used in their video was, "I'm interested in environmental science, because I look at life as one big question, and I'm constantly looking for the answers." They then wanted a CNN endorsement, so I said that airport CNN was useful while traveling. They used this quote in the video, because they did not know Minnie and I do not have cable TV, and sitting in airports was the only time I watched CNN. (The first time I saw the video was while seated in an airport.)

They wanted something much more enthusiastic, and the producer suggested a string of catchy CNN endorsements. But back then, CNN was becoming so biased in its reporting that I did not want to be too positive. After ten minutes or so of the producer's suggestions that I did not want to repeat, he finally suggested over the gurgling of the creek, "How about saying, 'I want to know the latest information, because I'm curious'?" So that is what I said, followed by a vigorous laugh. And that is what closed their thirty-second video, including the laugh.

The Japanese *Spaceship Earth* documentary was quite the opposite, for that producer never suggested what I should say during the two weeks they filmed me doing research at the Mauna Loa Observatory and carrying on life and doing science in Texas. While I did not laugh during the documentary, the US interpreter told me that some women who watched the documentary were moved to tears. I still wonder at which point, and why.

A *Scientific American* Reminder

By now, it was clear that Jonathan Piel's worry about the public relations nightmare that would have occurred if I continued writing The Amateur Scientist column had backfired. No other magazine or journal censored me, including *Nature*, the most prominent of them all, which published two of my formal papers cited previously and six of my letters after my evolution views were widely known.

As noted in chapter 2, *Nature* even published my letter teasing them for advocating the teaching of "all of Darwin" while failing to mention that Darwin himself wrote, "One Hand has surely worked throughout the universe"[10] and that he praised Christian principles.

10. Charles Darwin, *Voyage of the Beagle* (New York: Penguin, 1989), 325.

I disagreed with *Nature* editor John Maddox when he editorialized on April 21, 1994, "As the long (and soon to end) correspondence in Nature has shown, many professional scientists are deeply religious, often justifying their belief on the grounds that 'science cannot know everything.' . . . it may not be long before the practice of religion must be regarded as anti-science."[11]

In my reply, I asked, "Where does Maddox propose to draw the line? If Nature views religious scientists as anti-scientific, will this make all their scientific work suspect and therefore unworthy of publication? . . . If Nature plans to draw the line in religion, then those of us who faithfully subscribe and occasionally contribute to its pages, but who also happen to be religious, need to know precisely where the line is to be drawn."[12]

Maddox demonstrated his objectivity and fairness by publishing my letter. Even *Scientific American* saw the light after Jonathan Piel's departure. The magazine's Tim Beardsley reported on my 1997 findings in Brazil in "Smoke Alarm," a detailed news story published in December 1997. The magazine also published one of my letters to the editor. Ironically, *Scientific American*'s German owners eventually acquired N*ature*, and a search of my papers at the *Nature* website also lists some of my *Scientific American* publications.

"50 Best Brains in Science"

The cover of *Discover* magazine's December 2008 issue proclaimed, "50 Best Brains in Science." A major feature article inside was titled "The 50 Most Important, Influential, and Promising People in Science,"[13] and a media release led with the announcement that "Discover has identified 50 of the brightest people alive."

In a classic editorial exaggeration, *Discover* included me on their list, and this attracted the ire of some skeptics who have yet to accept the fact that I both do and publish serious science. One of them sent a protest letter to *Discover*, which was printed along with the magazine's response: "In our feature, we recognized Mims specifically for his contributions as an amateur scientist, and we stand by that assessment. His work on the Altair 8800 computer, on RadioShack's home electronics kit, and on The Citizen Scientist newsletter has been undeniably influential. Discover does not in any way endorse the Discovery Institute's views on

11. John Maddox, "Defending Science Against Anti-Science," *Nature* 368 (April 21, 1994), 185.
12. Forrest M. Mims, "Science Versus Anti-Science," *Nature* 368 (April 21, 1994), 682.
13. Corey S. Powell, "The 50 Most Important, Influential, and Promising People in Science," *Discover*, November 17, 2008, https://www.discovermagazine.com/planet-earth/the-50-most-important-influential-and-promising-people-in-science.

'intelligent design' At the same time, Mims's association with that group does not invalidate his role as a leading figure in the American amateur science community, any more than James Watson's dubious speculations about race take away from his groundbreaking research on DNA."[14]

14. Editors, *Discover*, February 2009, 6.

20

AIRPORT-SECURITY MISADVENTURES

Over the years, I have learned how to successfully pass through airport security with a carry-on bag loaded with 30 pounds of electronics. (I cannot tell you how.) But this success came only after many encounters with airport police, irate inspectors, and gate agents who had serious concerns about homemade electronic instruments. While I certainly respect and understand the need for responsible airport security, the downside is that today's security environment creates special problems for both inspectors and anyone carrying scientific instruments on airlines. Here, I shall relate just three of my many security misadventures.

A Chicago Chokehold

While checking in for a flight at Chicago's O'Hare International Airport in 1991, when an American Airlines security agent asked about the homemade near-infrared hygrometer in my carry-on bag, I explained I was planning to measure the water vapor outside the aircraft. That alarmed the woman, who called her supervisor. After I explained my intention to the equally alarmed supervisor, she called *her* supervisor. The three women then formed a huddle some distance away to decide what to do while I wondered whether I would miss the flight.

After a phone call, the supervisor's supervisor told me in a firm voice that the captain would have to approve my presence on his aircraft and ordered, "Follow me!" She then escorted me around the metal detector without completing the inspection of the many other instruments in my bag.

O'Hare was busy, and my gate was at the far end of a long concourse. When we finally reached the gate, the supervisor ordered, "Wait here while I find the captain!"

My flight was boarding, so several minutes after the security woman disappeared into the crowd, I decided to get in line. When I took one step forward, someone from behind threw an arm around my neck, placed me in a chokehold, and dragged me back toward the wall.

It is embarrassing to be choked in front of a hundred or more fellow passengers: What must they be thinking about me? Fortunately, a third of them did not notice. But another third did—and gasped. The final third glanced at the disturbance for a moment and continued to the gate. Maybe they were frequent flyers who thought it was just another routine passenger assault at the Chicago airport.

After my assailant dragged me to the wall, he loudly demanded, "You're gonna do what on *my airplane*?"

Suddenly, the people watching this attack on a fellow passenger became noticeably quiet. They also wanted to know what I planned to do on their plane.

While I could still breathe, it is difficult to speak with an arm around your neck. "If—you—let—go—of—my—neck—I'll—tell—you," I gurgled. The assailant said nothing as he pulled his arm away. I then turned around and saw an angry, frowning man with squinted eyes and his hands on his hips. He was not very tall, but he was built like a football player, and he was wearing the uniform of an American Airlines captain.

"I'm going to measure the water vapor outside the airplane," I said. The captain cocked his head slightly and sounded genuinely curious when he asked, "How are you going to do that?"

During the altercation, I had managed to hold onto my uninspected carry-on bag, which I placed on the floor and opened just enough to retrieve the water vapor instrument. I did not open it all the way lest the captain, the security supervisor, and the surrounding crowd see all the other instruments packed inside. The water vapor instrument was the size of a 1-inch-thick deck of cards. I handed it to the captain and explained how it measures water vapor when pointed at the sun through the plane's window.

"You're kidding," he said. I then told him how it has two light sensors: one that detects sunlight absorbed by water vapor and a second that detects sunlight unaffected by water vapor. A logarithmic amplifier calculates the ratio of the two measurements and sends the result to the device's display.

The captain might have had an engineering background, because he understood what I told him, so I decided to show him some other instruments in the bag. He was intrigued by them and said, "Great!" and "Neat!" He even called the first officer to have a look and repeated some of what I had explained. After a few minutes, the captain said it was time for us to board the flight. "Is there a way I can help you during the flight?" he asked.

I explained that the biggest problem in making measurements from airplanes is knowing the location and altitude. "It sure would help if you could announce the altitude when we pass over cities," I said. "I could then make a reading and know exactly where it was taken."

"OK," the captain responded. "I can do that." I thanked him, and we walked to the gate. By now, all the other passengers had entered the nearly full plane, and we were the last to board.

Not long after takeoff, the captain made a brief announcement over the intercom: "Peoria, 28,000." He made similar announcements every ten minutes or so during the two-hour flight across Illinois, Missouri, Arkansas, and Oklahoma before entering Texas.

Those announcements were more than worth the embarrassing chokehold. I could not believe this was happening, but neither could the others passengers. After the first few announcements, people asked questions like, "Why is he telling us this?" or "Who cares?" I cared, but I was too busy making measurements and entering the data in a notebook to explain why.

When the plane landed at Dallas–Fort Worth International Airport, I was one of the last passengers to exit. The smiling captain, waiting at the exit, asked, "How did I do?" I thanked him profusely and told him it was my best data-collection flight ever. And it was.

Two years later, Minnie and I were sitting in an interterminal Airtrans people-mover car at the DFW airport while returning home from an anniversary visit to Hawai'i when the car stopped between terminals. Suddenly, I recognized the captain seated directly across from us.

I carefully asked, "Are you the pilot who helped with my water vapor measurements during a flight from Chicago to DFW several years ago?" He smilingly said he was and that he remembered our encounter. I introduced him to Minnie, and there followed a few minutes of reminiscing.

My former assailant wanted an update on my science activities, and he agreed to write his name, address, and phone number in my notebook. When I said Minnie and I would likely miss our flight if the car did not start moving, the captain asked where we were headed. When I replied, "San Antonio," he said, "Don't worry—the plane won't leave without you. I'm the pilot."

This was a surprising coincidence, and I was glad Minnie was there that Sunday afternoon, June 20, 1993, to participate in the conclusion of the choke-and-drag story I had told her about two years before.

Saved by Rolex

In 1996, Rolex flew me to Geneva to give a short talk at a Rolex Awards ceremony. While departing from Zürich Airport, I requested that the security people hand-inspect the TOPS ozone instrument that led to the 1993 Rolex Award to avoid subjecting it to X-rays that, over time, might damage its UV sensors. They said that all carry-on things must be X-rayed, without exception. As I had done at other airports, I politely said they could x-ray my bag, but not TOPS.

The security people then called their supervisor, who immediately sent two uniformed soldiers armed with automatic rifles over their shoulders and pistols on their belts. The soldiers ordered me in German to place the bag and TOPS on the X-ray belt.

I asked to speak with the airport supervisor, but they persisted. Meanwhile, a crowd gathered to watch, and a Chinese woman who must have had a science background tried to help.

"Can't you understand he just wants you to inspect his instrument!" she shouted. She repeated this several times but was ignored.

Soon, a frowning man wearing a white shirt and tie arrived with two more armed soldiers. The odds were now four heavily armed soldiers and three grim-faced officials against the Chinese lady and me.

The white-shirted man sternly asked why I was not complying with the rules. He was the airport supervisor, and his manner reminded me of the flight attendant who had ordered me to give her my instruments during the flight from Geneva to New York after the 1993 Rolex Awards ceremony.

Suddenly, I remembered how the Rolex brochure had become a ticket to the cockpit of the plane, so I reached into the carry-on bag, pulled out the Rolex brochure, and handed it and TOPS to the supervisor. When he saw the photo of me holding the TOPS instrument on the cover of the brochure, his frown was replaced by a broad smile, and he loudly exclaimed, "Rolex Award!"

Suddenly, both X-ray people, the four soldiers, and some of the people watching looked at me with astonishment. Moments before, it had appeared that the airport supervisor might end the controversy by forming a firing squad. Now, he was excited to be holding the ozone instrument that had earned a Rolex Award.

After he examined TOPS and showed it to his team, the supervisor assured them that there was no problem and that he would personally escort me through security. It was a repeat of the Chicago experience in 1991, for he walked me around the metal detector, shook my hand, and wished me well. They never checked me for metal or inspected or x-rayed TOPS or the carry-on bag.

Brazilian Bullets

Brazil's São Paulo/Guarulhos–Governador André Franco Montoro International Airport is as busy as its sixty-seven-character name suggests. On September 9, 1997, on my way home from carrying out experiments described previously, I was accompanied by Brad White, a first-year student at Texas Lutheran University, as we waded through thick crowds at an American Airlines departure gate. The security people were unhappy about the electronics in our carry-on bags and asked detailed questions about what we had been doing in Brazil.

Fortunately, the plane's captain, and letters and emails from Dr. Paulo Artaxo and NASA about our three weeks in Alta Floresta, rescued us, and the security people became almost casual. They let our carry-on bags through with only another visual inspection and did not bother to open my laptop computer or a yellow box full of instruments.

We then advanced to the X-ray station, where the operators demanded that all the instruments be x-rayed. My trip log records that they said, "No problem."

We finally made it through the maze of security and were in line to exit the crowded terminal for the bus to the plane when two airline people pushed through the passengers, shouting, "Brad White! Brad White!"

Brad and I raised our hands, and they insisted that Brad go with them so they could search his checked baggage. I assumed that Brad's big knife looked suspicious when x-rayed. While I did not want us to be separated, they told me to get on the bus, and Brad would soon join me.

But Brad did not arrive, and I realized it must have been my bag that they wanted to search, so I left the bus and told the gate agent. She ran over to the security people and then ran back to me and confirmed that it was Brad's bag that concerned them.

After I returned to the bus, a gate agent jumped inside and shouted my name. I raised my hand, and he politely ordered me off the bus. After a few minutes, two armed security men arrived in a van with Brad's bag—which was my instrument case secured with a steel cable and lock to prevent theft of the expensive instruments. Brad's baggage tag was attached to the case. When I asked for Brad, they said he was already on the plane.

They asked me to unlock the bag, and I did. While I was showing them the instruments, the driver became concerned and told his partner the inspection should be done elsewhere.

While the driver rapidly drove us to a dark, remote area reserved for bomb inspections, the second agent said the X-ray inspection of one of the bags showed what appeared to be several large bullets. I immediately knew what caused their concern: Several of my homemade sunlight sensors are installed in brass compression fittings from a hardware store. They were perfect for their purpose (I still use them), but they resemble large bullets.

When we arrived at the bomb-inspection site, the agents jumped out and opened the back of the van. I told them I thought I knew what they were looking for, and they allowed me to reach inside the bag and retrieve the case holding the suspect light sensors. I plugged one of the sensors into the radiometer, switched on the power, and asked them to shine their flashlight into the tube installed in the bullet-shaped fitting. When they did, numbers appeared in the radiometer's display, and they realized the "bullets" were harmless. After resealing the bag, they quickly drove me to the plane.

I was the last person to board the plane. It was a relief to see that Brad was already aboard, having been taken there earlier by the same van. The other passengers had had to take a crowded bus to the plane, but we were escorted like VIPs. As the plane began to taxi, Brad and I agreed that our special trips to the plane were a fitting way to culminate our three weeks of demanding work in smoky Brazil. But that was a security experience I would not like to repeat.

A Closing Tip

If you carry instruments aboard a commercial aircraft, it is probably best to avoid showing them to fellow passengers. That includes Geiger counters, which are ideal for providing an estimate of the plane's altitude, for while the background count on the ground is around 11 per minute, it can exceed 400 per minute at 35,000 feet (10.7 km). Be sure the Geiger counter's sound is switched off, for the high-altitude count becomes a buzz that might alarm fellow passengers and flight attendants. (I speak with the voice of experience.)

On one occasion when this occurred, three female flight attendants asked to meet with me after the aircraft emptied. They wanted to know how cosmic radiation during their many long flights might affect their reproductive health. They were right to be concerned, and it's past time for aircrews to be fully informed about potential health consequences of years of flying.

21
MIMS FAMILY SCIENCE

When our children were growing up, I required that they keep their rooms neat, have supper with us every day, attend church on Sundays, and do a science fair project every year.

Science fair projects taught our children much more than science. They learned how to research and write their reports and include references to sources they used. They learned how to make attractive presentation boards and prepare slide shows or PowerPoint presentations. They learned how to fill out the bureaucratic paperwork required for science projects. And they learned how to respond to questions asked by judges and how to present their projects before an audience. Their projects earned many awards and scholarships.

Parental involvement in their children's science projects should be limited to providing only basic assistance. For example, parents often pay for some of the things used in a project, especially the display board. But parents must understand that their role should be limited to advising and assisting when an extra hand is needed. It is unfair to everyone when parents do most of the work. As a former science fair judge, it is easy for me to determine by asking a few questions whether a student's parents did more of the project than the student.

My policy was to provide ideas and help pay for their supplies and display boards. I also made sure they completed their projects on time, which is why I urged them to do most of their research during the summer. Some projects became so advanced that we sought advice from a professional. If a teacher, expert, or science fair judge expressed skepticism about their findings, I told them to follow the approach to doing science I had followed since high school and have mentioned before: If you are confident of your data, but a scientist is not, believe your data, not the scientist.

Minnie Discovers a Rare Mushroom

While Minnie never attended college, she was highly supportive of our children's science fair projects. After all, we met when she was a secretary in the Biophysics

Division of the Air Force's most important scientific laboratory. Perhaps Minnie's major contribution to our children's science fair achievements was her remarkable gift of observation that she passed on to them.

Shortly after moving to our place on Geronimo Creek, we were looking down into a secluded gully lined with tall trees when she noticed a tiny animal some 15 feet below us on the gully floor. It required half a minute for me to see what she saw; it was the first shrew that either of us had seen.

In November 1998, Minnie was standing on the back porch while I was talking to her from the yard when she suddenly exclaimed that she had seen a curl of smoke emerging from the ground. I turned around but saw nothing, for the "smoke" had already dissipated. We walked to the site to investigate and saw a star-shaped fungus unlike any we had ever seen. The fungus was not in any of my nature books, and I had no idea what it was, so it was quite a surprise to walk back inside the house, open the latest issue of *Texas Parks & Wildlife*, and find an article about what Minnie had just seen.

Minnie had discovered a Chorioactis geaster, one of the world's rarest fungi. Early Texas settlers called it the devil's cigar, for it emerges from the ground looking much like a cigar, opens into four to six cinnamon-hued segments, and emits occasional puffs of spores that look exactly like smoke. This rare fungus grows from the rotting roots of dead cedar elm trees. The fruiting bodies emerge during late fall, when the soil is moist. It is occasionally found in only half a dozen Texas counties and in Kyushu, Japan, and is one of only a dozen or so fungi that emits an audible hiss when it ejects its spores.

After I wrote several articles about what some naturalists prefer to call the Texas star, I sent some samples to Dr. Donald H. Pfister, curator of the Harvard University's Farlow Library and Herbarium and one of the very few experts on the devil's cigar. We heard from botanist Shuichi Kurogi, Japan's leading authority on this rare mushroom, which in Japan is classified as critically endangered.

Shuichi has been featured on several television programs in Japan about the mushroom, and during November 2007, he visited our place. During wet years, a dozen or more specimens appear in our woods. But 2007 was a dry year, and we found only a few.

Shuichi is a skilled photographer, and he put his equipment to good use as he lay flat on the ground, capturing images of his quarry. After photographing the most photogenic specimen, he was so moved by the experience that tears came to his eyes.

Minnie's keen observation of a Texas star pollen plume led directly to my newspaper articles about that rare fungus reminiscent of earlier articles by mycologists. Soon, a movement developed to name the Texas star the Texas state mushroom. Thanks to state senator Dr. Donna Campbell and other Texas legislators, in 2021, Governor Greg Abbott signed a resolution awarding the Texas star that designation.

As for our children's science projects, Minnie helped them acquire their poster materials and joined me in visiting them when their projects were displayed at their schools and various regional events. She was especially interested in Vicki's project that found ammonia in the bottled water she was giving baby Sarah.

Eric Detects Underground Nuclear Tests

Eric's most ambitious science fair project for the 1986–87 school year was a homemade seismometer. A professional seismometer expert said his project would be a failure, for our house is built atop soil and not rock. I told Eric that the concrete foundation under our house might substitute for rock, and he decided to ignore the expert and proceed with his plan.

While I suggested the general idea, Eric assembled the instrument entirely on his own. He suspended a lead weight from a 1-meter length of optical fiber directly over a red LED mounted under a pinhole. Seismic events caused the end of the fiber to oscillate back and forth across the pinhole. A light sensor at the opposite end of the optical fiber detected changes in light intensity, which were amplified by a circuit Eric built and sent to a printer that he programmed as a chart recorder. When the seismometer was quiet, the printer printed a straight string of dots on a moving paper strip. When movement was detected, the dots were printed from side to side like a professional seismometer.

Eric installed the fiber and the sensor inside a metal tube he bolted to the side of our house. However, the slightest breeze caused so much seismic noise that it was necessary to peel back the carpet in a corner of his bedroom and bolt the contraption directly onto the slab (without telling Minnie but receiving her after-the-fact approval). The instrument responded to doors opening and closing and toilets flushing. It also responded to stormy conditions with enough wind to slightly move the house.

Once these noise sources were recognized, it was easy to detect trains passing by a mile away and large trucks crossing a bridge. Eric also detected distant earthquakes. He kept the seismometer on twenty-four hours a day, and Minnie and I wondered how he could sleep with the printer's constant *clack-clack-clack* all night.

The ultimate test was whether Eric's seismometer could detect an underground nuclear test in Nevada some 1,200 miles (1,930 km) from his bedroom. This was shortly before such tests were banned by treaty, and the Nuclear Test Site was required to provide notice before each test.

I will never forget watching Eric in his bedroom when he was waiting to see whether his seismometer would detect a scheduled nuclear test. The scheduled time arrived, and Eric began clocking the estimated arrival of the first shock wave. As we watched, the noisy printer plotted the vibrations from the shock within seconds of the predicted time. It then indicated the arrival of the slower second and third waves. A second nuclear test a few weeks later also indicated the three waves from the explosion.

Eric's project earned first place at Seguin High School and the most awards ever given at the Alamo Regional Science and Engineering Fair. He also won a trip to the International Science and Engineering Fair. A museum in Canada exhibited an interactive seismometer inspired by Eric's project. Eric graduated from Texas A&M University with a degree in earth science, and his extensive computer experience during his science fair years led to his position as an IT security director within a major university system.

Vicki Measures the Sun's Rotation and Solar Flares

When Sarah was a baby, Vicki's science project was testing the quality of water using an aquarium test kit. We were quite surprised when she detected ammonia in the bottled water we were giving Sarah, for this suggested possible bacteria contamination. Minnie tossed the unused water and found a source that passed Vicki's tests.

Vicki's eighth-grade project was measuring the rotation of the sun, a project she completed when she was twelve during the summer before school started. She did this by projecting the image of the sun against a sheet of paper held fast by a clipboard. She then used a pencil to mark the presence of sunspots. She repeated this procedure each day for a few weeks until sunspots she marked had moved across the solar disk. She soon learned that the sun rotates once about every twenty-eight days. This project was so elegantly simple that I selected it for my first installment of The Amateur Scientist in *Scientific American*. The column reported this about Vicki's project:

> *If you have a small telescope, you can easily assemble a permanent solar projection observatory. Two years ago Vicki Rae Mims, my teenage daughter, did just that. Vicki constructed her system from scrap lumber, a clipboard, a cardboard shade and a small 10-power telescope. . . . For four months Vicki made almost daily observations of sunspots using this system. She measured the rotation of the sun by tracking several spots and groups completely across the solar disk. Vicki observed that some spots moved across the solar disk faster than*

others, a difference that occurs because the gases at the solar equator rotate more rapidly than the gases toward the poles.[1]

Vicki's most significant project was detecting X-class solar X-ray flares using a Geiger counter during 1989, a year with major solar activity. Similar to the negative advice Eric received from a scientist about his seismometer project, a NASA scientist advised Vicki that a Geiger counter would probably not detect solar flares. I told her that meant it probably would, and she detected twelve X-class solar X-ray flares by the increase in the background radiation that accompanied them.

Vicki's solar-flare project failed to earn the recognition it might have received when one of the five judges could not find her project and never evaluated it. Space was limited, and her project was placed on the floor above with middle school projects. Since the total number of judging points determined awards, her project was outscored a few points by another project and she came in second. Nothing could be done to correct the problem, but Vicki had the last word when her project became a chapter in a book by science author Joseph J. Carr.[2]

Vicki graduated from Texas A&M with a degree in English. Her public speaking experience during her science fair years served her well during her nineteen-year career with Mothers Against Drunk Driving.

Sarah's World-Class Discovery

In July 1995, Minnie, Sarah, and I flew to Colorado on a science vacation. Vicki stayed home to take a Spanish course at Texas State University and to operate the Brewer automated ozone instrument, which the University of Georgia, under contract to the EPA, had placed on my property. She even learned how to measure the ozone layer at night by directing the Brewer to look at the moon, something only a handful of people have done.

Vicki had developed altitude sickness during our trip to the summit of Mauna Kea in Hawai'i in 1992, when she was sixteen. Sarah was only eight years old, and she experienced worse symptoms when we drove to the 14,265-foot (4,348 m) summit of Colorado's Mount Evans on the highest paved road in the US.

Just as Vicki did not complain during the Mauna Kea trip, Sarah said nothing at Mount Evans. But she was suffering while seated in snow as she did her best to measure the sun's ultraviolet for her fourth-grade science fair project. She recently

---

1. Forrest M. Mims, "Sunspots and How to Observe Them Safely," *Scientific American* 262, no. 6 (June 1990), 130–3.
2. Joseph J. Carr, *Radio Science Observing*, vol. 1 (Independence, KY: Delmar Learning, 1998).

told me, "Even though I was sick, it was still a great experience."

During the Colorado trip, I left Minnie and Sarah at our motel while I drove to the summit of Pikes Peak to make measurements. While I hoped that Pikes Peak would provide a simpler alternative than the Mauna Loa Observatory, I was wrong. The many cars arriving and leaving the summit created considerable dust. The air was also filled with plant debris blown up from the forest below. The updraft was so strong that paragliders jumping from the summit rose in the sky before descending. This experience, and the fact that Mount Evans was inaccessible during winter, persuaded me that the Mauna Loa Observatory was a far better atmospheric monitoring site.

The next day, we visited the Florissant Fossil Beds National Monument and the adjacent Florissant Fossil Quarry, where for a small fee, visitors can search for fossil insects, fish, and leaves.

We then visited a waterfall on the Roaring River in Rocky Mountain National Park. I found a long pole and attached to one end the same UV sensor and data logger I had used to measure the UV reflected from whitewater off the Kona Coast in Hawai'i. I then leaned out over the rushing water and held the pole so that the sensor was looking straight down while it recorded data. I expected to detect significant UV reflected from the mass of white foam, as in Hawai'i, but this did not occur. Perhaps the microorganisms in ocean surf cause far more reflectance than pure mountain water, a hypothesis I have not published until now.

This thought about microorganisms brings me back to Sarah's world-class discovery. Perhaps because she observed how well Eric and Vicki had done with their many science fair projects, Sarah was especially enthusiastic about her projects. They began in first grade with a project to make a pH indicator from purple cabbage juice.

For another elementary school project, Sarah raised monarch butterflies from larvae collected from milkweed plants in our field. When they emerged from their chrysalises, she tagged and released them.

When Sarah reached high school, I was concerned that I might have made a mistake requiring Eric and Vicki to do science projects every year, so I called them to seek their advice about what to do in Sarah's case. They both emphasized how much they had learned from their projects and recommended that Sarah should continue hers.

Sarah's ninth-grade project was measuring heat islands during a family vacation from Texas to New Mexico by mounting a temperature sensor outside her back seat window. (One of her judges did not know that a heat island is a natural or artificial object or environment warmer than its surroundings.)

Sarah's tenth-grade research was using microscope slides placed outdoors to capture dust arriving in Texas from the Sahara, a project that earned many awards despite the science fair official who mocked her $29.95 toy microscope. While driving back home from the Texas State Science Fair, she asked me for new ideas. I told her that bacteria had been found in Saharan dust, and maybe she could find out if bacteria are associated with Asian dust that sometimes arrives in Texas during spring.

When spring arrived, Sarah was ready with her microscope slides. We also bought her a package of Petrifilms, square plastic films coated with a thin layer of nutrient that stimulates individual bacteria or fungal spores to form visible colonies. She began exposing the film in our field much like she had earlier exposed microscope slides to collect Sahara dust. Soon, her films were covered by colonies of spores and bacteria.

Previously, Minnie, Sarah, and I had met Dr. Thomas E. Gill at Texas Tech University to discuss Sarah's African-dust project, so I notified Tom to tell him Sarah had detected spores and bacteria in Asian dust. Tom replied that there had been no Asian dust over Texas when Sarah exposed her Petrifilms, but there had been substantial smoke from fires in Yucatan in Mexico.

How spores and bacteria could survive fire did not make sense, so I told Sarah she should collect some dry native grass from the field, place it in a metal trash can, and ignite it while holding Petrifilms attached to a pole in the smoke. She did this and followed up with a second set of Petrifilms that were not exposed to the smoke. After a few days of incubation, the Petrifilms held in the smoke were loaded with colonies, while the other Petrifilms had only a few.

In 1997, I had captured spores and bacteria on agar trays in Brazil during major smoke events. These results confirmed my hypothesis that there would be more nonpigmented bacteria (white colonies) than pigmented bacteria (orange colonies) on smoky days due to the suppression of solar UV by smoke.

My hypothesis assumed that pigmented bacteria are better protected against UV than nonpigmented bacteria. The significance of this finding is that infectious bacteria are generally nonpigmented, and this could explain the increase in respiratory distress during the burning season. Sarah's finding was a major surprise, for it had never occurred to me that the overall number of bacteria would be higher on smoky days.

There was nothing in the scientific literature about mold spores and bacteria in smoke from biomass fires, so it was clear that Sarah had made an important discovery. But a well-known biologist at a major university denied this was possible. He said that spores would be destroyed by the fire that produces smoke. This was reminiscent of the negative advice from professional scientists about Eric's seismometer, Vicki's solar flare project, and some of my projects. I advised Sarah not to acknowledge his erroneous response in her report, for he would be embarrassed.

How do microbes survive the fire that produces the smoke in which they are entrained? The answer became obvious when Sarah burned a patch of dead ragweed in our field. Some of the air rushing in to feed the fire was warmed and rose skyward before entering the flames. Presumably, living microbes were carried up with the smoke plume.

While her science fair project, "Smoke Bugs," earned major awards, some of the judges asked whether it was possible that the spores and bacteria she was detecting on Petrifilms did not arrive from Yucatan but, rather, might have been blown into the sky somewhere between the Texas coast and our field. To resolve the contamination possibility, Minnie and I drove Sarah to Padre Island, off the coast of Texas, on a day with considerable smoke arriving from Yucatan and a few months later, when the wind was from the same direction but there was no burning in Yucatan.

On both trips, Sarah flew from a kite a homemade air sampler she had made from a plastic cup with a microscope slide clipped inside. On the smoky day, the slide collected carbon particles and spores. On the clear day, there were few spores and no carbon. These findings proved that she had discovered that biomass smoke carries living spores and can do so over a considerable distance.

Meanwhile, I helped transform Sarah's science fair report into "Fungal Spores Are Transported Long Distances in Smoke from Biomass Fires," a formal paper we submitted to *Atmospheric Environment*, the leading atmospheric science journal. The paper was assigned fast-track status and quickly reviewed by two experts. Its conclusion included this hypothesis: "A beneficial role for spores carried skyward with forest fire smoke could be reinoculation of fire-sterilized soil with mycorrhizal fungi and the dispersal of such symbiotic fungi to new locations."[3]

This was Mark Hartwig's idea, and we included him in the acknowledgments: "Mark Hartwig suggested the design hypothesis that dispersal in smoke of mycorrhizal fungal spores might be beneficial."[4]

Mark's design hypothesis raised a red flag in the mind of one reviewer, who objected. We simply told the editor that Mark's idea was a valid product of his worldview and should remain despite its intelligent design connection. The editor

3. Sarah A. Mims and Forrest M. Mims III, "Fungal Spores Are Transported Long Distances in Smoke from Biomass Fires," *Atmospheric Environment* 38, no. 5 (February 2004), 651–5.
4. Mims and Mims.

agreed, and Sarah's first scholarly paper was accepted after only twenty-one days and published in February 2004. As of this writing, the paper has received 111 citations in the scholarly literature. (As this manuscript was being typed, a request for a copy of the paper arrived from a zoologist in Austria.)

Meanwhile, Sarah's second science fair project affirming her discovery earned many awards. She also received a $1,000 ($1,594 in 2023) Young Achievers Award and a trip to New York City from *Popular Mechanics* magazine. Minnie and I were also included. NASA produced "Smoke's Surprising Secret," a four-page website about Sarah's discovery.[5]

It also displayed a poster about Sarah's discovery at the Smithsonian Museum of Natural History in Washington and profiled her on a Smithsonian website with the title "Atmosphere Explorers — Sarah Mims: Up in Smoke": "You don't have to be a Smithsonian scientist to know a lot about the atmosphere! Teenage high school science student, Sarah Mims, made an unexpected discovery while working on a science fair project to see if spores of fungi could travel on air-borne dust. With a homemade, 35-cent smoke detector attached to a kite, she found that smoke particles from major fires in [Yucatan] had carried live fungi spores across the Gulf of Mexico to her home in Texas. Since some fungi can spread plant diseases, Sarah's finding suggests that the atmosphere can transport smoke-riding, disease-causing microbes long distances."[6]

Sarah also received a congratulatory message from Dr. Yoram J. Kaufman, one of NASA's top experts on atmospheric chemistry, who was a part of the 1995 SCAR-B contingent in Brazil:

Date:10/27/2003

Dear Ms. Sarah Mims

Your paper on the survival of bacteria in fires and long-range transport to Texas is very interesting and innovative. Your measurements are very exciting in their simplicity and outcome. You were very smart not to discard the "unrelated" measurements of bacteria that were not associated with dust and recognize the importance of the measurements regarding the survival of the bacteria in the fire. The implications to agricultural activity are very important. Once again we

5. "A Scientific Family," Earth Observatory, January 7, 2004, https://earthobservatory.nasa.gov/features/SmokeSecret.
6. "Sarah Mims: Up in Smoke," *Forces of Change*, Smithsonian Environmental Research Center, https://forces.si.edu/Atmosphere/03_00_02.html.

see that in our global village what one farmer does may affect another farmer an ocean away.

Your work shows the highest standards of science work and intellectual activity. I am sure that you will be one of the best candidates, with the highest intellectual ability and curiosity, in any university you may apply for.

Regarding experimental science, once I read an interesting observation that a Nobel prize recipient mentioned regarding his finding. He asked the question: "what is the difference between the Nobel prize and the trash can". Then he answered that "it is the degree to which you believe in your instrumental procedures". Apparently you had enough confidence in the quality of your experimental work that you did not discard the information about bacteria without dust and got your exciting new discovery.

Good luck with your studies and future work. If you land in atmospheric sciences, maybe you will consider joining us as a scientist here in NASA GSFC.

With best regards
Dr. Yoram J. Kaufman
NASA/GSFC Senior Fellow, Climate and Radiation Branch

Note his endorsement of the principle I taught all our children and have espoused elsewhere herein: If you are confident of your experimental protocol, always give your data a higher priority than the opinions of experts.

During one of her appearances, Sarah and I met Nobel Laureate Jack Kilby, who received the award for his invention of the integrated circuit when he was an engineer at Texas Instruments. He autographed the back of a TI calculator I brought for the occasion.

Sarah echoed Eric and Vicki when asked whether her science projects played a role in her working life after college: "Yes, especially with public speaking. I became very nervous when asked to speak in public. But because of having to defend my research before many panels of judges, I know how to properly prepare for public speaking. All the nervousness goes away soon after I begin to speak. Science fair experience has also helped me with technical writing and organizing various projects at work. I know how to outline a project and make everything come together."

When asked her advice for students who want to do serious science projects, Sarah again echoed Eric and Vicki, "If someone tells you it can't be done, that often means it can be done." One writer used this as the title for her interview with Sarah.

Recently, I learned about the smoke research of Dr. Leda Kobziar of the University of Idaho and notified her about Sarah's findings. She already knew, for she replied, "That was a critical pioneering study indeed and is cited in each of my papers . . . . I've told the story of your paper many times (the 'genius father-daughter team') and in many interviews . . . . We can only imagine what we would know now had Sarah's work been continued 20 years ago. It has always baffled me (as you know) that others (especially aerobiologists) didn't continue the work back then. It is a great story indeed to inspire young scientists to follow their curiosity. It certainly has inspired me!"[7]

Sarah's "Smoke Bugs" discovery, a classic example of serendipity, has become a new field of science, which Kobziar has named *pyroaerobiology*.

Sarah graduated from Texas A&M with a degree in rangeland ecology and works as an environmental remediation and disposal expert. You can learn more about her "Smoke Bugs" project by searching online (keywords: "Sarah Mims smoke"). Ksanna Mazhurina interviewed Sarah about her "Smoke Bugs" project for GlobalLab.[8]

---

7. Leda Kobziar, email to Forrest M. Mims, April 24, 2021.

8. Ksanna Mazhurina, "If Someone Tells You It Can't Be Done, That Often Means It Can Be Done," GlobalLab, April 2, 2014, https://discovery.globallab.org/en/blog/message/62a894fe-ba67-11e3-8fde-08606e697fd7.html.

22 ROLEX REDUX

The 2016 Rolex Awards were presented during the fortieth-anniversary celebration of the awards November 14–15 in Los Angeles, and Rolex arranged travel for former laureates. During the reception, a sharp-looking gentleman rapidly approached me and introduced himself. He was Dr. Andrew McGonigle, a Scottish physicist turned volcanologist and a reader in the Geography Department at the University of Sheffield. Andrew had received a 2008 award for his pioneering measurements of volcanic emissions using an instrumented drone.

I was surprised when Andrew said I had inspired him to enter the competition. I was equally surprised when he said Rolex would soon begin to fund research by collaborations of two or more laureates.

One of my key research goals had long been to conduct an organized UV-B survey of Hawai'i Island, a project I had begun in 1994, and I proposed this to Andrew. He agreed, and we soon began work on "Into the UV," a 6,000-word proposal to Rolex. The proposal mentioned that the Microtops II instruments we planned to use sprang directly from my 1993 award. I had also reserved $1,000 ($2,093 in 2023) from the award to make some two dozen light and ultraviolet sensors, several of which we planned to use during the project.

On August 25, 2017, Rolex notified us that our project had been approved. We had plenty of time to prepare, because our proposed schedule would not begin until July 2018. The long delay was necessary, for we wanted to make UV measurements when the sun is straight overhead at the zenith, which occurs only between latitudes of 23.5 degrees north and south of the Equator. At our planned measurement sites on Hawai'i Island, the sun would be at the zenith on July 24 and within a few degrees of the zenith the week before and after.

Sunny Sam

A key objective was measuring the UV on a human face, but we didn't know the best approach. While visiting a hobby store, I noticed a *Phantom of the Opera* half-mask. Reasoning that UV sensors could be attached to the mask, I bought one and fitted UV sensors on the cheek and forehead. The result was a working setup that stimulated either ridicule, laughter, or both from Minnie and those who saw it when I posted it on Facebook.

I returned to the hobby shop and found a foam plastic mannequin head that would work even better, for it could be mounted on a tripod and its position easily controlled. I soldered ultrathin wires to each of seven calibrated UV sensors and mounted them on the eyes, ears, cheeks, and forehead of the mannequin head. Phone plugs soldered to the sensor wires were inserted into the input ports of a pair of Onset data loggers fastened on each side of the head's neck with Velcro.

I mounted the instrumented head on a $25 battery-powered turntable that could be mounted on a tripod. The head rotated 100 times per hour. When the data loggers were programmed to store a UV measurement from all seven sensors every second, the loggers collected 420 samples each minute and 252,000 samples per ten-hour day. After I posted a photo of the head on Facebook and asked for a name, sixty-three suggestions arrived; I selected Sunny Sam.

Kilauea, the world's most active volcano, had been in eruption since 1988. The large volume of sulfur dioxide it emitted formed considerable vog. (That term is the popular name for volcanic haze.) On May 3, less than three months before our arrival, Kilauea began a major eruption from the Leilani Estates subdivision near Pahoa, Hawai'i. The massive plume of sulfur dioxide (SO_2) gas from this eruption would provide unexpected measurement opportunities for Andrew and his doctoral student, Thomas Wilkes.

Hawai'i Island Field Research

Sunny Sam and I arrived at Kona on July 17. The following day, I drove to the Mauna Loa Observatory and installed the Colorado State University instruments I manage on the solar deck. Sky conditions were ideal, so I calibrated four of Sunny Sam's UV sensors against a professional Solar Light UV sensor.

On July 19, Andrew and Tom arrived. Tom brought two ultraviolet cameras he had helped develop from a smartphone camera sensor. I brought three Microtops IIs and a host of UV measuring instruments, cameras, data loggers, and two laptops. I also brought two Mavic Pro camera drones.

We found an ideal site for our sea-level UV measurements at the Old Kona Airport Park, only a few miles from our hotel. On July 21–24, we spent the day at these sites while Sunny Sam was recording UV at the beach. Andrew and Tom took UV photos of us and various plants and flowers. When the sun was straight overhead at Lahaina noon on July 24, we photographed vertical posts without shadows.

On July 29, we visited MLO. I needed to check the Colorado State radiometers, and Sunny Sam wanted to make some UV measurements. We were surprised to see Dr. Emma Liu and her team collecting samples at MLO. Emma is a Leverhulme Early Career Research Fellow in Volcanology in the Department of Earth Sciences at the University of Cambridge. She's following Andrew's lead in using drones to measure volcanic gases and was in Hawai'i to sample the air around the island for emissions from the ongoing Kilauea eruption.

July 31 was Volcano Day. The pastor of Puna Baptist Church led us to the Pahoa High and Intermediate School, where he introduced us to the principal. She permitted Andrew and Tom to set up Tom's UV camera and spectrometer on a second-floor balcony with an ideal view of the volcanic plume. While I took visible photos of the huge plume, they took dozens of UV photos and spectrometer measurements of the plume, which Andrew described as the largest he had seen in his years studying active volcanoes.

The thick, gray plume rose a few thousand feet (600 m), where it formed a large cumulus cloud that streamed toward the south. Later that afternoon, we drove under the plume along Highway 130 while Andrew held Tom's UV spectrometer out his window to measure the SO_2 in the plume.

Some 700 nearby homes had been destroyed by lava. Highway 130 was threatened by large cracks across the pavement that emitted clouds of steam, but the highway was reopened after steel plates were placed across the cracks.

On August 2, we made a second visit to MLO to calibrate our instruments. We then drove to Kona International Airport and parted ways after fifteen highly productive days. Performing field research with Andrew and Tom was a highlight of my science career. Thanks to the Rolex grant, we collected a huge amount of data and many hundreds of visible and UV photographs. We did so while spending every night in hotels with a breakfast buffet every morning, a restaurant meal for supper, and a shower every evening. That was a first for my many years of Hawai'i research, during which I had spent some 235 days and nights at the observatory eating grocery store food and Subway sandwiches during trips to the coast for a fast swim and groceries.

Sunny Sam's Major Finding

While I was testing Sunny Sam in Texas before the Hawai'i field trip, it soon became obvious that his UV sensors received more UV at midmorning and midafternoon than at noon. This was confirmed repeatedly in Hawai'i, and on August 10, I posted the following on Facebook: "Sunny Sam is working hard in Hawaii—turning more than 1,000 times every day he's deployed. He's providing a new way to evaluate human exposure to the sun's UV rays that might result in a revision of the EPA's

UV Index. While the UVI is very helpful if you're lying flat on a beach, your face can receive far more UV at mid-morning and mid-afternoon anywhere from the Equator to at least 30 degrees north/south. (Sam approves this message.)"

Kilauea Stops Erupting

On August 4, I drove to Kalapana Seaview Estates Park, where residents allowed me to place Sunny Sam in operation. I then used Microtops II to measure the ozone layer through occasional gaps in the volcanic cloud overhead. This instrument measured up to 410 Dobson units (DU) of ozone, which was much too high, for the average of ninety-nine total ozone measurements from several sites far away from the eruption cloud was 270 DU.

Volcanoes do not emit ozone, but as Andrew and Tom had affirmed several days before, they emit considerable SO_2. This gas absorbs some of the same UV wavelengths as ozone, and that's why Microtops II gave ozone amounts up to 52 percent too high.

When I returned to the park the next day, the thick volcanic cloud cover was replaced by ordinary cumulus clouds, and the sky was much more open. The average of 128 ozone measurements by Microtops II was only 2 percent over the background ozone. Something had happened to the volcano's SO_2, but I did not know what it was until I was packing Sunny Sam and the instruments in the rental car and turning on the radio. A news program reported that Kilauea had nearly stopped erupting that morning. The *Hawaii Tribune* later reported that the Hawaiian Volcano Observatory described the dramatic drop in SO_2 emissions that followed: Emissions on August 3 indicated tens of thousands of tons of SO_2 coming from the fissure 8 vent, but just two days later, the emission rate was only about 200 tons per day.

The wonderful irony in the two days of measurements from the Kalapana Seaview Estates Park was that Sunny Sam and I had unknowingly made measurements during the final day of the eruption and the first day thereafter.

Moving to the Mauna Loa Observatory

The end of the 1983 Kilauea eruption fit perfectly with my schedule, and the next morning, I checked out of my Hilo hotel after a final visit to its breakfast buffet and moved to the tiny bedroom at the lonely Mauna Loa Observatory. The following morning, I put Sunny Sam to work measuring UV at the emergency helicopter landing area while I calibrated my instruments.

While checking on Sunny Sam shortly before noon, I saw what I had never seen in twenty-six years of annual MLO calibration visits: a large brown cloud of smoke was slowly drifting over the observatory! The smoke was from the Keauhou Fire (August 5–15, 2018), a major brush and forest fire on the southeast slope of Mauna Loa that covered most of the sky over MLO from 11:45 a.m. to 12:25 p.m. Brush fires caused by lightning, human activity, and military exercises are relatively common on Hawai'i Island during drought years, but smoke rarely reaches the observatory.

That solar-noon smoke event was a blessing in disguise, for UV is strongly absorbed by smoke. This smoke event provided an ideal comparison with the very thick smoke over Brazil during my 1995 and 1997 field campaigns. My original LED sun photometer used in both Brazil campaigns and the Microtops II used in the 1997 Brazil campaign were also in use during the 2018 MLO smoke event. The smoke event also provided comparison data for all the MLO sunlight instruments, including the Colorado State radiometers I had installed at MLO the day after I arrived in Hawai'i. The smoke event and the measurements made on the last day of the Kilauea eruption and the following day were among the most exciting of my science career.

The Hawai'i UV Survey Paper

Shortly after returning home, I began work as lead author on a lengthy paper about our findings in Hawai'i. In addition to Andrew McGonigle and Thomas Wilkes, the authors included Dr. Alfio Parisi, a noted Australian ultraviolet expert, and Dr. William Grant, the former NASA ozone and UV scientist introduced earlier. Parisi assisted with technical UV issues, and Grant wrote the section on UV and disease. We also included Dr. Joseph M. Cook, who was conducting parallel UV measurements in Greenland while we were in Hawai'i, and Dr. Tom D. Pering, who played a key role in developing the UV smartphone camera and spectrometer.

The paper's key point was its confirmation of findings by others that the EPA's UV Index needs revision. Because the UVI applies to horizontal surfaces, like people lying flat on a beach, people in tropical and temperate latitudes should be informed that their upright face is more vulnerable to UV at midmorning and midafternoon than at noon.

Preparing the paper was significantly slowed by the giant spreadsheets containing Sunny Sam's UV data. After many sixteen-hour days and important revisions by the coauthors, on January 14, 2019, "Measuring and Visualizing Solar UV for a Wide Range of Atmospheric Conditions on Hawai'i Island" was submitted to the *International Journal of Environmental Research and Public Health.*[1] The

1. Forrest M. Mims III et al., "Measuring and Visualizing Solar UV for a Wide Range of Atmospheric Conditions on Hawai'i Island," *International Journal of Environmental Research and Public Health* 16, no. 6 (March 2019), 997, https://www.mdpi.com/1660-4601/16/6/997.

11,200-word paper was then sent to three anonymous referees, and responding to their many concerns required a full month. The revised paper was accepted on March 12, 2019. As of this writing, 3,983 people have read some or all the paper.

23

MAVERICKS DO NOT RETIRE

During a lifetime pursuing science, I've learned many lessons that might assist readers who are thinking about doing a science project. Seven of those lessons stand out:

- **A formal education in mathematics and science is not required to design and assemble scientific instruments and make publishable discoveries with them.** As I have learned many times, hands-on experience trumps the classroom. I have met many scientists who know this well, for they specialize in fields far away from what they studied during college. When your project requires a skill or technique you were never taught, spend time in libraries and online to learn everything you need to know. That is what I did when I began making atmospheric measurements and when I needed to understand how to use basic calculus to analyze the data from my twilight instruments.
- **Investing money in a carefully planned science project might provide more intellectual growth than a semester of college courses.** College is expensive, especially if you lack assistance. The money required for a three-hour science course might be better spent acquiring the materials needed for a science project. Thanks to the internet, you can find used and even new tools, parts, instruments, and other scientific supplies for very reasonable prices.
- **When a scientist says your carefully planned project will not succeed, maybe it will.** I have learned this lesson many times, and so did all three of my children when they were doing their science fair projects.
- **Persistence is important.** Wishing my travel aid for the blind could be fitted onto eyeglasses was not enough. I had to keep trying. An especially significant value of persistence is in measuring natural phenomena as well as or even better than the professionals. This will give you far more credibility than a string of college degrees.
- **Pursue projects you enjoy.** I so much enjoyed doing projects for science fairs during middle and high school that I have never stopped doing them. Science projects are my lifelong hobby.

- **Share your projects and results with the world.** I began this approach in 1969 with my first magazine article about how to make a model-rocket light flasher. I followed through with a few thousand articles in newspapers and electronics and computer magazines. Since 2008, I have described many of my science projects in *Make* magazine, the most creative space for makers in the known universe. Then there's the internet, where you can find NASA, NOAA, and many professional scientific organizations, along with other amateur scientists who post their projects on websites.
- **If you can do serious science without a degree, don't be concerned if you are ridiculed for your lack of professional qualifications or your personal beliefs.** Earlier I related how the famous atheist Richard Dawkins ignored my science while ridiculing my personal beliefs in a string of emails he has not allowed me to share. Others who share Dawkins's beliefs have been much more courteous. Consider Michael Shermer, the well-known skeptic who cofounded the Skeptics Society and edits *Skeptic* magazine. I met Shermer during a meeting of the Society for Amateur Scientists organized by my friend Shawn Carlson, who is both a friend and a skeptic. While Shermer and I hold opposite views on evolution, we are both bicycle fanatics. He was as intrigued by the 1,000-mile bike trip I organized for teens in my church in 1974 as I was in his role as a professional long-distance bicycle racer. During our meeting and in the emails that followed, not once did Shermer ridicule my Christian faith.

These lessons are on my mind as I continue the daily atmospheric measurements begun in 1990 while also pursuing other projects. Since this memoir is being completed during my twilight years, I shall conclude by describing some of my recent and ongoing science projects.

The 30-Year Paper

As reported in prior chapters, on February 4, 1990, I began a one-year project to prove that a self-taught amateur can do serious science with the homemade LED sun photometer that impressed the editor of *Scientific American* in 1989. That initial year has yet to end, and my measurements of total water vapor, aerosol optical depth, and the ozone layer during this ongoing research have passed 32.5 years. The water vapor and optical depth readings have become the longest-running series of such data compiled in the US since the Smithsonian Astrophysical Observatory conducted long-term measurements at Table Mountain, California (1926–1957). These measurements are why the Intergovernmental Panel on Climate Change (IPCC) designated me an expert reviewer of several sections of its 5th (October 2014) and 6th (March 2023) Assessment Reports.

Thirty years is a climatological mean, the length of time weather observers employ to define the average temperature, precipitation, and humidity at specific weather stations. Therefore, I proposed to *Bulletin of the American Meteorological Society* (*BAMS*), the leading weather-related journal, a paper to be entitled "A 30-Year Climatology (1990–2020) of Aerosol Optical Depth and Total Column Water Vapor and Ozone Over Texas." After the editors approved my suggestion, I spent several months documenting the data, establishing its accuracy, and preparing graphs. The completed paper was then sent to three independent reviewers, all of whom were curious about my background as an amateur scientist.

Their major concern was the 30-years of total water vapor, which clearly showed the ups and downs of El Niños and La Niñas. But the 30-year trend was perfectly flat. This was unexpected, for global warming science assumes that total water vapor should increase as CO_2 increases, for warm air can hold more water vapor than cool air.

My first homemade LED sun photometer that impressed the editor of *Scientific American* in 1989 provided all 30 years of the water vapor and optical depth measurements. It also survived prior peer review, for I described its design in a leading journal ("Sun photometer with light-emitting diodes as spectrally selective detectors," *Applied Optics*, 1992, 6965-6967). I had also published the first 12 years of its water vapor measurements in another leading journal ("An inexpensive and stable LED Sun photometer for measuring the water vapor column over South Texas from 1990 to 2001," *Geophysical Research Letters*, 2002, 29, 20-1 to 20-4.)

Nevertheless, the reviewers wanted the paper to include total water vapor measurements from an independent nearby instrument, for long-term stability of water vapor suggests the same for temperature. I called Brent Holben, manager of NASA's AERONET to ask if any of their robotic sun photometers had lengthy water vapor records. Brent suggested I check the 20 years of data from an AERONET instrument in Oklahoma. I did so and was surprised that it showed a decline in total water vapor. So did some satellite measurements over the central US. I added these facts to the paper, the reviewers approved, and the paper was published in *BAMS* on January 1, 2022. (The Covid pandemic delayed the print version until late 2023.)

While the data are only valid for my site, it illustrates how climate is regional in nature. Total water vapor over my site, much of the central US, northern China, and elsewhere is flat or even declining while elsewhere, especially over oceans, it is increasing. The implication is that warming and cooling trends should be derived from regional measurements rather than only the IPCC's global models that lump all regions together.

The Twilight Project

No time of day is more visually appealing than twilight, especially when the glow is enhanced by a semipermanent mist of sulfuric acid droplets in the lower stratosphere around 10 to 15 miles (16 to 24 km) high. This aerosol layer is created and refreshed by volcanic eruptions and emissions from major forest fires. Major volcano eruptions can greatly inflate the stratospheric aerosol layer, and this leads to extended, colorful twilights.

My twilight studies have benefitted from *Sunsets, Twilights, and Evening Skies*, by Aden and Marjorie Meinel, a wonderful book based on the authors' lifetime of sky observations. The dedication page states in Hebrew and English, "The heavens declare the glory of God and the vault of heaven shows the work of his hands. Psalm 19:1." The book closes with, "O God of our salvation! . . . thou dost make the dawn and the sunset shout for joy!" The friendly narrative and superb photographs between these quotations make this book a keeper, which is why my copy is beginning to fall apart.

Pinatubo Twilights

As recounted previously, during a cruise to the Gulf of California to observe and measure the ozone layer during the solar eclipse of July 11, 1991, Minnie and I were treated to a brilliant red twilight glow seventeen minutes after sunset the evening of the eclipse. This stunning twilight was caused by volcanic aerosols injected high into the stratosphere by the eruption of Mount Pinatubo twenty-seven days earlier half the world away.

Brilliant, long-lasting twilight glows appeared over Texas after we returned home. Trees block the western horizon at our place, and for two years, I often rode my bicycle to a vantage point three miles away to take more than 500 photographs of Pinatubo twilights.

The book by the Meinels includes charts that estimate the height of volcanic aerosols by measuring the duration of the twilight from sunset to glowset, when the illuminated portion of the sky slips below the horizon. This qualitative approach works reasonably well, and in August 1999, I created a spreadsheet that estimated the height of volcanic aerosols based on the formula the Meinels had used. Based on the Meinels's equations, the highest height of the Pinatubo aerosols I measured was around 26 miles (42 km). That was very close to what professionals estimated. But I wanted a quantitative method based on physical measurements instead of visual estimates.

Scientists use aircraft and instrumented balloons to directly measure the altitude of atmospheric aerosol layers. However, these methods are expensive and require teams of people.

In another method, which I described in chapter 16, a laser beam is pointed straight overhead when the sky is dark. Some of the photons scattered downward from aerosol layers are captured by a telescope and focused onto a sensitive light detector. The speed of light is about 1 foot (0.3 m) per nanosecond (a billionth of a second), so the distance in feet between the laser and the aerosol layer is half the time in nanoseconds between the laser pulse and its return. This method, called lidar—which I have discussed previously—resembles radar, which uses radio waves.

Former Mauna Loa Observatory station chief John Barnes designed and assembled the powerful MLO lidar, and many nights, I've photographed its brilliant green beam shining up to the stratosphere. As I have mentioned before, John taught me how to operate his lidar, which I have done a dozen times when I was alone at the observatory.

Twilight Photometry

Those Pinatubo twilights made a major impression on me, and I became determined to learn how to measure the altitude of aerosols in the atmosphere. A homemade lidar system would be expensive. And because it might affect the eyes of pilots flying overhead, I might also need permission from the Federal Aviation Administration to use it. Therefore, I turned to an inexpensive alternative in the most expensive book I have ever purchased, *Polarization and Intensity of Light in the Atmosphere*, by former MLO director Dr. Kinsell L. Coulson.[1] I bought this book for $100 ($192 in 2023) in August 1996 at the International Radiation Symposium (IRS) at the University of Alaska Fairbanks. It and the Meinels's book are the best science books in my modest collection.

Coulson's book explains how he and a few other scientists have estimated the height of aerosol layers by measuring the intensity of the twilight glow from sunset to an hour or so thereafter. The method, which dates to the 1920s, is called twilight photometry.

Australian scientist Dr. E. K. Bigg was among the leading advocates of twilight measurements, and he described some of his results in "The Detection of Atmospheric Dust and Temperature Inversions by Twilight Scattering."[2]
Dr. Frederick Volz, another pioneer of the method, was a professional atmospheric scientist who invented homemade sun photometers. When I met Volz at his home in Massachusetts in the late 1990s, he gave me one of his homemade instruments.

1. Kinsell L. Coulson, *Polarization and Intensity of Light in the Atmosphere* (Hampton, VA: Deepak, 1988).
2. E.K. Bigg, "The Detection of Atmospheric Dust and Temperature Inversions by Twilight Scattering," *Journal of the Atmospheric Sciences* 13, no. 3 (June 1956), 262–8.

My Homemade Twilight Photometers

When pointed straight up at the zenith sky during twilight, a light-sensitive photometer can detect layers of smoke, dust, and sulfuric acid aerosols caused by forest fires, volcano eruptions, and even meteor smoke. A set of equations can then extract the altitude of those aerosols based on the time they are detected after sunset or before sunrise.

The small number of scientists who have employed the twilight method use instruments equipped with telescopes and sophisticated electronics. I wanted a much simpler method that could be published in my *Make* magazine column. All my previous sun photometers used a light-emitting diode as a light sensor, but they were not sensitive enough to detect the twilight glow.

An external lens to collect more light would have helped, but I wanted a photometer that needed no optics and could be easily built. The solution was to increase the amplification (gain) of the integrated circuit connected to the LED light sensor by ten billion, which is far more than that recommended by the manufacturer. Electronics engineers said this would not work, which motivated me to prove it would. And I did so during tests at the Mauna Loa Observatory June 27–July 8, 2013.

My ultrasimple twilight photometers require no lens, and they use inexpensive red LEDs like those in vehicle taillights and near-infrared LEDs like those in TV remote controls. Their construction is described in *Make:* magazine.[3]

My inexpensive twilight photometers have measured the altitude of smoke from faraway fires, haze from distant power plants, African dust that arrives over Texas every summer, volcanic aerosols in the stratosphere, and meteor smoke in the mesosphere.

Dr. Thomas E. Gill, who advised my daughter Sarah about her discovery of living microbes in biomass smoke, moved from Texas Tech to the University of Texas at El Paso. He and Dr. Joseph Prospero, professor emeritus at the University of Miami, provided a valuable sounding board for my twilight charts depicting the altitude of dust layers. Both men are well-known atmospheric scientists, and Prospero is famous for his research on the transport of enormous clouds of dust from the Sahara desert to North America.

During a trip to Washington, DC, in October 2013, an hour before sunrise, I placed a twilight photometer on a table in the alcove between the Wall Street Deli and the entrance to the headquarters of the US Federal Emergency Management

---

3. Forrest M. Mims III, "Build a Twilight Photometer to Detect Stratospheric Particles," *Make:*, March 10, 2015, https://makezine.com/projects/twilight-photometer.

Agency (FEMA). Minnie was sleeping at the nearby Holiday Inn.

Because of the city's crime rate, the proximity to an emergency agency seemed ideal. It did not occur to me that my experiment would be considered an emergency by a FEMA police officer, who soon arrived to check out me and the twilight instrument. Minnie was right; I should not have installed the instrument in an easily carried ammunition box.

When I showed the police officer a graph that depicted how the instrument measured the altitude of aerosols in the sky, he allowed me to continue while keeping an eye on me from inside the FEMA building's front entrance. After the shift change, a second police officer arrived, and an exact copy of the first encounter occurred. However, I left immediately after sunrise with some wonderful data that showed my instruments could provide usable data from a brightly illuminated city.

How the Twilight Method Works

The next time you are watching a beautiful sunset, turn around and look east. If the sky is clear, you will see a pink band forming a wide arc just over the horizon. This antitwilight arch is called the Belt of Venus. The gray or purplish sky below the arc is the earth's shadow. As the sun continues to slip farther below the western horizon, the shadow continues rising under the pink band over the eastern horizon as the sun's rays illuminate progressively higher angles in the sky. Eventually, the visible pink band disappears, but sunlight is still passing higher and higher overhead.

Air molecules scatter sunlight, and their abundance falls with altitude. Therefore, the brightest part of the sky directly overhead is just above the earth's shadow as it rises higher in the sky after sunset. This means a photometer looking straight overhead receives sunlight scattered mainly from aerosols just above the shadow, the height of which can be calculated. If you plot the signal from a twilight photometer second by second during twilight, the resulting chart will show a smoothly curved, declining arc as the earth's shadow rises overhead. The heights of any aerosol layers present are hidden in this curve.

I learned to determine the height of the earth's shadow at any point along the curve by using an equation by Indian twilight scientist Dr. B. Padma Kumari and her colleagues.[4] But where are the aerosol layers along the relatively smooth line in the chart? I could not understand how twilight scientists were extracting aerosol layers from their twilight photometers until I discovered that Kumari's paper explained,

4. B. Padma Kumari et al., "Exploring Atmospheric Aerosols by Twilight Photometry," *Journal of Atmospheric and Oceanic Technology* 25, no. 9 (September 1, 2008), 1604.

"The most effective way of retrieving the aerosol layer is by the logarithmic gradient of intensity . . . ."[5]

The paper includes an equation that invokes the derivative of the data to tease out subtle changes in the twilight intensity. But this is calculus, and I had no idea what a derivative is until John Barnes explained it to me late one night during a lidar session at the Mauna Loa Observatory.

I then studied derivatives online and devised a spreadsheet that transformed the light measured by my twilight photometer into a chart that depicted elevation on the *y*-axis (left vertical) and the twilight intensity gradient on the *x*-axis (bottom horizontal). The revised spreadsheet transforms the smooth curve of twilight intensity into a curve punctuated by spikes and bumps where aerosol layers occur. This was fantastic!

Meteor Smoke and Noctilucent Clouds

What is especially fascinating about the twilight photometry paper by E. K. Bigg is its emphasis on detecting dust from disintegrating meteors entering the atmosphere 50 to 56 miles (80 to 90 km) above the surface. Although my interest in twilight photometry was initially restricted to dust and smoke in the lower atmosphere, I began noticing bumps and spikes in the data much higher up when I allowed the instruments to collect data for an hour or more. Could those bumps be caused by meteor smoke?

This reminded me of a visit with Dr. Donald Pettit, NASA's oldest astronaut and a veteran of several space flights, including two lengthy stays aboard the International Space Station. In July 2006, Pettit and I were guest speakers at a NASA-sponsored science teacher workshop at the University of Alaska Anchorage. Pettit was also there to meet with Dr. James M. Russell III, principal investigator of the AIM Team at Hampton University's Center for Atmospheric Sciences. *AIM* (*Aeronomy of Ice in the Mesosphere*) was a satellite planned to observe noctilucent clouds (NLCs) that form 51.6 miles (83 km) above sea level, which is one of the altitudes where I was detecting meteor smoke. *AIM*, launched in 2007, is still providing data.

In February 2003, Pettit photographed NLCs from the International Space Station, and when Pettit and Russell were discussing NLCs, I was there to listen. I had no idea that a decade later, my $25 twilight photometers could measure the height of meteor dust that forms an essential ingredient of NLCs: a tiny nucleus on which water vapor freezes.

5. Kumari.

One evening, some of the teachers and I drove to the top of a hill near Anchorage after sunset. There, we soon observed a wispy, bluish NLC atop a bright orange twilight glow. It was associated with the Aquarids, a meteor shower that had peaked the day before.

A decade later, two of my homemade twilight photometers were detecting meteor smoke like those that form NLCs. Especially fast meteors leave smoky trails as high as 56 to 75 miles (90 to 120 km), which my instruments have detected. For example, on October 22, 2019, one of my photometers detected several layers of meteor smoke from 56 to 65 miles (90 to 105 km) overhead. This occurred during the Orionids meteor shower. Orionids travel much faster than most meteors; thus, they begin to burn much higher where there are fewer air molecules.

The Volcanic Eruption of Raikoke

On June 22, 2019, a major eruption of the volcano Raikoke occurred. Raikoke, 1.5 mile (2.4 km) wide, is a Russian island in the Pacific Ocean, 230 miles (370 km) south-southwest of the southern tip of the Kamchatka Peninsula. The uninhabited island is dominated by a crater a half-mile (0.8 km) wide from which the 2019 eruption propelled the most ash and sulfur dioxide (SO_2) into the lower stratosphere since the 1991 eruption of Pinatubo. By August, the Raikoke cloud was causing brilliant twilight glows across North America, Europe, and Asia.

From July 23, 2019, to March 6, 2020, I made fifty twilight measurements of the Raikoke cloud. Each set of measurements was accompanied by twilight photos from a drone flying at 100 feet (30 m) to avoid trees that block the horizon at my site. These measurements provided an average elevation of the SO_2 cloud of 11.25 miles (18 km), with a few peaks of 16 miles (26 km).

The Raikoke aerosols I measured were consistently higher than measurements by the Cloud-Aerosol Lidar and Infrared Pathfinder Satellite Observation (*CALIPSO*) satellite, which consistently indicated a maximum Raikoke plume elevation of only about 9 miles (15 km).

These altitude differences were significant, for volcanic SO_2 can deplete the ozone layer, which peaks between 12 and 15.5 miles (20 km and 25 km). Therefore, it is important to know when volcanic aerosols reach the ozone layer. Pinatubo aerosols penetrated most of the ozone layer, where they destroyed enough ozone to cause large increases in solar UV-B at the surface.

John Barnes informed me that his MLO lidar had measured Raikoke aerosols as high as 15.5 miles (25 km), which matched my peak measurement. Another lidar measured heights of as much as 16 miles (26 km). These lidar measurements, which supported my twilight measurements, meant that Raikoke injected more aerosols into the ozone layer than CALIPSO indicated. I later learned that CALIPSO has a sensitivity problem, which I had suspected all along.

The Tree Ring Project

For many years, I have attempted to extract climate clues from the annual growth rings of trees. If you examine the cross-section of a branch from a bald cypress or another conifer, you will notice that the lower side of the branch is much darker than the upper side. This is caused by tannin, and I assume it is related to sunlight exposure on the branch's leaves. Is this extra tannin a defense against light-sensitive insects that might otherwise bore into the bottom side of a branch? Do the rings in individual branches provide higher-resolution sunlight data than trunks, which essentially integrate the sunlight illuminating the entire tree? I have been working on these and other questions for thirty-five years and have yet to complete a paper on this project.

Another aspect of the tree ring study is the identification of two variants I have identified of the common bald cypress (*Taxodium distichum*). Neither variety produces knees, the vertical extensions of the roots around the base of the common bald cypress. The growth rings and the spread of the branches of one variety are very different from that of the common bald cypress. I have documented these variants in detail in many photographs, and a bald cypress expert will join me on a paper if we can find funds for a DNA study.

While Minnie enjoys looking at my tree ring collection, this project annoys her when we visit the lumber section of hardware stores. The reduced tannin on one side of most knot holes points toward the top of the tree from which the board was sawed. So, when 2x4s are displayed upright, I feel compelled to rotate upside-down boards into their proper position.

The UV-B Monitoring and Research Program

Since 1992, the US Department of Agriculture (USDA) has sponsored the UV-B Monitoring and Research Program at Colorado State University. Dr. James Slusser was director of the program, which manages thirty-five sites across the mainland US and in Alaska, Hawai'i, Canada, and New Zealand, from 1999 to 2007.

Jim was well known for his ultraviolet research, and he was aware of my atmospheric research and my Radio Shack books. He often called to discuss topics of mutual interest. In 2001, Jim asked me to give a talk at an ultraviolet conference he cochaired sponsored by the International Society of Optical Engineering and NASA in San Diego. My paper was titled "Solar UV-B Measured at the Surface and Inferred by Satellite at a Rural Texas Site: 1994–2001." I felt like an amateur when surrounded by the experts at the meeting, but Jim treated me as if I were one of them.

After the San Diego conference, Jim asked if I would like to manage one of his program's sunlight-monitoring sites that he would like to place in Texas. This would provide an ideal way to compare my sunlight measurements with those by one of the UV-B Network stations. Texas Lutheran University is in Seguin, and I contacted John T. Sieben, dean of the College of Natural Science and Mathematics, to ask whether TLU could host the proposed site. Sieben wrote Jim, "Texas Lutheran University will be happy to become part of the USDA UV-B Monitoring and Research Program . . . ."[6]

On March 15, 2004, Bill Durham and Roger Tree arrived from CSU to install the instruments on the roof of TLU's Moody Science Hall, which began sending data to CSU every evening, which was then posted online.[7]

Most rooftop science instruments do not appear very busy, but two of the USDA instruments certainly do! They are called radiometers. Every fifteen seconds, curved metal bands rotate over the light sensors of each instrument, pausing briefly when the light sensors are shaded before rotating out of sight. This method allows the sensors to measure light scattered from the full sky with and without direct sunlight.

One of the photometers measures UV, from which the ozone amount can be determined. The other measures seven colors of sunlight, from which the amount of haze and water vapor can be extracted. Other instruments measure the UV-B, which causes sunburn and the blue and red colors of sunlight that make plants grow. Temperature and relative humidity are also measured.

I should add that these instruments include a third shadowband radiometer, in which all seven sensors are LEDs. A local power plant provided this instrument for my research. It is the only one to use LEDs as sunlight sensors.

Once a week, I climb a steel ladder to the roof of the Moody Science building to inspect the instruments and clean their optical surfaces. I have taken the instruments to the Mauna Loa Observatory for calibration during four of my stays there. Eventually, I hope to write a paper covering their results.

---

6. John T. Sieben, email to James Slusser, February 3, 2004.
7. UV-B Monitoring and Research Program at Colorado State University, http://carhenge.nrel.colostate.edu/UVB/uvb-dataAccess.jsf.

Return to Summit (RTS)

One of my newest projects is Return to Summit (RTS). As noted earlier, the Mauna Loa Observatory is widely acclaimed as the world's most important atmospheric monitoring station. The original 1951 building on the 13,468-foot (4,105 m) summit of Mauna Loa, abandoned in 1953 due to the poor condition of the road, was replaced in 1956 by the present MLO at 11,141 feet (3,397 m).

MLO founder Dr. Robert Simpson often recommended that at least some key MLO measurements should be resumed at the summit, and today's unmanned aircraft systems (drone) technology provides a convenient means to fly instruments from the present MLO to the same elevation as the original observatory.

In 2018, I conducted a preliminary pilot study of twenty instrumented UAS flights up to 400 feet (122 m) over MLO during the Rolex-sponsored UV survey of Hawai'i Island. Results from this simple study justify a much more organized series of test flights in which data will be collected under a wide variety of conditions. I hope to expand this study during a future MLO visit approved by NOAA and the observatory but delayed by the 2022 eruption of Mauna Loa. One mile of the MLO road is now under a thick layer of lava, which also destroyed the power line to the observatory.

Ongoing Atmospheric Studies

The 30-year paper reports only measurements made by instruments pointed directly at the sun. In 1995, I used $1,000 from the Rolex Award to build twenty light sensors that measure the intensity of total sunlight from the full sky and the sky when the sun is blocked. The full-sky sensors show that at noon on a typical summer day, about half the ultraviolet arriving at the ground is from the direct sun and the other half is scattered from the sky. This explains why people can be sunburned in the shade. I hope to write "A 30-Year Climatology of Full and Diffuse Sky Measurements at Multiple Wavelengths" in 2026.

After making the daily sun measurements, I photograph the sky. These photographs nicely visualize the color and condition of the sky and the solar aureole. I began with a Pentax Spotmatic film camera, which, in 1991–1993, I used to record some 500 images of Pinatubo-caused twilight glows. Since September 28, 1998, I have made 25,859 digital photos of the sky, and it is time to determine how best to publish them.

A thirty-year (1998–2028) climatology of sky photos accompanied by updated charts of physical measurements will provide an engaging visual history of the sky over my Texas site. These photos will clearly show the effects of Saharan dust, volcanic aerosols, and reduced sulfate pollution from fewer coal-fired power plants.

Some rare winter photos will show spectacular solar coronas caused by thick layers of pollen from juniper trees in the Texas Hill Country. (I published some of these photos in a paper for *Applied Optics.*)

In 1998, I bought a Fuji MX-700 digital camera and used it to photograph the solar aureole, the bright disk around the sun caused by stuff in the air, by blocking the sun with a disk on a wire holder. Twenty years later, the Fuji failed, and in 2007 I replaced it with a Canon G9 equipped with a simple fixture to block the solar disk so the solar aureole around the disk could be captured. As noted earlier, that became my first project in *Make* magazine.

In 2000, I began using a Nikon CoolPix 990 digital fish-eye camera to photograph the entire sky. That camera failed in 2018 and was replaced with a high-resolution Sony α6000 with a fish-eye lens. The many thousands of photos taken by this array of cameras are immensely helpful when analyzing sunlight measurements during dust, smoke, and power plant pollution events.

A Simple Method for Measuring Total Water Vapor

I like simple instruments, and that's what led me to develop a simple method for using a noncontact infrared thermometer to measure total column water vapor. The method exploits the fact that water vapor is the key greenhouse gas. Thus, when the thermometer is pointed straight up at the sky, its measurements of infrared are directly proportional to the amount of water vapor in the sky. This method is simpler than using a sun photometer, and it can be used at night.

I invested two years perfecting this methodology, and my GLOBE partner, Dr. David R. Brooks, and NASA scientist Dr. Lin Chambers joined me in writing "Measuring Total Column Water Vapor by Pointing an Infrared Thermometer at the Sky," a formal paper for *Bulletin of the American Meteorological Society.*[8] Besides their advice, their participation provided the page fee for the paper. Dr. Francis J. Merceret and Dr. Lisa L. Huddleston of NASA's John F. Kennedy Space Center evaluated my method and concluded in a 2014 report: "This consistency across geographic region and time suggests that the methodology is robust and stable enough to be applied in a variety of operational circumstances where inexpensive, field-deployable estimates of IPW are required."[9]

8. Forrest M. Mims III et al., "Measuring Total Column Water Vapor by Pointing an Infrared Thermometer at the Sky," *Bulletin of the American Meteorological Society* 19, no. 10 (October 2011), 1311–20.
9. Francis J. Merceret and Lisa L. Huddleston, "Assessment of a Technique for Estimating Total Column Water Vapor Using Measurements of the Infrared Sky Temperature," NASA/TM—2014–218368, June 2014, https://ntrs.nasa.gov/api/citations/20140012644/downloads/20140012644.pdf.

Because I make so many measurements at solar noon, I stopped the IR thermometer measurements after the paper was published. I plan to restart the project by mounting a suitable IR thermometer and data logger on a fence post outside my office to measure total water vapor every minute.

Monitoring the Historic Hunga Tonga Eruption

On January 15, 2022, Hunga Tonga–Hunga Ha'apai, a sleeping volcano under the South Pacific, exploded. In what some scientists described as a once-in-a-lifetime event, the eruption propelled ash and water 36 miles (58 km) into the sky, the highest eruption plume ever measured. Its sonic boom twice encircled the earth. The historic event was viewed by satellites, and this permitted the height of the eruption plume to be accurately measured.

The 1991 Pinatubo eruption blocked enough sunlight to cause the earth to cool a few degrees. The Tonga eruption was hugely different, and in a paper in *Geophysical Research Letters*, NASA scientist Dr. Luis Millán and his team reported that they found that Hunga Tonga had injected the equivalent of the water from about 58,000 Olympic-size swimming pools into the stratosphere, which increased the abundance of stratospheric water vapor by around 10 percent.

Water vapor is the chief greenhouse gas, so scientists have predicted that the Hunga Tonga water vapor will cause warming. They have also predicted that Hunga Tonga water vapor could remain in the stratosphere for a decade, where it could cause some destruction of the ozone layer.

During early 2022, I began checking my twilight photometer data for any sign of the Hunga Tonga water vapor over Texas. On May 6, 2022, one of my photometers detected unprecedented water vapor in the lower mesosphere 36 miles (58 km) over my field. I sent a chart showing the water vapor to the authors of a Hunga Tonga paper in *Geophysical Research Letters*.

One of the authors was Dr. Dong Wu, a well-known atmospheric scientist at NASA's Goddard Space Flight Center. Wu expressed serious interest in my finding and arranged for me to design and build five high-quality twilight photometers through an assignment from a Goddard Space Flight Center subcontractor, Science Systems and Applications.

The SSAI contract lasted June 16–August 31, 2022. I quickly called Scott Hagerup, my friend who had designed and built the first Microtops back in 1994, and he agreed to design and build the photometers based on a basic design I provided. By the end of August, Scott had assembled all five photometers, but I had barely begun to use them.

Wu wanted data, so he extended the contract to February 28, 2023. I then began collecting high-quality Hunga Tonga aerosol and water vapor data from the five photometers, each of which was equipped with a small telescope that looked at an angle of the sky of only 0.27 degrees, slightly larger than half the sun's diameter. After I continued to collect high-quality data over my site and during four field trips to Padre Island National Seashore, Wu extended the contract to February 29, 2024.

Thus far, I have made more than 200 twilight observations with the five photometers Scott assembled for the GSFC, and I have made 140 with a four-channel instrument Scott built for me.

A Hunga Tonga Discovery

On May 16–17, 2023, Wu and I participated in a Hunga Tonga online workshop sponsored by Stratosphere-Troposphere Processes and Their Role in Climate (SPARC), a project of the World Climate Research Programme. My contribution was a poster coauthored by Scott and Wu. Because our poster was posted online, during the two minutes allotted for poster presenters, I showed a slide that illustrated how the twilight method provides water vapor profiles. The slide indicated Hunga Tonga water vapor at an altitude range of about 12–25 miles (20–40 km). It also showed a layer at roughly 20–37 miles (50–60 km). Later, Dr. Gerald Nedoluha of the US Naval Research Laboratory informed me that the higher layer matches the strongest water vapor layer consistently observed in the thirty-year record of NRL microwave measurements of atmospheric water vapor.

Included in that one slide were two photos of Hunga Tonga twilight glows. Those and one in my poster were the only ones shown during the entire meeting. Twilight glows should be an important aspect of volcanic-eruption studies, which is why I made more than 500 twilight photos after the 1991 Pinatubo eruption and about 50 after the Raikoke eruption in 2019. I was glad to see that my twilight photo suggestion led those in the summary of suggestions from the second day. The Hunga Tonga paper I will write with Wu and Hagerup will include both altitude profiles of volcanic aerosols and water vapor together with stunning twilight photos.

Where It All Began

Childhood visits to see Papaw, my blind, paternal great-grandfather, and my great-grandmother in Lufkin, Texas, made a permanent impression on me. I remember the front porch swing and the record player in their tiny living room, where he listened to recordings of the Bible on vinyl disks provided by the National Library Service's "talking books" program at the Library of Congress. As I mentioned earlier, during one visit, I accompanied him on a walk and was amazed how he could count the power poles along the rural road.

Today, one of Papaw's canes hangs by our front door, where it reminds me of his strong work ethic and faith. It also reminds me of the childhood visits, when the old man with empty eye sockets under his snow-white hair called us to his chair and carefully felt our faces with his big hands to see how we had grown. Looking back, it was as if his sturdy fingers were sculpting my future, for he had inspired a dream that led to my science career.

The Explosion

For many years, I had wondered how Papaw had been blinded. I knew that the accident had occurred in 1906, while he was working on the track bed for a new railroad, and a few years ago, I found memoirs by Papaw and Nanaw, my grandmother, that included more details. The key detail was that Papaw's father-in-law, J. Norries, had a contract with the Kansas City, Mexico, and Orient Railway to prepare the track bed for a half-mile segment 15 miles (24 km) south of Sweetwater, Texas.

During October 1905, Papaw, his brother, and his father-in-law drove three wagons to Sweetwater, where they hired five men as helpers. Within a few months, Papaw's wife, Malissie, arrived with their two young daughters, Letha, who was only four at the time (she became my grandmother), and her baby sister, Pearl. Papaw's mother-in-law also arrived with her daughters. My grandmother's memoir describes how the women did the cooking.

The men were only 100 feet (30 m) from the end of the half-mile contract when they reached a rise that would require explosives to remove. The afternoon of May 3, the fuse Papaw ignited burned much faster than expected, and a massive explosion stopped his watch at 4:20 p.m. Letha, who we called Nanaw, wrote that her father was blown into thick brush that saved his life. She remembered how he was "a mass of blood" and that "one eye & part of his face was torn away and destroyed. The remaining eye was damaged and had to be removed without sedation." After sixteen days in the railroad hospital in Fort Worth, he returned to the worksite to help finish the job.

The Texas and Oklahoma Railroad

Details in Papaw's memoir provided enough information to look for the accident site on Google Earth, and I soon found a likely location the correct distance from Sweetwater, where the railroad track passes through a cut in a ridge. I then contacted Larry Locke, who owns the only train on the Texas and Oklahoma Railroad (TXOR) track between Sweetwater and the Buzzi Unicem USA cement plant at Maryneal.

I told Larry about Papaw and asked permission to hike to the site along the railroad track with my cousin, Don Mims. Instead, I was surprised when Larry offered to take Don and me to the site on his train!

The morning of October 31, 2017, Don and I arrived at the pickup point to wait for the train. We soon heard a train whistle, and a minute later, a beautiful blue engine with three bright headlights appeared. I launched my Mavic Pro camera drone and took photos and a video of the train's two engines and seventy-one cars as it approached.

Engineer trainee Kenneth Ballenger and conductor trainee Austin Kamer helped us carry our gear up the steps to the cab. There, we met engineer Lonnie Clowers, who has been driving the train on the TXOR short line to the cement plant for nineteen years. We then stood on the platform at the front of the locomotive as we headed south at 10 miles per hour. An hour and a half later, Lonnie checked the Google Earth photo I gave him and said we were nearly there. Suddenly, we recognized a pattern of juniper trees that matched the Google Earth image, and Lonnie stopped the train.

The Accident Site

While Austin watched me make aerial photos of the site with the drone, Don and Kenneth hiked to the top of the ridge about 15 feet above the west side of the tracks. After I landed the drone, Don and Kenneth arrived with a crushed, rusted-out blasting powder keg they had found on the ridge's crest. "DuPont" was embossed around the bottom of the keg, and "25 pounds" was embossed in the center. This unexpected find provided significant evidence that the accident had happened where we stood. The "brush" where Papaw landed was probably the juniper trees (*Juniperus ashei*) along the east side of the railroad track. Based on the diameter of their trunks, some of those trees had been there in 1906.

Drone Photos Reveal the Campsite

After our trip, I downloaded the drone photos and was surprised to see what we had missed at the site: a large stone circle adjacent to the railroad track and several smaller, partial circles. Nanaw had written in her memoir, "My first memories are from living in a tent about fifteen miles from Sweetwater, Texas, alongside the road bed which my Grandfather Norries had contracted to build for the Orient and Santa Fe Railroad." The stone circles were just 200 feet (61 m) from the end of the half-mile section of roadbed, and a creek was only 400 feet (122 m) away. Could those stone circles be their campsite?

We contacted Larry Locke to inform him about what the aerial photos revealed, and on May 1, 2018, Larry took us to the site in his hi-rail, a pickup truck with railroad wheels that can be lowered onto railroad tracks. Don and I spent several hours carefully exploring the stone-circle campsite while taking many photos from the ground and the drone. I photographed fourteen species of pink, yellow, blue, and white wildflowers scattered among the limestone rocks and slabs.

Our visit was only three days before the anniversary of the accident 112 years before. The women and girls back then must have enjoyed seeing the same flowers that were the last Papaw saw. We then explored the ridge where Don had found the blasting powder keg during the October visit. We found a dozen more kegs, including some crushed under large boulders.

Thanks to Larry and his TXOR crew, we discovered facts about our family history that had long been unknown. And we found and explored the campsite where our ancestors had spent five chilly months while the men and their crew worked on the roadbed.

The personal computer era might have been delayed several years or more if Intel had not developed the 8080 microprocessor. The Altair 8800 microcomputer would never have been developed at MITS if Ed Roberts had not been determined to build a computer. Microsoft would have never begun in Albuquerque if the Altair 8800 had not been the cover story of the January 1975 issue of *Popular Electronics*. And MITS would have never begun if Papaw's blindness had not inspired the travel aid for the blind that led directly to my model-rocket light flasher article in *Model Rocketry*.

That is why I brought something special to where Papaw lost his eyes: one of the LEDs Texas Instruments had given me in 1966 for the prototype travel aid for the blind. I photographed the LED lying on limestone shards blown apart where the explosion occurred 112 years before. Papaw's tragic accident inspired my science and electronics career, and that remote site south of Sweetwater is where it all began.

EPILOGUE

The preface related how the man banging on the door at the Mauna Loa Observatory smiled and politely asked, "Are you a *real* scientist?" No one had ever asked me that, and I did not want to disappoint a visitor who had gone to considerable trouble to drive what was then a miserable road to MLO to meet a real scientist. After all, I am just a self-taught scientist. So I paused several seconds before providing the four words that required this book to explain: "I just do science."

This brings to mind J. R. R. Tolkien's *The Lord of the Rings*, which closes with Frodo giving Sam the unfinished memoir by Bilbo and Frodo. Frodo then tells Sam that the book ends with several blank pages that he should use to describe his personal reflections.

I suggest you do likewise by acquiring a notebook to record your observations of the natural world and your personal science-related projects and activities. If you enjoy experimenting while cooking, you are already doing science. You are doing likewise if you take regular photographs of the sky and sunsets, keep a record of precipitation where you live, or take photographs of birds that appear at feeders and wildflowers that bloom in a nearby park.

If you are a student, you might learn more from science projects, possibly much more, than from the courses you take. Electronic instruments described in my books and those by other experimenters can provide you with the means to pursue more advanced projects.

When Dr. Jonathan Witt, a close friend and executive editor at the Discovery Institute Press, read this manuscript, he suggested the title and wrote that this memoir should "inspire up-and-coming scientists, particularly those who may feel themselves outside the standard pathways to scientific accomplishment (e.g., persons like you who lack a tenured academic post, complete with a lab and graduate students, at a top university). Your memoir is a call to action—get out there, stop making excuses, and discover something. A great big wondrous world awaits."[1]

Jonathan is right. I hope my adventures in the world of science will inspire you to join me in making inventions and exploring, measuring, and photographing the

1. Jonathan Witt, email to Forrest M. Mims, May 3, 2023.

natural world. Perhaps your personal scientific adventures can provide a sense of accomplishment, benefit humanity, and advance our knowledge of our planet and the universe.

Several years before the thirtieth year of my near-daily atmospheric measurements arrived, I mentioned to my brother, Milo, that thirty years of measurements are considered a climatological mean. Milo's enthusiastic response was, "Why don't you set a goal of 50 years?"

That had never occurred to me, but that is the goal I have set, plus a year to write the summary paper. I will need that extra year, for I will be ninety-five years old when the fiftieth year arrives. By then, I hope to have learned that many others have joined me in conducting long-term monitoring of the environmental world that sustains us.

ACKNOWLEDGMENTS

This book would have been impossible without the enthusiastic support of Dale Dougherty and his superb staff at Make: Community. Kevin Toyama and Mark Nichol carefully edited and fact checked the book and inserted its many references. The pages and illustrations were then expertly arranged by Juliann Brown, who has laid out my columns in *Make* magazine for more than a decade.

Prolific writer John Ash McCormick, who has written twenty-three books and thousands of articles, arranged for me to address the National Press Club about the *Scientific American* affair in 1990. When I sent John the first draft of the book, he kindly offered many helpful suggestions, as did his wife, Beth Goldie, a former employee of the Central Intelligence Agency, author, and newspaper copy editor.

Dr. Jonathan Witt, executive editor and senior fellow of the Discovery Institute, provided significant suggestions about the organization of the text and its overall content. Jerry Cooper, editor of the *Texas Aggie* magazine for thirty-one years, reviewed the text and made many helpful suggestions.

Inflation-corrected dollar amounts throughout are derived from the US Bureau of Labor Statistics CPI Inflation Calculator (bls.gov/data/inflation_calculator.htm).

The author is at 433 Twin Oak Road, Seguin, TX 78155, USA.
Email: fmims@aol.com.

INDEX

Note: Photos are indicated by an *i*, followed by the photo #.

C

H

I

J

K

L

M

N

O

P

Q

R

S

T

U

V

W

X

Y

Z

Mark Langford

Forrest M. Mims III is a scientist, inventor, and researcher. He has written for such publications as *Scientific American*, *Nature*, and *Science*, and his books have sold more than 7.5 million copies. In 1993, Forrest earned a Rolex Award, given to those whose endeavors have made a significant contribution to improving life and protecting the planet. He also served as a consultant to NASA's Goddard Space Flight Center, National Geographic Society, and National Science Teachers Association. Forrest lives in Seguin, Texas, with his wife Minnie.

ALSO BY FORREST M. MIMS III

Introduction to Electronics, Radio Shack, 1972.

Light Emitting Diodes, Howard W. Sams Co., 1973.

LED Circuits & Projects, Howard W. Sams Co., 1973.

Light-Beam Communications, Howard W. Sams Co., 1975.

Optoelectronics, Howard W. Sams Co., 1975.

Lasers: The Incredible Light Machines, David McKay Company, 1977.

Number Machines, David McKay Company, 1977.

Understanding Digital Computers, Radio Shack, 1978.

The Forrest Mims Circuit Scrapbook, McGraw-Hill, 1983.

Getting Started in Electronics, Radio Shack, 1983.

Forrest Mims's Computer Projects, Osborne McGraw-Hill, 1985.

Forrest Mims Engineer's Notebook, Radio Shack and High Text, 1985.

Science & Communication Circuits & Projects, Radio Shack, 1986.

Electronic Sensor Circuits & Projects, Radio Shack, 1986.

Electronic Formulas, Symbols & Circuits, Radio Shack, 1986.

Fundamentals of Surface Mount Technology, Heath Company, 1989.

Timer, Op Amp & Optoelectronic Circuits & Projects, Radio Shack, 2000.

Sun and Sky Monitoring Station, Radio Shack, 2003.

Hawai'i's Mauna Loa Observatory: Fifty Years of Monitoring the Atmosphere, University of Hawaii Press, 2011.

Make: Forrest Mims' Science Experiments, Maker Media, 2016.

Environmental Science: An Explorer's Guide, Intelligent Education, 2018.

Milton Keynes UK
Ingram Content Group UK Ltd.
UKHW021110290924
448960UK00005B/16